MOVING QUESTIONS

American Physiological Society

People and Ideas Series

Circulation of the Blood: Men and Ideas
Edited by Alfred P. Fishman and Dickinson W. Richards
1982

Renal Physiology: People and Ideas
Edited by Carl W. Gottschalk, Robert W. Berliner, and Gerhard H. Giebisch
1987

Endocrinology: People and Ideas
Edited by S.M. McCann
1988

Membrane Transport: People and Ideas
Edited by Daniel C. Tosteson
1989

August & Marie Krogh: Lives in Science
Bodil Schmidt-Neilsen
1995

Respiratory Physiology: People and Ideas
Edited by John B. West
1996

MOVING QUESTIONS

A History of Membrane Transport and Bioenergetics

Joseph D. Robinson, M.D.
Department of Pharmacology
State University of New York Health Science Center
Syracuse, New York

New York   Oxford
Published for the American Physiological Society
by Oxford University Press
1997

Oxford University Press

Oxford   New York
Athens   Auckland   Bangkok   Bogota
Bombay   Buenos Aires   Calcutta   Cape Town
Dar es Salaam   Delhi   Florence   Hong Kong   Istanbul
Karachi   Kuala Lumpur   Madras   Madrid
Melbourne   Mexico City   Nairobi   Paris
Singapore   Taipei   Tokyo   Toronto

and associated companies in
Berlin   Ibadan

Published by Oxford University Press, Inc.
198 Madison Avenue, New York, New York 10016

Oxford is a registered trademark of Oxford University Press

Library of Congress Cataloging-in-Publication Data
Robinson, Joseph D.
Moving questions : a history of membrane transport and
bioenergetics / Joseph D. Robinson.
p. cm.   (People and ideas series)
Includes bibliographical references and index.
ISBN 0-19-510564-8
1. Biological transport.   2. Bioenergetics.   3. Membranes
(Biology)   I. Title.   II. Series.
QH509.R63   1997   96-48139
571.6'4—dc21

9 8 7 6 5 4 3 2 1

Printed in the United States of America
on acid-free paper

This book is dedicated to the memory of my grandfather
  Heath Carrier
and my father
  Douglass Robinson
who taught me about the world and words

PREFACE

This book describes a half century of research on cellular membrane transport and on metabolic energy capture and utilization. During this time—which begins in the late 1930s—the effort and imagination of various scientists overthrew reigning formulations, created novel explanatory models, and unified previously distinct experimental fields. My primary goal is to display the course of that research, showing how new experiments defined novel entities and processes, and how an encompassing field, bioenergetics, then emerged.

A secondary goal is to present examples of mainstream biological research that illustrate how experimental results—seen as refutations, confirmations, and elaborations—can sway opinion toward a solid consensus. This interpretation differs from the currently fashionable view of some commentators that stresses instead the central roles of power, prestige, gender, class, and ethnicity. In any case, the scientific practices exhibited here deserve proper philosophical scrutiny. Although constraints of space have squeezed any analysis from this draft, brief mention of salient issues does appear in relevant chapters and in the final conclusions. (Oddly, historians and philosophers seem reluctant to deal with this science. Those who do consider biological topics tend to focus on the theory of evolution, even though the bulk of biological research in this century, in terms of papers published and technology influenced, has dealt not with evolution per se but with what may be termed physiology and biochemistry. And these endeavors, which are the aims, efforts, and accomplishments of the vast majority of biologists, have been largely ignored.)

Histories written by scientists are almost inevitably "internal" accounts, relating the course of scientific toils that culminate in our knowledge at a particular time. This is my aim here. But occasional reference will also be made to such topics as scientific fashion, availability of material and intellectual resources, personal and social characteristics influencing productivity, ability to recruit allies, and public willingness to finance scientific research. These factors surely matter.

Identifying success and error raises the specter of "whig history" written from the vantage of current opinion. Although I endorse admonitions to judge efforts within their contemporary setting, I believe that scientific formulations must also be viewed in the context of discoveries verified and mistakes corrected. When past results and interpretations have subsequently been deemed faulty, such later judgments need to be acknowledged as well as the earlier results reported.

To these ends I rely predominantly on published scientific papers, on their described results and argued discussions. Since in the era considered here published papers represented the preeminent channel for communicating scientific results and were the major means for persuading others, these papers must command attention. Scientific papers have been faulted as unreliable sources for establishing the origin and development of particular experiments, and that is demonstrably true in certain cases. In a few instances, published memoirs and reminiscences have recounted origins and odysseys, sometimes describing routes far from a planned assault on the problem solved; these have been noted here. But in this book inspiration is not often the central issue. Instead, my chief concern is with the experiments performed and the interpretations proffered, and that is what the published accounts display.

Another illuminating source has been formal interviews with many of the participants, whose recollections and opinions (and hospitality) have been extremely helpful: Wayne Albers, Amir Askari, Richard Cross, Philip Dunham, Ian Glynn, Joseph Hoffman, Leon Heppel, Peter Hinkle, Lowell Hokin, Mabel Hokin, Andrew Huxley, André Jagendorf, Bernard Katz, Richard Keynes, Henry Lardy, Jerry Lingrel, Peter Mitchell, Jens Nørby, Harvey Penefsky, Robert Post, Kurt Repke, Aser Rothstein, Arnold Schwartz, Amar Sen, Jens Skou, E. C. Slater, A. K. Solomon, Wilfred Stein, Daniel Tosteson, and Hans Ussing.

Richard Cross, Philip Dunham, Jeffrey Freedman, Peter Hinkle, Robert Post, and Lisa Robinson read drafts of various chapters, and I am deeply grateful for their valuable criticisms, necessary corrections, and useful suggestions. With characteristic generosity, Jens Nørby shared his trove of notes and reprints on the history of membrane transport that he had been accumulating for a similar project.

I am also indebted for assistance, advice, and information to: Douglas Allchin, Rhoda Blostein, Paul Boyer, Waldo Cohn, Setsuro Ebashi, Steven Grassl, John Harris, Alan Hodgkin, Peter Holohan, Patricia Kane, Steven Karlish, Anthony Martonosi, Helen Mitchell, Ruth Nadelhaft, Makoto Nakao, Gerda Pandit-Hovenkamp, Alfred Pope, John Prebble, Priscilla Roslansky, H. J. Schatzmann, Vladimir Skulachev, Carolyn Slayman, Joseph Steinbach, Mikulas Teich, Bruce Weber, and Ronald Whittam.

My wife Carol provided the abundant encouragement, consolation, and succor that was so continuously required.

I appreciate the sponsorship of the American Physiological society and the editorial assistance of Oxford University Press. And without financial support from the National Science Foundation this work would not have been possible.

Finally, I must apologize for the obvious clash between the author's enthusiasm for breadth and detail and the publisher's draconian strictures on length.

CONTENTS

1. Introduction, 3
   Cells and Their Structure, 4
   Cellular Metabolism, 4
   Membrane Transport, 10
   Information Transfer, 11
   Experimental and Explanatory Hierarchies, 12

2. Views in the 1930s, 13
   Cells and their Structure, 13
   Cell Constituents, 18
   Enzymes and Metabolism, 21
   Methods, 23

3. Accounting for Asymmetric Distributions of Na^+ and K^+ in Muscle, 26
   Fenn's View in 1936, 26
   Conway's "Double Donnan" Model of 1939/1941, 29
   Establishing Permeability to Na^+, 34
   Krogh's Summation, 40
   Responses and Rebuttals, 41
   Ussing's Measurement of ^{24}Na Outflux, 44
   Commentary, 47

4. Accounting for Asymmetric Distributions of Na^+ and K^+ in Red Blood Cells, 53
   Establishing Permeability of Red Blood Cells, 53
   Demonstrating Net K^+ Influx and Na^+ Outflux, 56
   Criticisms and Continuations, 59
   Progressing Further, 61
   Commentary, 63

5. Ion Gradients and Ion Movements in Excitable Tissues, 68

Beginnings, 68
Giant Axons, 70
The Hodgkin-Huxley Model, 70
Conflicts and Criticisms, 78
Direct Measurements of Ion Fluxes, 79
Maintaining Ionic Asymmetries, 80
Generalizations, 81
Commentary, 83

6. Epithelial Transport by Frog Skin, 89

Early Studies on Frog Skin, 90
Ussing's Frog Skin Model, 91
Commentary, 95

7. Contemporary Events: 1939–1952, 98

Structure, 98
Metabolism, 99
Methods, 101

8. Characterizing the Na^+/K^+ Pump, 103

Cardiotonic Steroids Inhibit the Pump, 103
Coupling Na^+ and K^+ Transport: A Na^+/K^+ Pump, 107
Maintaining the Steady State: Matching Pump Flux to Leak Flux, 112
Energy Sources for Active Transport, 113
Mechanisms Depicted for Active Transport, 118
Commentary, 119

9. Identifying the Na^+/K^+-ATPase, 126

Skou Discovers the Na^+/K^+-ATPase, 127
Precursors, 130
Successors, 132
Developments through 1965, 134
Commentary, 138

10. Contemporary Events: 1953–1965, 147

Membrane Structure and Composition, 147
Protein Structure and Dynamics, 149
Other Scientific Advances, 151
Sponsors, 152

11. Characterizing the Na^+/K^+-ATPase, 156

Multiple Transport Modes, 156
Purification and Reconstitution, 159
Reaction Mechanism, 162
Reaction Sequence, 168
Commentary, 168

12. Structure and Relatives of the Na$^+$/K$^+$-ATPase, 183
    Chemical Identity, 183
    Isoforms, 186
    Three-dimensional Structure, 187
    Reaction Mechanism, 188
    The Family of P-type ATPases, 188
    Commentary, 198

13. Alternatives, 205
    The Hokins' Phosphatidic Acid Cycle, 205
    Ling's Association–Induction Hypothesis, 212
    Commentary, 216

14. Using the Transmembrane Cation Gradients: Transporters and Channels, 220
    Na$^+$-coupled Transport Systems, 220
    Signalling by Ion Fluxes through Channels, 228
    Commentary, 233

15. Contemporary Events: 1966–1985, 238
    Membrane Structure, 238
    Proteins, 240
    Transporters, 242
    Protein Synthesis, 244

16. Oxidative Phosphorylation: Chemical-coupling Hypothesis, 247
    Slater's Formulation, 247
    Attempts to Identify Intermediates, 249
    Other Developments: 1953–1965, 253
    Commentary, 256

17. Oxidative Phosphorylation: Chemiosmotic Coupling Hypothesis, 261
    Mitchell's 1961 Formulation, 261
    Mitchell's 1966 Reformulation, 266
    Development, 268
    Acceptance and Further Revisions, 275
    Generalization and Integration, 277
    Commentary, 277

18. Oxidative Phosphorylation: F_1, F_oF_1, and ATP Synthase, 283
    Conformational Mechanisms and Binding-change models, 283
    Structure and Mechanism, 288
    Na$^+$-driven ATP Synthase, 292
    Families of Cation-transporting ATPases, 292
    Commentary, 293

19. Conclusions, 300

Histories, 301
Assumptions, 301
Goals and Approaches, 302
Means, 303
Course, 304
Characteristics of Scientific Practice, 304
Resolving Conflicts Rationally, 306
Generality and Quantitation, 309
Recommendation, 310

Appendix I: Units of Measurement, 312
Appendix II: Amino Acids and Proteins, 313
References, 315
Index, 365

MOVING
QUESTIONS

chapter 1

INTRODUCTION

B Y the 1930s, chemical analyses of vertebrate tissues—chiefly muscle—and of isolated cells—chiefly red blood cells—had demonstrated a peculiar and puzzling asymmetry. Tissues and cells generally contained high concentrations of potassium ions (K^+) relative to sodium ions (Na^+); by contrast, the environments of these tissues and cells, such as the blood plasma, contained the opposite concentration ratio: high Na^+/low K^+. Meanwhile, studies on the changeable contents of tissues and cells suggested that a membrane, although not visible by the microscopy of that time, surrounded each cell. Such a membrane could then have different permeabilities to solutes such as Na^+ and K^+. And investigations of excitable tissues, such as nerve and muscle, also focused on Na^+ and K^+ contents, with the electrical activity explained by some in terms of varying permeabilities to ions.

While permeability and excitability were major interests of physiologists in the 1930s, metabolism was a major interest of biochemists. In that decade they linked the capture of metabolic energy to the synthesis of adenosine triphosphate (ATP). Biochemists also described a synthesis of ATP dependent on oxidative metabolism, thereby identifying the process that became known as oxidative phosphorylation.

An account of the development, transformation, and ultimate unification of these issues and interests fills the following chapters. But first I will provide a present-day sketch of the relevant aspects of cells and their function to emphasize the pertinent aspects and to provide non-biologists with a contemporary understanding as well as the necessary vocabulary. Placing such a survey here specifies the end to which this tale is unfolding, but knowing the destination need not spoil a journey. And since this ultimate picture informs any biologist's reading, it seems only fair to grant others that foreknowledge. (Appendix I catalogs scientific units employed; Appendix II lists pertinent amino acids and notes how they are linked to form proteins.)

CELLS AND THEIR STRUCTURE

Two essential flows sustain the living world. One is the flow of information—from generation to generation, between cells and environment, and within cells. The other is the flow of energy, which is extracted from the environment, then transformed, stored, and used to drive all vital processes: growth, reproduction, movement, and communication. This book is concerned with ways this flow of energy is channeled by cells to power those workings and to signal that information. The first topic must then be cells, which are the basic functional units performing a range of essential activities, whether from single-celled or multicellular organisms. (In most multicellular organisms, various cells are specialized for particular chores, grouped then into tissues and organs, and with control of one cell's workings often modified through interactions with other cells.)

The schematized mammalian cell (Fig. 1.1) illustrates pertinent structures and components. Separating it from all else is an outer "plasma membrane;" exterior to this in plants and bacteria may be rigid cell walls providing mechanical strength, but it is the plasma membrane that not only defines out and in but controls what crosses out of and into the cell. Underlying these roles is a characteristic structure, a bilayer of lipid molecules oriented with their hydrophilic heads toward the surface and hydrophobic tails toward the interior, thus forming a barrier to polar, hydrophilic molecules. Modifying that barrier are proteins that lie within, and extend across, the bilayer, as well as proteins that lie on the bilayer surfaces. Some proteins mediate transport functions, which we will discuss shortly; some combine with particular chemicals in the extracellular environment, serving as receptors of the information signaled by those chemicals; and still others are connected to the cytoskeleton, mediating mechanical roles.

Inside the cell are assorted organelles, including mitochondria, which are centrally involved in energy transformation, and the endoplasmic reticulum, which forms an interconnected system of cisterns and tubules. These organelles are bounded by one or two membranes similar to the plasma membrane and that separate the organelles from the cytoplasm. The cytoplasm itself is far from structureless, containing rods and filaments of the cytoskeleton responsible for maintaining shape and for governing motility of cells and of their organelles within. The cytoplasm is filled also with "soluble proteins" (those which are not "membrane-bound").

CELLULAR METABOLISM

Glycolysis

The pertinent outline of energy processing begins with the metabolism of glucose through a succession of catalyzed changes termed "glycolysis" (Fig. 1.2). Details of these steps are not essential here, but five points deserve notice.

1. That multitude of steps allows a given product (pyruvate) to be synthesized over a specific route; this entails the formation and cleavage of particular

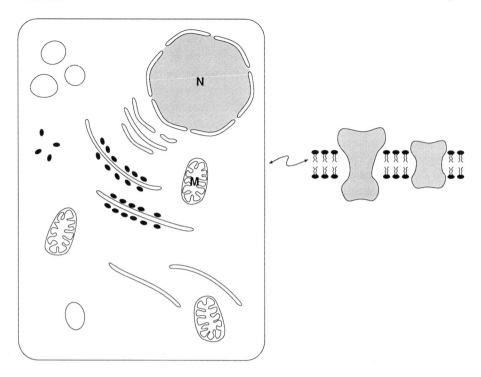

Fig. 1.1 Schematic view of a mammalian cell. Within the outer plasma membrane is the cytoplasm and the nucleus (N) surrounded by its membranes. Other organelles include mitochondria (M) with their outer and inner membranes (the latter folded into cristae and enclosing the inner matrix space), tubules of the endoplasmic reticulum (some of which have attached ribosomes), and various other vesicles also bounded by membranes. To the right is an enlarged, schematic cross section of the plasma membrane, indicating the bilayer of phospholipids (each with polar headgroups and two nonpolar fatty acid sidechains) and containing intrinsic protein molecules extending across the bilayer.

covalent bonds. Enzymes catalyze distinct reactions by binding certain reactants (e.g., glucose and ATP) and, through guiding the reaction mechanism, favor from among legions of potential products one set (e.g., forming glucose with phosphate esterified to the hydroxyl on its sixth carbon, glucose-6-phosphate, plus adenosine diphosphate, ADP).

2. Enzymes can "couple" the catalysis of two reactions, as in ATP cleavage/glucose phosphorylation. This is a crucial means for driving one reaction at the energetic expense of another.

3. Natural selection established ATP as the preeminent molecule for coupling various sorts of cellular work flexibly to metabolism, thereby creating a common currency for energy storage and utilization. That establishment required the development of enzymes catalyzing ATP synthesis and enzymes coupling energy-liberating ATP hydrolysis to energy-requiring needs. ("Hydrolysis"

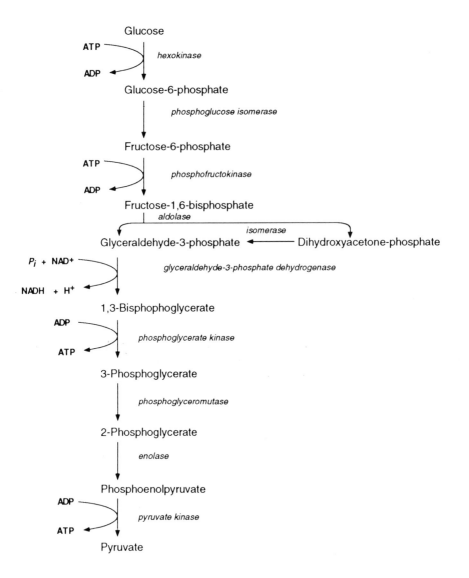

$$\text{Glucose} + 2\,P_i + 2\,\text{ADP} + 2\,\text{NAD}^+ \Rightarrow 2\,\text{pyruvate} + 2\,\text{ATP} + 2\,\text{NADH} + 2\,\text{H}^+ + 2\,\text{H}_2\text{O}$$

Fig. 1.2 The glycolytic pathway. Shown are the sequence of intermediates and the responsible enzymes. Six-carbon fructose-1,6-*bis*phosphate is split into two 3-carbon fragments, both of which (after isomerization of one to the other) are then metabolized further. The essential steps for energy capture are those coupled to forming ATP and NADH.

refers to cleavages where water is involved; for ATP hydrolysis to ADP plus inorganic phosphate, OH^- from H_2O is transferred during the cleavage of ATP to one fragment and H^+ to the other.) ATP is often termed an "energy-rich" compound, referring to its "free energy" of hydrolysis, the energy available for useful work. Under physiological conditions this is suitably large, 10–12 kcal/mol.

4. When one molecule of glucose is metabolized to two pyruvates, two ATP are formed through a process called "substrate-level phosphorylation" (because it is catalyzed by the same enzyme complex on which a substrate is being metabolized).

5. An intrinsic problem in metabolizing glucose to pyruvate is the net consumption of nicotinamide adenine dinucleotide (NAD^+), a biochemical cofactor that must be recycled to keep glycolysis flowing. This is achieved by converting the product, reduced NAD (NADH), back to NAD^+ by oxidizing NADH. (Conversion of NAD^+ to NADH is reduction, a gain of electrons; in this case two electrons, e^-, plus a hydrogen ion, H^+, participate:

$$NAD^+ + H^+ + 2e^- \Rightarrow NADH$$

Whenever one species is reduced—gains electrons—another is oxidized—loses electrons. Oxidation-reduction reactions thus come as couples and are refered to as "redox" reactions. In glycolysis, when NAD^+ is reduced, glyceraldehyde-3-phosphate is oxidized to 1,3-*bis*phosphoglycerate: each molecule of glyceraldehyde-3-phosphate donates two electrons to NAD^+.) Thus, the difficulty in converting glucose to pyruvate is finding an acceptor for the two electrons of each NADH formed.

Three pathways can accomplish this. The most favorable energetically is oxidizing NADH to NAD^+ coupled to reducing oxygen to water, thus taking electrons from NADH and donating them to oxygen:

$$4\,e^- + 4\,H^+ + O_2 \Rightarrow 2\,H_2O$$

Under physiological conditions, this reaction can provide about 50 kcal of energy: it represents the major energy-capturing system in animal metabolism, oxidative phosphorylation. When oxygen is absent two alternatives are available. Pyruvate may be reduced to lactate (CH_3COCOO^- to $CH_3CHOHCOO^-$), with this reduction coupled to the oxidation of NADH back to NAD^+ (this is the major solution for mammalian cells when oxygen is insufficient). Or pyruvate may be reductively decarboxylated to ethanol (CH_3CH_2OH) and carbon dioxide (CO_2) (as when yeast form alcohol and carbon dioxide anaerobically by fermentation).

Krebs Cycle and Oxidative Phosphorylation

Vastly greater energy capture results from converting pyruvate to carbon dioxide and water over the Krebs cycle (Fig. 1.3). Here, 32 ATP (or equivalents) can be

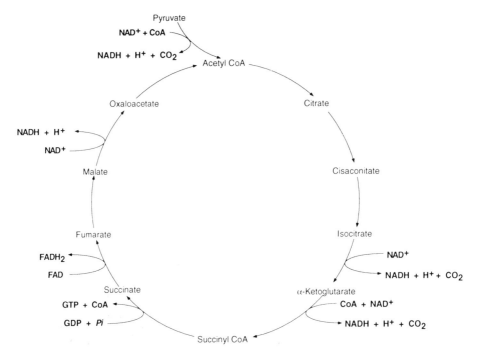

Fig. 1.3 The Krebs citric acid cycle. Shown are the sequential steps in which three-carbon pyruvate is converted into three molecules of carbon dioxide, with the associated energy-capturing steps (forming NADH, GTP, and $FADH_2$).

formed from each glucose. Details of the cycle are not essential here, but two points deserve notice.

1. Guanosine triphosphate (GTP, which is akin to ATP and interconvertible with it) is formed by substrate-level phosphorylation.
2. Most energy, however, is captured through oxidizing NADH, formed during the cycle, back to NAD^+: coupling the oxidation of NADH to the reduction of oxygen, forming water. From two turns of the cycle (one glucose forms two pyruvate), 20 ATP are generated from oxidizing this NADH. Including the NADH produced during glycolysis raises the yield to 25. Analogous oxidation of two reduced flavin adenine dinucleotides ($FADH_2$) produced in two turns of the cycle generates three more ATP. Two ATP, net, are formed during glycolysis, and two GTP are formed with two turns of the Krebs cycle.

Electrons are conveyed from NADH to oxygen by carriers embedded in the inner membrane of mitochondria; these carriers are collectively termed the "respiratory chain." Details of these redox reactions are not crucial here, but the mode of energy transformation is: free energy liberated in redox reactions is used to pump H^+ out of the inner mitochondrial compartment (Fig. 1.4). That transport thus creates an electrochemical gradient of H^+ across the inner mitochondrial membrane.

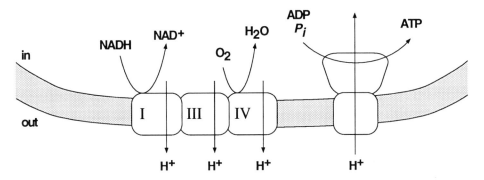

Fig. 1.4 Scheme for mitochondrial oxidative phosphorylation. Three protein-lipid complexes (I, III, and IV) in the inner mitochondrial membrane transfer electrons from NADH to oxygen, forming water. Coupled to the redox reactions is the extrusion of H^+ from the matrix space, creating a transmembrane H^+ gradient. Flow of H^+ down the electrochemical gradient into the matrix space—through the enzyme ATP synthase—then drives the formation of ATP.

These H^+ gradients require energy to form; conversely, these gradients can be converted back into usable energy. Tapping such potential energy reserves can occur by several processes, but the most general one involves synthesizing ATP. Potential energy stored as an asymmetric distribution of ions is converted into a compound with a high free energy of hydrolysis by coupling the downhill flow of H^+ through the synthesizing enzyme, ATP synthase (Fig. 1.4). The synthesized ATP, after transport out of the mitochondria, is then available for powering activities throughout the cell.

Similar transport of H^+ across plasma membranes of bacteria, also driven by sequences of redox reactions, creates electrochemical gradients that drive ATP synthesis for these organisms too. A similar system operates in chloroplasts of green plants. Light energy absorbed by chlorophyll drives sequences of redox reactions coupled to H^+ transport across chloroplast membranes, thereby creating an electrochemical potential that is then consumed to synthesize ATP.

Transmembrane ion gradients thus represent a second common currency for the cellular economy.

MEMBRANE TRANSPORT

Transport across cell membranes encompasses far more than redox-driven H^+ pumping and H^+-driven ATP synthesis. Membranes inherently create problems of access. Accepting and accumulating the good while rejecting and extruding the bad pose fundamental challenges—not only of selective discrimination but also of maintaining osmotic balance against the risk of catastrophic rupture.

A solution providing the requisite flexibility involves adjusting the concentrations of key solutes through energy-consuming active transport systems. Common inorganic cations found in and around cells—H^+, Na^+, K^+, calcium ions (Ca^{2+}), and magnesium ions (Mg^{2+})—are all maintained at particular nonequilibrium values by continuous energy-consuming pumping. Moreover, essential nutrients such as sugars and amino acids are accumulated by transport systems and waste products such as bicarbonate are expelled.

(For the moment, "active transport" may be considered as transport coupled to a consumption of cellular energy reserves. This contrasts with "passive transport", which is driven by random thermal energy and can achieve only equilibrium distributions of the transported substances. Passive transport is further subdivided into "facilitated diffusion," which uses protein transporters in the membrane to speed the approach to equilibrium, and "simple diffusion," which occurs as if by simple passage through the membrane, although in many cases protein channels are involved.)

Again, the converse of using energy to establish transmembrane distributions away from their equilibrium values is using those gradients to regain energy—just as mitochondria consume redox energy to create H^+ gradients and then tap those gradients to make ATP. Thus, cells of many organisms create asymmetric distributions of Na^+ across their plasma membranes by active transport systems, and then tap those Na^+ gradients to drive the transport of such other solutes as Ca^{2+}, H^+, glucose, amino acids, and neurotransmitters. These solute-driven transport systems are termed "secondary active" transport systems, the "primary active" transport systems are those established by metabolic activity, such as redox reactions or ATP hydrolysis (Fig. 1.5).

In most mammalian cells the primary means for establishing these Na^+ gradients is an enzyme in the plasma membrane, the Na^+/K^+-ATPase, which pumps Na^+ out of and K^+ into cells at the expense of ATP hydrolysis (Fig. 1.5). This enzyme thus maintains transmembrane Na^+ and K^+ gradients in the face of inherent leakiness as well as the consumption of these gradients to power cellular needs; in performing these duties it consumes a major fraction of all ATP produced.

At other sites within mammalian cells H^+ gradients are established across organelle membranes by other ATPases (primary active transport), and these gradients also are used to drive transport across the organelle membranes (secondary active transport).

INFORMATION TRANSFER

Signalling between and within cells also occurs through both making/breaking covalent bonds and creating/dissipating transmembrane ion gradients. For example, chemicals ("neurotransmitters," "hormones") released by one cell can bind to and interact with specific receptors on the surface of other cells. This process can occur through two ways. (1) By occupying receptors containing transmembrane channels for certain ions ("ligand-gated ion channels"), these chemicals can trigger electrical signals by initiating ion fluxes. The consequent change in transmembrane voltage

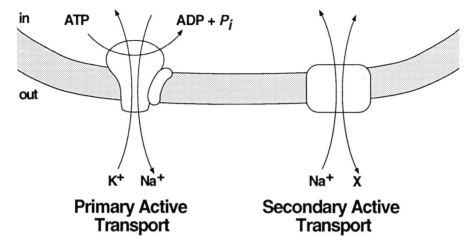

in   ATP   ADP + P_i

out

K⁺   Na⁺   Na⁺   X

Primary Active   **Secondary Active**
Transport   **Transport**

Fig. 1.5 Primary and secondary active transport systems. The enzyme $Na^+/$ K^+-ATPase in the plasma membrane couples ATP hydrolysis to pumping Na^+ out of and K^+ into the cell (primary active transport). Transporters in the plasma membrane can then pump other solutes into or out of the cell by coupling their movements to the influx of Na^+ down its electrochemical gradient (secondary active transport).

can then propagate as an "action potential": a wave of ion fluxes and resultant voltage changes that spreads by sequential activation of transmembrane channels sensitive to voltage fluctuations ("voltage-gated ion channels"). (2) By occupying surface receptors these neurotransmitters and hormones ("first messengers") can produce "second messengers": through stimulating synthesis (e.g., forming cyclic adenosine monophosphate, cAMP, from ATP) or release into the cytoplasm (e.g., liberating Ca^{2+} from stores within the endoplasmic reticulum). The second messengers can in turn alter cellular processes by activating protein kinases; these catalyze transfers of phosphate from ATP to enzymes and channels, thereby modifying their capabilities.

All these changes are transient. Phosphorylayed enzymes and channels are dephosphorylated and cAMP is cleaved enzymatically. Electrochemical gradients are restored by ATP-driven membrane pumps transporting those ions. And ATP is regenerated by consuming mitochondrial H^+ gradients formed during oxidative metabolism.

EXPERIMENTAL AND EXPLANATORY HIERARCHIES

Levels of experimental approach include the following.

(1) *Whole organisms* are obviously the most meaningful representative of the intact individual but the most difficult to analyze in terms of identifying and manipulating hordes of variables.

(2) *Whole organs* also present problems for analysis, including controlling variables and attaining experimental access. These problems are compounded by difficulties in ensuring adequate oxygenation (which can be minimized by perfusing through the blood vessels of the isolated organ or by using very thin organs, such as frog sartorius muscle, that allow diffusion from the surface).

(3) *Tissue slices* of a few tenths of a millimeter thickness allow better diffusion of oxygen and added reagents, and although cells on the surfaces are cut, many others within are unharmed. Tissue slices are frequently considered "whole cell preparations."

(4) *Isolated cells* that live free, such as bacteria, or that are obtained from organisms, such as red blood cells, or that are grown in tissue culture provide still simpler systems for study.

(5) *Homogenates*, tissue ground in salt of sugar solutions, circumvent problems of access at the expense of losing structure and suffering dilution; they are termed "broken cell preparations."

(6) *Isolated organelles*, which are separated from homogenates usually by differential centrifugation, provide further simplification.

(7) *Isolated molecules*, such as purified enzymes, afford the ultimate specificity and the ultimate loss of interactions with other constituents.

Preceding sections in this chapter displayed biological explanation from organism to cell to molecule, roughly paralleling these approaches. The final exploratory chains then rested on the entities and processes of chemistry. But explanations were achieved at multiple levels, often concurrently at descriptive and conceptual levels. Progress has been slow, however, in constructing *quantitative* models of highly organized systems from the properties of isolated constituents: the multitudes of entities, of structural relationships, and of interactive capabilities combine to form a daunting complexity.

chapter 2

VIEWS IN THE 1930s

A S background for later developments, this chapter will describe some of the reigning views from the latter 1930s concerning cells, membranes, proteins, and metabolic systems. A brief account of some of the methods then available is also included.

CELLS AND THEIR STRUCTURE[1]

By the 1930s cell theory had attained the orthodoxy of textbook pronouncements that identified cells as both the structural and functional unit of life. But how did structure relate to function?

Components

Microscopy had by this time revealed various structures within cells, although some parts required specific staining to be seen. Generally there was just one nucleus, and its importance in reproduction was firmly established. Mitochondria, inclusions staining distinctively with dyes such as Janus Green, were detectable in both plant and animal cells, although their function was then unknown. Chloroplasts were seen as green granules in plant cells. Additional structures were less certainly identified through available microscopes. On the other hand, early experimenters, by teasing apart assorted organisms and tissues, found a jelly-like substance filling the interior that became known by the mid-nineteenth century as "protoplasm"; to many biologists this seemed the ultimate living substance. (Whereas protoplasm referred to the entire content of the cell, "cytoplasm" was introduced to identify protoplasm outside of the nucleus.)

Plasma Membrane

In 1962 Homer Smith called the concept of plasma membranes the second most significant generalization in cellular biology—following only the cell theory itself—but noted that it had not achieved the status of an index entry in histories of biology then available.[2] Since the plasma membrane is a cellular structure fundamental to this book, a quick synopsis of its earlier history is warranted.

Early microscopists saw walls outlining—and thus defining—plant cells, but these cell walls, to the confusion of today's readers, were often termed "membranes." By modern electron microscopy, the cell wall—a thick, rigid, but porous structure of cellulose—lies just outside the plasma membrane (Fig. 2.1). This figure also illustrates another characteristic difference from conventional animal cells (Fig. 1.1), an internal vacuole containing watery cell sap and bounded by the vacuolar membrane, analogous structurally and functionally to the plasma membrane. Cytoplasm is thus confined to a thin layer called the "protoplast."

From observing that pigments diffused within the protoplast but could neither escape nor enter it from the exterior or the vacuole, Carl Nägeli in 1855 inferred the presence of membranes surrounding the protoplast—one within the cellulose wall and another surrounding the vacuole. In studies that were extended profitably by Wilhelm Pfeffer and Hugo de Vries in following decades, Nägeli found that immersing plant cells in high concentrations of sugars or salts caused both vacuole and protoplast to shrink (Fig. 2.1). Conversely, replacing bathing media with more dilute solutions caused the protoplast and vacuole to swell back to their original volumes. Moreover, solutions producing equivalent degrees of shrinkage were subsequently shown to have equal osmotic pressures; indeed, the data of Pfeffer and de Vries were used by J. H. van't Hoff to support his formulation of osmotic pressure.

Particularly significant was Pfeffer's observation that protoplasts and vacuoles of plant cells immersed in certain solutions first shrank rapidly but then returned slowly to their original volumes. The explanation centered on selectively-permeable membranes that allowed water to pass freely but solutes only slowly. Initially, water flowed rapidly across the selectively-permeable membranes from vacuoles and protoplasts to bathing media, from regions of lower solute concentration to those of higher concentration. This water flow was associated with shrinkage of vacuoles and protoplasts. But as solutes slowly crossed the selectively-permeable membranes into protoplasts and vacuoles, moving down their concentration gradients, water followed: protoplasts and vacuoles then swelled back toward their original volumes. That process thus permitted the relative rates of solute entry to be evaluated by measuring rates of swelling after shrinkage, a calculation used extensively to categorize membrane permeabilities.

In notable experiments at the turn of the century, Ernest Overton inferred a characteristic composition of the permeability barrier from analogous experiments. He found that solutions of alcohols, ethers, and related substances did not induce shrinkage at concentrations having the same osmotic pressure as solutions of sugars and salts that did; he concluded that these nonpolar substances crossed plasma and

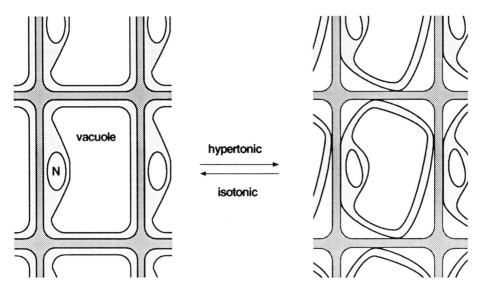

Fig. 2.1 Schematic view of plant cells and osmotic changes. Within the plant cell wall is a plasma membrane enclosing the cell, which contains a nucleus (N) and a large vacuole (also bounded by a membrane). When cells are immersed in hypertonic media, water moves out of the vacuole and the cell, and both shrink. When shrunken cells are returned to isotonic media, water moves back in, and the cell and vacuole swell back toward their original volumes.

vacuolar membranes so rapidly that there was then no osmotic flow of water. And since these rapidly permeating substances dissolved lipids, he inferred that the membrane barrier was lipoidal. Overton did not specify that the membrane was solely lipid but suggested that it was impregnated with cholesterol and phospholipid.

Experiments showing osmotic swelling and shrinking of animal cells lagged appreciably, but around the turn of the century a number of investigators—notably H. J. Hamburger, G. Gryns, C. Eykman, and S. G. Hedin—applied the methods of Pfeffer, de Vries, and Overton to red blood cells. These cells undergo characteristic and reversible changes in shape and volume: in plasma or in isotonic media they are biconcave discs (Fig. 2.2*C*); in hypertonic media they shrink to a wrinkled, crenated form (Fig. 2.2*D*); in mildly hypotonic media they swell to a sphere of equal surface area (Fig. 2.2*B*). But in still more dilute media they swell further and rupture, undergoing "hemolysis," a process by which hemoglobin and other cytoplasmic constituents are released (Fig. 2.2*A*). Through such approaches, permeabilities both to lipid-soluble compounds and to certain small polar substances were established quantitatively, thus demonstrating the lipoidal nature of the barrier and the selective permeability to water and to anions such as chloride (Cl^-). To cations, however, red blood cells appeared essentially impermeable. By the turn of the century techniques for osmotic rupture and extraction of red blood cells also permitted observation and characterization of the remaining membranous "ghosts."

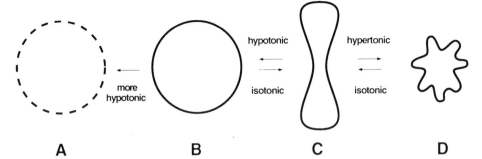

Fig. 2.2 Osmotic changes in red blood cells. In isotonic media (C), red blood cells are biconcave discs. In hypertonic media (D), water leaves and the cells shrink to smaller volumes without a change in surface area. In hypotonic media (B), water enters and the cells swell without a change in surface area. Further swelling in hypotonic media (A) causes rupture of the membrane and loss of internal contents (hemolysis).

Since red blood cells lack internal organelles such as nuclei and mitochondria, the only membranes are plasma membranes. This allowed E. Gorter and F. Grendel, with laudable insight, to perform in 1925 one of the more significant studies in this field.[3] They extracted the lipid from a counted number of cells and measured the area that lipid occupied when spread as a monolayer on water: with red blood cells from six species they found the area of the monolayer to be twice the calculated surface areas of the cells.[4] They concluded that the membrane is composed of a bilayer of lipids, with their polar heads forming the surfaces and their nonpolar tails filling the interior (Fig. 2.3A). They did not explicitly generalize this structure to other cell types but their successors did, and Gorter and Grendel's formulation is a crucial feature of most subsequent models, including the current favorite.

Although unaware of Gorter and Grendel's proposal,[5] James Danielli elaborated a similar structure in 1935 that became the standard for decades to follow.[6] Danielli based his model (Fig. 2.3B) on five lines of evidence. (1) Membranes were demonstrably permeable to nonpolar substances and generally impermeable to polar ones, which implied a lipoidal barrier. (2) Monolayers of amphiphilic lipids (polar on one end and nonpolar on the other, as are phospholipids) would be unstable between the two aqueous environments of cell interior and exterior. Moreover, earlier measurements of electrical capacitance were interpreted as indicating a thin membrane on the order of 30 Å (Gorter and Grendel had calculated their bilayer to be 30 Å; Danielli hedged the possibility of thicker membranes by allowing unspecified amounts of additional lipid in the middle). (3) The measured interfacial tension between aqueous media and cells (< 1 dyne/cm²) differed markedly from that between water and monolayers of certain lipids, such as "oil" from mackerel eggs (9 dyne/cm²). Danielli found, however, that the aqueous part of the mackerel egg decreased the surface tension of the water–oil interface; since chicken egg proteins behaved similarly, he concluded that protein must coat lipid membranes to reduce the interfacial tension.[7]

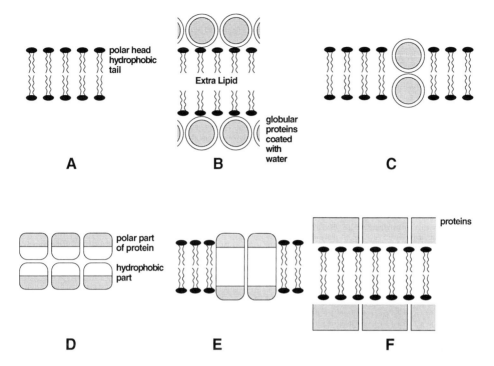

Fig. 2.3 Membrane models. *A:* Gorter and Grendel (1925) proposed a bilayer of lipid, with polar heads projecting out and nonpolar tails within. *B:* Danielli (Danielli and Davson, 1935) added globular proteins on the surface, including the possibility of additional lipid to make the membranes thicker. *C:* Including such proteins within bilayers would be unstable since their hydrated surfaces would contact nonpolar lipid tails. *D:* Danielli (1936) drew membrane models composed of lipoproteins, each extending halfway across the membrane, but he did not represent a lipoprotein extending across the full bilayer (*E*). *F:* Harvey and Danielli (1938) were later less explicit about the state of membrane proteins, but they still drew them only on the bilayer surface.

(*4*) X-ray crystalographic studies on the protein pepsin were interpreted as showing a globular mass with a hydrated surface layer.[8] (5) Danielli concluded that proteins at an oil–water interface did not denature.[9] Thus, Danielli's model (Fig. 2.3*B*) is Gorter and Grendel's bilayer with globular, hydrated proteins on both surfaces.

To allow passage of small polar molecules, Danielli proposed that spaces between globular proteins served as sieves; he invoked the thinness of the lipid layer and fluctuations in packing to allow diffusion across that phase.[10]

An alternative to both Gorter and Grendel's and Danielli's models is a mosaic model in which patches of pores (through or between protein molecules) are interspersed with patches of lipid (Fig. 2.3*C*). Mosaic models had been proposed since the turn of the century, and a vision of 1936 reads remarkably like one 50 years later:

There coexist side by side porous areas, where permeability is controlled by the relation between the pore diameter and the diameter of the solute molecule, and "lipoid" areas, where—independent of the molecular size—the relative lipoid solubility is responsible for the rate of penetration.[11]

But Danielli rejected such mosaics since he felt lateral interfaces between lipid tails and proteins (viewed as globular structures with hydrated surfaces) would be unstable.[12] Imagining various alternatives, he admitted that membranes composed of lipoproteins would be "satisfactory, provided that the lipoid part of the molecule is not smaller in cross sectional area than the protein part"[13] (Fig. 2.3D). The structure of lipoproteins was then even more uncertain than that of other proteins (they were defined simply as lipid-containing proteins), so it should not have been inconceivable for single lipoproteins to stretch across the whole membrane; such lipoproteins could associate hydrophobically with lipids laterally, as well as with themselves, providing stable mosaics (Fig. 2.3E).

Further studies on proteins at oil–water interfaces indicated that proteins "unrolled" under such stress,[14] and subsequent drawings were less explicit about the state of the membrane protein (Fig. 2.3F). Francis Schmitt interpreted optical polarization studies of red blood cell membranes as showing lipids with axes perpendicular to the membrane plane and proteins with axes tangential.[15] And A. K. Parpart extracted protein from red blood cell ghosts, concluding that ghost protein was a *homogeneous* substance, akin to collagen, and arranged "like spaghetti" on the surface.[16]

Whereas a consensus in favor of the Danielli model was thus forming, adapting the model to certain aspects of permeability was less straightforward. Particularly puzzling was how essential nutrients like polar sugars and amino acids could gain entry; moreover, extracellular glucose entered red blood cells by a process kinetically different from simple diffusion. By the late 1930s such phenomena were increasingly apparent—particularly with regard to absorption by the intestine and kidney where materials were accumulated against apparent concentration gradients—but plausible mechanistic models were not forthcoming. Danielli depicted only three ways for ions to cross membranes: (1) "through a rigid pore . . . as in [artificial] collodion membrane[s]", then popular experimental models; (2) "by combination with a constituent of the . . . membrane, followed by diffusion of the ionic doublet" accross the membrane; or (3) "penetration by simple diffusion . . . across the relatively homogeneous lipoid membrane."[17]

CELL CONSTITUENTS

Comparisons of protoplasm with proteinacious materials such as egg white were prominent in the nineteenth century, and attention to colloidal phenomena as explanatory models of cellular processes—which accompanied those observations—persisted into the 1930s. Ultimately more successful, however, were approaches based on characterizing identified constituents.

Proteins[18]

By the beginning of this century concepts of proteins as cellular constituents were refined through considerations of composition, structure, and size. At that time Emil Fischer turned to separating and identifying amino acids liberated from proteins, although it was not until 1940 that his successors recognized the last of the 21 common amino acids. Concurrently, Fischer also launched a synthetic approach, linking amino acids together: by 1907 he coupled a string of 18 amino acids before constraints of technique and cost terminated the effort. In 1932 Max Bergmann and L. Zervas developed an alternative method allowing the addition of further amino acids, but throughout the 1930s the synthetic approach was still severely limited.

These syntheses were founded on Fischer's proposal for a particular type of linkage between amino acids of the protein chain: amide bonds—termed in this context "peptide bonds"—between carboxyl and amine groups of successive amino acids. This linkage formed a "polypeptide" chain, but Fischer cautioned against generalizing the peptide structure to all linkages within proteins. And in 1924 Emil Abderhalden proposed diketopiperazine complexes joined by "partial valency." Another proposal arose from X-ray diffraction patterns of fibrous proteins: in the 1930s W. T. Astbury considered that α-keratin contained cyclical loops. More elaborate was Dorothy Wrinch's model featuring cyclol links that joined amino acids into sheets, with these then folded into polyhedra of definite molecular weight. Yet another ill-fated proposal was Bergmann and Carl Niemann's conclusion that the total number of amino acids in a protein—as well as the number of occurences of each amino acid in that protein—is invariably given by the formula 2^m3^n, where m and n are integers.

In the first successful X-ray study of a globular protein, J. D. Bernal and Dorothy Crowfoot in 1934 concluded from examining crystals of pepsin that "peptide chains in the ordinary sense may exist only in the more highly condensed or fibrous proteins, while the molecules of the primary soluble proteins may have their constituent parts grouped more symmetrically about a prosthetic nucleus"; they admitted, however, that "such ideas are merely speculative."[19] Two years later, A. E. Mirsky and Linus Pauling argued instead that a native protein molecule "consists of one polypeptide chain which continues without interruption throughout the molecule . . . [but that] this chain is folded into a uniquely defined configuration, in which it is held by hydrogen bonds."[20]

At the turn of the century the question was whether proteins were amorphous colloids of no definite molecular weight or distinct molecules each of a characteristic size. The analytical centrifuge, developed by Theodor Svedberg in the 1920s, revealed a range of molecular weights: from 17,000 for myoglobin to 6,700,000 for snail hemocyanin. At least some of the larger species were apparently in equilibrium with smaller components. Some proteins thus seemed to be composed of subunits, and a consequent interpretation—again in the quest for comprehensible, organizing principles—was that all proteins were assembled from a characteristic building block of 17,600.

In his 1938 textbook of biochemistry, Meyer Bodansky listed 21 amino acids as constituents of proteins (listing two now excluded and lacking two now included); noted that a "protein molecule is not merely a single large polypeptide"; devoted less space to hydrogen bonding in proteins than to Wrinch's cyclol model or Bergmann and Niemann's 2^m3^n proposal; considered that proteins are simple multiples of a fundamental weight unit; and cited Astbury's X-ray studies on keratin but not Bernal and Crowfoot's study of pepsin.[21]

Bodansky also presented a standard scheme for classifying proteins. "Simple" proteins were subdivided largely according to solubility and physical properties: albumins were "soluble in pure water and coagulable by heat," globulins were "insoluble in pure water but soluble in neutral solutions of salts," etc.[22] "Conjugated" proteins were those "united to some other molecule or molecules" such as glycoproteins containing carbohydrate and lecithoproteins containing the phospholipid lecithin (the only example of a lipoprotein given).[23]

Carbohydrates and Lipids

In 1827 William Prout separated foodstuffs into proteins, carbohydrates, and lipids. Carbohydrates are of minimal interest to this account beyond their essential role as metabolic fuels—chiefly as glucose and its storage polymer glycogen—and in conjugation with proteins to form glycoproteins. Although lipids are centrally involved in membrane structure, the details of their biochemistry are not important to this discussion, except to note that by the latter 1930s neutral lipids were known to include cholesterol and triglycerides, and phospholipids to include lecithin and cephalin.

Salts

Measuring inorganic salts was a laborious, tedious, and technically demanding enterprise since the chemical similarities challenged current capabilities. Moreover, the contents changed with time after the death of an animal, making extrapolations back to the living state necessary yet uncertain. Even more troublesome was evaluating *cellular* contents from measurements on *tissues*, which required that the sizes of the intracellular and extracellular compartments in those tissues also be evaluated.

A conspicuous emphasis of such studies was on comparing one type of cell or tissue among various species. For example, in 1876 G. Bunge found that blood serum of pigs, horses, cattle, and dogs all contained high contents of Na^+ and low contents of K^+, whereas red blood cells of the first two species had high K^+ and low Na^+ content while the reverse was true for the other two species. He reported no measurements on other cells or tissues. Twenty years later, Julius Katz measured Na^+ and K^+ contents of muscle from 12 species (but of no other tissues), reporting that in all cases the ratio by weight of K^+ to Na^+ was high, ranging from 1.6 in pig muscle to 14.2 in pike. In 1898 Abderhalden extended Bunge's results to eight species, in five of which red blood cells contained more Na^+ than K^+. This tradition of studying *either* muscle *or* red blood cells—and only rarely examining other animal cells or tissues to determine which was the more commonplace distribution throughout the organism—continued into the 1930s, when Stanley Kerr extended Abderhalden's approach to red blood cells of 21 species, again measuring no other cells or tissues.[24] A few descriptions of other tissues were reported, but in one study brain, liver, kidney, and lung were analyzed for K^+ in one species and for Na^+ in another.[25] As August Krogh complained in 1946, only "a ridiculously small number [of cell types] . . . have been studied."[26] Still, John Peters and Donald Van Slyke assumed that measurements on muscle and red blood cells "are fair examples of the relative distribution of [ions] between the cells and the extracellular fluids."[27]

ENZYMES AND METABOLISM[28]

Enzymes

Bodansky's 1938 textbook defined an enzyme as "a catalyst, organic in nature, produced by living cells."[29] This characterization represents a clear-cut aspect of enzyme function. It also reflects the research program of that era, which was founded on the "enzyme theory of biochemistry"[30] and its assumption that "sooner or later a particular specific [enzyme] will be discovered for every vital function."[31]

The relationship between enzymes and proteins was noticed early, but that obvious relationship was confounded by a technical difficulty. Analytical methods for detecting proteins were then quite insensitive, whereas intrinsic activities of many enzymes are quite high: enzymatic activity could be demonstrated when no protein could be detected (equated by some to no protein being present). Resolution came with the purification of enzymes through crystalization: of urease (1926) and, more definitively, of pepsin (1930). Those studies showed an ultimate, fixed association between enzymatic activity and protein content after purification, a ratio maintained through alternative modes of recrystallization. Bodansky's 1938 textbook reported that seven enzymes had been crystallized.

Another recognized attribute of enzymes was an observed specificity: the ability to catalyze a particular reaction by accepting only certain reactants. In an oft-quoted analogy, Fischer likened the fit between enzyme and reactant to a lock and key. The binding of reactant to enzyme was also implicit in kinetic formulations devised to account for how the rates of enzymatic activity varied with the concentrations of reactant ("substrate"). Leonor Michaelis and Maud Menten's equation, which describes a "saturable," hyperbolic increase in enzyme velocity with substrate concentration, is based on a model in which the substrate must first bind to the enzyme and the consequent enzyme–substrate complex then breaks down into enzyme and product. Also in accord with this model and with the principle of intrinsic specificity was the discovery of "competitive inhibition," in which structurally similar chemicals vied for occupancy of the substrate-binding region of an enzyme.

Wilhelm Ostwald had elaborated in 1900 the concept of chemical coupling, the discrete and causal linkage of one chemical reaction to another, but the concept of one enzyme coupling two distinct chemical reactions was still a rudimentary idea in the 1930s.

Metabolism

At the start of this century glucose was acknowledged as a preeminent biological fuel, its ultimate products (in the presence of oxygen) being carbon dioxide and water:

$$C_6H_{12}O_6 + 6\,O_2 \Rightarrow 6\,CO_2 + 6\,H_2O.$$

By the late 1930s the first third of that process, the conversion of glucose to pyruvate (or to lactate or to ethanol plus carbon dioxide), was known, as far as writing down the sequence of products, in forms close to that shown in Fig. 1.2. Textbook presentations differed in some details, but more serious was the absence of recent information about the sources and fates of the phosphates joined to several of the intermediate compounds.

In 1929 Karl Lohmann and, independently, C. H. Fiske and Y. Subbarow identified ATP. J. K. Parnas then demonstrated that tissues could form ATP in the presence of added 3-phosphoglycerate, and he surmised that such phosphate transfer was not linked to the *overall* process of glycolysis but to specific *steps* in the reaction chain. In 1935 Otto Meyerhof showed that the phosphate of phosphoenolpyruvate could be transferred to glucose, carried by ATP, i.e., that there were two phosphate transfers: from phosphoenolpyruvate to ADP, making ATP, and from ATP to glucose, making glucose-6-phosphate. Such interpretations underlay the textbook characterization of ATP as a "phosphate carrier and donator."[32] Three years later, Meyerhof showed that oxidation of glyceraldehyde-3-phosphate to 3-phosphoglycerate was linked to the formation of ATP, indicating a second point for ATP synthesis. The reaction was clarified further by Otto Warburg and Walter Christian in 1939, who used a purified, crystalline enzyme preparation. This was a notable achievement, for while the successive products of the glycolytic pathway had by then been fairly well identified, the characteristics and often even the identities of the responsible enzymes were unknown. As Joseph Fruton pointed out, Warburg's approach, which contrasted with much of the previous work using minced or extracted tissue as crude enzyme sources, "made it clear that the chemical dissection of complex biochemical processes depended on the isolation and characterization of individual enzyme proteins," and set a precedent for studying as chemical entities "many enzymes whose activity had been thought to be associated with the integrity of cellular structures."[33]

The second third of the process, converting the carbon and oxygen of glucose to carbon dioxide, was also left out of the textbooks of the late 1930s, again due to the newness of the explication. In 1937 Hans Krebs described the "citric acid cycle" (also known subsequently as the "Krebs cycle"), which accounted for the conversion of a two-carbon unit condensing with four-carbon oxaloacetate to form six-carbon citrate; the citrate then lost two carbon dioxides to form, successively, five-carbon α-ketoglutarate and four-carbon succinate; and the remnant finally reverted to oxaloacetate (Fig. 1.3).

The last third of the process, uniting with oxygen the hydrogens[34] extracted down the glycolytic pathway and around the Krebs cycle, required far more time for understanding. By the beginning of the 1930s, H. Wieland and T. Thunberg had demonstrated the extraction of hydrogens by "dehydrogenases," and Warburg the action of "oxidases." By the end of that decade David Keilin had showed how these processes were linked, distinguishing, by their optical spectra, three cytochromes (a, b, c) that underwent cyclical oxidation/reduction. Keilin also identified the terminal cytochrome oxidase of this respiratory chain with Warburg's Atmungsferment. In 1937 Hermann Kalckar described the coupling of oxygen consumption to phosphorylation of pyruvate (while glycolysis was inhibited); subsequently he demonstrated an oxidative phosphorylation forming ATP. And in 1939 V. A. Belitser and E. T. Tsybakova extended these experiments significantly, calculating that more than one phosphate was incorporated per atom of oxygen consumed (P/O > 1): "not only primary but also some intermediate oxidation-reductions are coupled with phosphorylation."[35]

Finally, in studies initiating a new area of inquiry, W. A. Engelhardt and M. N. Ljubimowa described in 1939 the ability of the muscle protein myosin to hydrolyze ATP.[36] This accorded with previous studies indicating that ATP was the proximate energy source for muscle contraction. It also was the first description of an ATPase activity, foreshadowing enzymatic mechanisms that transform the energy from ATP hydrolysis into the performance of various types of cellular work.

METHODS

Experimental methods available in the late 1930s constrained approaches, although for electrophysiological studies electronic amplifiers and stimulators were available as well as oscilloscopes for recording. The resolution attainable by light microscopy, roughly 1000 Å, was an order of magnitude too poor to detect plasma membranes. For biochemical studies tissues were sliced, minced, or ground; further separation was achieved by centrifugation (although techniques were recognized as crude), extraction with various solvents, and precipitation. Chemical separations relied on extraction, crystalization, precipitation, distillation, and liquid electrophoresis. Analytical techniques involved forming derivatives, weighing products, titration (using various indicators), and colorimetric determinations. Oxygen consumption was measured as changes in gas pressure, using manometers. Permeability was assessed by volume changes (from secondary flows of water) that were detected optically or as the loss of cell contents following osmotic rupture. Net transport could be determined by measuring differences in composition. As noted above, assaying cellular content of Na^+ and K^+ was an exhausting as well as an exacting chore.[37]

NOTES TO CHAPTER 2

1. For relevant historical accounts see Baker (1948, 1949, 1952); Davson (1989); Jacobs (1962); Kleinzeller (1995); and Smith (1962).

2. Smith (1962).

3. Gorter and Grendel (1925).

4. Bar et al. (1966) concluded that sufficient lipid is in red blood cell membranes to form a bilayer, but that Gorter and Grendel both extracted the lipid incompletely and underestimated the surface area.

5. Danielli (1962) reported that he "did not encounter the work of Gorter and Grendel until 1939" (p. 1165) but later recalled seeing the Gorter and Grendel paper in 1935 (Danielli, 1982).

6. Although the model appears in Danielli and Davson (1935), it is in one of four sections stated in a footnote to be the work "of JFD alone."

7. Danielli and Harvey (1935).

8. Bernal and Crowfoot (1934).

9. Danielli and Harvey (1935). Nevertheless, the interfacial tension of the lipid/protein system increased over time (see their Fig. 1).

10. Danielli and Davson (1935); Harvey and Danielli (1938).

11. Höber (1936), p. 367. Collander (1937) proposed a sieve mechanism in lipoidal membranes.

12. Danielli (1936).

13. Ibid., p. 402.

14. Danielli (1938).

15. Schmitt et al. (1936).

16. Parpart and Dziemian (1940), p. 22.

17. Harvey and Danielli (1938), p. 335.

18. For relevant historical accounts see Fruton (1972); Leicester (1974); Srinivasan et al. (1979); and Teich (1992).

19. Bernal and Crowfoot (1934), p. 795.

20. Mirsky and Pauling (1936), p. 442.

21. Bodansky (1938), p. 101.

22. Ibid., p. 83.

23. Ibid., pp. 83–84.

24. Kerr (1937).

25. Manery and Hastings (1939).

26. Krogh (1946), p. 163.

27. Peters and Van Slyke (1931), pp. 758–759.

28. For relevant historical accounts see Florkin (1972); Fruton (1972); Kohler (1973); Leicester (1974); and Teich (1992).

29. Bodansky (1938), p. 126.

30. Kohler (1973).

31. F. Hofmeister (1901), quoted in Kohler (1973), p. 185.

32. Best and Taylor (1939), p. 989.

33. Fruton (1972), p. 373.

34. See Chapter 1 for a description of redox reactions in terms of electron gain and loss. Redox reactions of organic molecules often involve gain and loss of a hydrogen *atom*, i.e., an electron *plus* H^+. Thus, oxidation of glyceraldehyde-3-phosphate involves loss of two electrons plus two H^+, with the incorporation of two electrons plus one H^+ into NAD^+ to form NADH (with one H^+ left over).

35. Translated in Kalckar (1969), p. 225. The implication is that redox reactions along the respiratory chain, as well as the initial dehydrogenation, are coupled to ATP formation.

36. Engelhardt and Ljubimowa (1939).

37. For example, measuring Na^+ by the procedure of Ball and Sadusk (1936) involved the following procedures: 1. Heat the sample with sulfuric acid overnight at 600°; 2. Add to the ash one drop of 2 N sulfuric acid and transfer that residue to a new tube with two 0.5 mL portions of water; 3. Add 10 mL of uranyl zinc acetate reagent and stir vigorously for 10 minutes; 4. Centrifuge this mixture and then decant the supernatant fluid, allowing the inverted tube to drain for 5 minutes; 5. Add to the residue 10 mL of glacial acetic acid and then stir; 6. Centrifuge and then decant again; 7. Dissolve the residue in 15 mL of 2 N sulfuric acid and transfer to a "Jones reductor" (a column filled with amalgamated zinc, which is prepared from mercury, nitric acid, and granulated zinc and then washed with sulfuric acid), drawing it through the reductor in 45 seconds; 8. Flush the reductor successively with four 20-mL portions of 2 N sulfuric acid at a rate of 40 mL/minute; 9. Draw air through the solution for 5 minutes; 10. Add 5 mL of a 5% ferric sulfate solution, then 5 mL of 85% phosphoric acid, and finally five drops of barium diphenylaminesulfonate indicator, swirling the flask after each addition; 11. Titrate with 0.025 N potassium dichromate until a violet color persists for at least 30 seconds.

chapter 3

ACCOUNTING FOR ASYMMETRIC DISTRIBUTIONS OF Na$^+$ AND K$^+$ IN MUSCLE

H OW can cells maintain high K$^+$/low Na$^+$ interiors while exposed to low K$^+$/ high Na$^+$ environments? A major concern from the late 1930s and through the following decade—the period considered here—was to distinguish between alternative models for explaining that perplexing asymmetry. This chapter describes studies on muscle. The next two chapters deal with studies carried out simultaneously on red blood cells and nerve cells—in some cases by the same individuals and for similar purposes, in some cases by other investigators with quite different interests.

FENN'S VIEW IN 1936

Wallace Fenn (Fig. 3.1) arrived at the new medical school in Rochester in 1924, equipped with degrees from Harvard University and experience in muscle physiology expanded by a visit with A. V. Hill in England. As founding chairman of the physiology department, Fenn spent the next decades developing his studies on muscle function—including investigations of how ions affect muscle activity—as well as building a strong research group. And in 1936 he published a detailed review of ionic distributions in muscle that is a revealing summation of the successes and failures to that date.[1] Fenn's explanation for asymmetry, a restatement of formulations by Rudolf Mond and Hans Netter, was notably simple. Plasma membranes were endowed with three sets of permeabilities:

1. Membranes were freely permeable to small[2] cations such as K$^+$ and H$^+$ but impermeable to larger cations such as Na$^+$ and Ca^{2+}. Selection by size was

Fig. 3.1 Wallace O. Fenn. (Photograph courtesy of Priscilla Fenn Roslansky.)

intuitively appealing and reflected the ability of porous artificial membranes then being studied to discriminate similarly.

2. Membranes were impermeable to anions (imaginable as negatively-charged pores accepting positively-charged ions but rejecting negatively-charged ones). This postulate thus barred Cl^-, the major extracellular anion, from entering or leaving cells; Fenn referred to "the general agreement that . . . muscle is normally impermeable to chloride."[3] The calculated volume of distribution of Cl^- in frog muscle was 14.7% of the total muscle fluid volume, which was close to the extracellular volume estimated from microscopic examination of muscle. Measures of "chloride space" were also similar to measures of the extracellular volume according to other indicators common at that time. Cl^- and Na^+ thus seemed to inhabit similar compartments in muscle. But Fenn

acknowledged problems with this tidy analysis: calculated volumes for Na^+ were by some reports larger than those for Cl^-.[4] Mond and Netter suggested that "excess" Na^+ might be bound to muscle cell surfaces.[5] Fenn was willing to admit some Na^+ inside the cell, but just how Na^+ could enter the cell if membranes were impermeable to it he did not explain.

3. Membranes were freely permeable to water. This stipulation acknowledged the swelling and shrinkage of muscle cells in response to changes in osmotic strength of the bathing media. Correspondingly, a comparison of the intracellular concentrations of solutes, in which K^+ is by far the most common cation, to the osmotic strength of the extracellular fluid required that K^+ in cells be free—neither bound nor complexed.

How can these properties account for the asymmetric distribution of ions? Since anions cannot leave muscle cells (stipulation 2), they must be balanced electrically by an equal charge of cations. Na^+ cannot enter muscle cells but K^+ and H^+ can (stupulation 1). Consequently, both will enter to neutralize anions confined within the cells; since the concentration of K^+ is 10,000-fold greater than that of H^+, K^+ will predominate. And water will move (stipulation 3) until the osmotic strengths inside and out are equal.

The model thus accounted economically for asymmetric distributions of Na^+ and K^+, but it left a residue of troubling concerns. That same year Fenn reported experiments showing that muscles gained Na^+ and lost K^+ during stimulation but then lost Na^+ and regained K^+ when allowed to recover.[6] How could impermeable muscle first gain and then lose Na^+? Fenn argued that with stimulation "only the surface of the fiber breaks down and exchanges its [cations,] i.e. the effective membrane surface moves inward [so that] sodium never really gets inside the membrane."[7] Moreover, the model placed a fundamental dependence on cellular anion content, which could be increased or decreased plausibly only by synthesis or degradation of organic anions within the cell, an awkward restriction. Related concerns arose from reports that muscle cells were in fact permeable to some anions, such as lactate and phosphate. The model required that these anions partition analogously to permeant cations, in disagreement with observed values. Fenn was compelled to admit: "It seems . . . likely that the theory is fundamentally inadequate in its present form."[8]

Why then did Fenn consider this model at such length—one that by some criteria was already refuted? It seems he had no alternative that he deemed more plausible. Early in his review he specified three ways osmotic balance could be achieved: (1) by a fixed total content of ions requiring total impermeability to at least some ions; (2) by strong cell walls that could resist osmotic pressures arising from a free permeability to ions; or (3) by "some force involving a continuous expenditure of energy (as in secretion processes)."[9] He then stated: "Other possibilities must certainly be exhausted before secretory properties are ascribed to muscle membranes in general."[10]

Why was this conclusion "certainly" the choice? There were two linked rationales. One was the quest for simplicity, and, in terms of current knowledge of membranes, impermeability was thought to be a less demanding requirement. The other rationale was an inability to imagine how secretory mechanisms could propel ions through lipid barriers while discriminating between closely similar K^+ and Na^+.[11]

In any case, the route to advocating alternative #3 (above) was, as we shall see, through fostering disenchantment with alternative #1. A strong cell wall, the second alternative, was incompatible with data on animal cells; Fenn included it because of the established role for plant cell walls in sustaining the elevated osmotic pressures within them.

Finally, it is worth emphasizing that the alternatives available to Fenn are *only* three. Similarly, *only* three alternatives could account for asymmetric distributions of ions: (1) equilibrium distributions with membranes absolutely impermeable to at least certain major ions; (2) equilibrium distributions but with some ions largely bound; and (3) nonequilibrium states maintained by energy-consuming transport ("secretion") of at least some major ions. Within the constraints of chemical laws, no other alternatives are available. We shall consider (and dismiss) alternative #2 in Chapter 13; here the conflict is between the first and third alternatives.

CONWAY'S "DOUBLE DONNAN" MODEL OF 1939/1941.

E. J. Conway (Fig. 3.2) formulated the ultimate model based on ionic impermeability, a similar but more precise and rigorous model than Fenn's and one arrived at from a different perspective. A year younger than Fenn, Conway received both his medical and D.Sc. degrees from University College, Dublin, and in 1932 he was awarded the first chair in biochemistry and pharmacology at that institution. There he, like Fenn, established a strong research program, although initially he studied kidney metabolism and only later membrane transport. But before considering his two definitive papers, an essential bit of chemistry must be introduced.

The Donnan Equilibrium

In 1911 F. G. Donnan described the equilibrium distribution of ions across membranes permeable to water and to some but not all of these ions. The distribution is most easily illustrated for membranes permeable to a single species of diffusible cations (C^+) and anions (A^-) but impermeable to a single species of nondiffusible anions (P^-): Fig. 3.3A. Donnan calculated from electrostatic and thermodynamic principles that at equilibrium (1) in each compartment the sum of the positive and negative charges must be equal, and (2) the product of the concentrations of diffusible ions in one compartment must equal the product of the concentrations of diffusible ions in the other:

$$[C^+]_{in} \times [A^-]_{in} = [C^+]_{out} \times [A^-]_{out}$$

Fig. 3.2 E. J. Conway. (Photograph © Godfrey Argent Studio, London.)

(the brackets indicate concentration and subscripts the compartment). The total number of particles in each compartment is then unequal (the compartment containing impermeant ions having more), so water will flow osmotically unless balanced by an opposing force, such as hydrostatic pressure.

For biological cells the consequences are obvious. If cells contain nondiffusible ions, such as proteins, in addition to diffusible ions, there are *only* four ways to avoid rupture from osmotic swelling—three equilibrium ways and one that is not.

1. Membranes can be mechanically strong enough to sustain hydrostatic pressures sufficient to halt further osmotic entry of water.
2. Membranes can be impermeable to water so osmotic concerns are irrelevant.
3. Impermeant ions in one compartment can balance impermeant ions in the other; the number of osmotically active solutes on each side would then be equal. This "double Donnan" distribution was the solution Conway favored, with impermeant Na^+ outside balancing impermeant anions inside cells (Fig. 3.3B).
4. Energy-consuming mechanisms can maintain concentrations of ions and/or water away from equilibrium distributions.

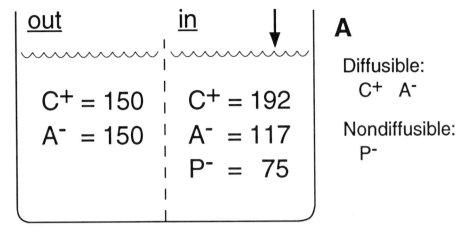

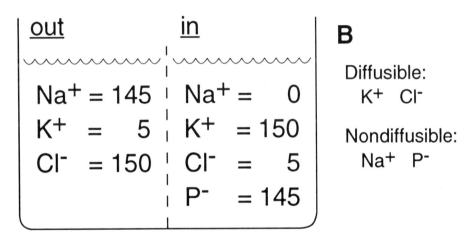

Fig. 3.3 Donnan distributions of diffusible ions. Two compartments are shown, "out" and "in," separated by a selectively permeable membrane. In *A*, the equilibrium concentrations of diffusible cation C^+ and anion A^- are shown in the presence of a given concentration of nondiffusible anion P^-. The arrow indicates that the "in" compartment contains more solutes than the "out" compartment, and thus a hydrostatic pressure would be required to prevent the flow of water from "out" to "in." In *B*, a "double Donnan" distribution is shown with a nondiffusible ion in each compartment. In this case, an equal number of solutes is in each compartment and no osmotic flow of water will occur.

Conway's Formulation

Conway saw in muscle the realization of mechanism 3. This formulation he published with P. J. Boyle[12] in a severely abridged form in 1939 and in a lengthy exposition in 1941; the latter will be considered here.

Membranes, according to this model, were permeable to K^+ and H^+, although they were impermeable to larger cations like Na^+. But they were permeable also to smaller anions like Cl^-. This permeability to Cl^- represented a major break with previous assumptions, but it was founded on direct evaluation by Conway as well as on descriptions for *some* tissues of Cl^- volumes that were larger than Na^+ volumes.[13] With these characteristics Conway derived equations, which were based on the Donnan equilibrium and standard chemical principles, specifying how K^+ and Cl^- would be distributed; he then compared calculated values with those from experiments on frog sartorius muscles. These muscles he soaked in salt solutions of various composition and then measured their K^+ and Cl^- contents, allocating the total contents between extracellular and intracellular volumes. Three examples can represent his successes and failures.

The Donnan equilibrium specifies that the product of the concentrations of diffusible ions in one compartment must equal the product in the other compartment. The experiments, however, did not satisfy this requirement at extracellular concentration below 12 mM K^+ (Table 3.1), and Conway admitted that muscle fibers lost K^+ when soaked at room temperature in concentrations of K^+ less than 29 mM. Since the K^+ concentration in frog plasma is roughly 3 mM, the model thus failed at physiological concentrations. To explain away this contradiction, he pointed out that membrane electrical potentials increase as extracellular K^+ is reduced (for a given concentration of intracellular K^+), and he attributed the discrepancy to the higher membrane potentials altering permeability. Without demonstrating such changes he merely stated: "muscle can stand only a certain concentration difference across it [and] when this is exceeded the membrane system breaks down."[14] Why it should break down under physiological conditions was not addressed.

But another prediction of the model—that as long as extracellular Na^+ was held constant the intracellular volume would not change, despite increases in extra-

Table 3.1 Donnan Relationship between K^+ and Cl^- inside and outside frog muscle cells*

$[K^+]_{out}$	$[K^+]_{out} \times [Cl^-]_{out}$	$[K^+]_{in} \times [Cl^-]_{in}$
3	240	680
6	490	660
12	1050	1000
120	23500	24200
300	112800	118700

*Frog sartorius muscles were soaked at 2–3° in media containing the concentrations of K^+ listed (K^+ was substituted for Na^+). K^+ and Cl^- contents were then measured and the concentrations calculated (listed as mmol/L of extracellular fluid or of cell water). The product of $[K^+]_{out} \times [Cl^-]_{out}$ is compared with $[K^+]_{in} \times [Cl^-]_{in}$; according to the Donnan equilibrium model these should be equal. [From Boyle and Conway, 1941, Table 8. Reprinted by permission of the Physiological Society.]

Table 3.2 Effects of Adding KCl on Intracellular K^+ Concentration and Intracellular Volume*

Extracellular	Intracellular			
$[K^+]_{out}$	$[K^+]_{in}$ measured	$[K^+]_{in}$ calculated	Volume measured	Volume calculated
3	91	93	1.14	1.17
6	92	96	1.14	1.17
12	101	102	1.26	1.17
120	212	210	1.12	1.16
300	353	390	1.07	1.15

*Extracellular K^+ was varied by adding KCl in the presence of a constant concentration of Na^+ at 2–3°. Experiments were otherwise performed as in Table 3.1. The volume is the intracellular water content, expressed relative to that of fresh muscle. [From Boyle and Conway, 1941, Table 6. Reprinted by permission of the Physiological Society.]

cellular K^+—was satisfied impressively across a wide range of K^+ concentrations (Table 3.2).

Still another prediction—that when K^+ was substituted for extracellular Na^+ the intracellular volume would increase—agreed quite well with data at lower concentrations of K^+ but not at higher levels (Table 3.3). This failure is more easily excused for it occurred far from physiological conditions, where various other perturbations might ensue.

The shortcomings should not overshadow the genuine achievement: a quantitative model in good agreement with experimental results, derived in accord with accepted chemical principles. The model also satisfied a proclaimed mechanistic value, by interpreting asymmetric distributions as a "true equilibrium [without any] continual expenditure of energy."[15] The cost to the model, however, was the necessity for *absolute* impermeability to Na^+. Since Conway acknowledged that at least certain cells contain Na^+, explaining how *some* Na^+ got in was a lurking challenge.

Table 3.3 Effects of Substituting K^+ for Na^+ on Intracellular Volume*

Extracellular $[K^+]_{out}$	Intracellular Volume Measured	Intracellular Volume Calculated
10	1.00	0.95
20	1.08	1.04
40	1.30	1.29
80	2.12	2.48
100	2.52	4.69

*The experiments were performed as in Table 3.2 except that K^+ was varied by substituting it for Na^+. [From Boyle and Conway, 1941, Table 5. Reprinted by permission of the Physiological Society.]

Fig. 3.4 Leon A. Heppel. (Photograph courtesy of Leon Heppel.)

ESTABLISHING PERMEABILTY TO Na^+

By the end of 1941, when Conway and Boyle's paper appeared, published evidence contradicted that fundamental requirement of impermeability to Na^+. And if that requirement were not universally satisfied, then mechanisms dependent on it could not be the sole process for establishing ionic asymmetries.

Heppel's Demonstration of Cellular Na^+ Uptake

Leon Heppel (Fig. 3.4) arrived in Rochester as a medical student in 1937, having just received a Ph.D. in biochemistry from the University of California. Heppel shifted to medicine because academic positions in biochemistry during the Great Depression were discouragingly rare; he came to Rochester because George Whipple, when he departed from California to become founding dean at Rochester, left a request that bright and research-oriented undergraduates be directed eastward. Heppel's thesis was on the consequences for rats of diets deficient in K^+. Whipple referred Heppel to Fenn, who recognized the potential of Heppel and Heppel's approach and offered him a stipend.

Heppel worked part-time during his first 2 years of medical school, extending his earlier studies. He fed young rats for 6 to 7 weeks on K^+-deficient diets. At that point the rats had failed to gain weight (control rats fed K^+-supplemented diets gained 20–30 grams per week); otherwise the deficient rats were not grossly abnormal. Rat tissues were then analyzed: K^+-deprived rats had lost over a third of their total muscle K^+ and gained an equivalent amount of Na^+ (Table 3.4).[16]

This was a straightforward experiment with little room for equivocation. Indeed, the data were not challenged in print. Continuing uncertainties about the size of extracellular and intracellular volumes were less pressing here because of the magnitude of the changes: extracellular volume could not have increased enough to account for all the gain in Na^+ and loss in K^+ (and the Cl^- content, then commonly taken to represent extracellular volume, did not change appreciably). Thus, the gain in Na^+ either represented binding to new sites on muscle cell surfaces (induced somehow by the diet) or Na^+ entering muscle cells. That Na^+ was free in the cytoplasm was indicated by the lack of osmotic swelling in the face of decreased muscle K^+: Na^+ was serving as an osmotic replacement for intracellular K^+.

Heppel reported two other indicators that muscles were not severely affected. A "preliminary" microscopic examination revealed no pathological changes (gross alterations in extracellular volume would have been apparent). And muscle removed from K^+-deficient rats exerted nearly as much tension as did muscle from control animals.

Surprisingly absent, however, were experiments to show that the gain in Na^+ and loss of K^+ was reversible: i.e., feeding deprived rats normal diets and determining what happened to muscle cations then. Heppel's brief account of his thesis research reported that K^+-deprived rats gained weight when fed K^+-supplemented diets, but the failure to demonstrate a restoration in K^+ content was unfortunate.[17]

Table 3.4 Effect of a K^+-deficient Diet on Rat Muscle Electrolyte Content*

| Rat group | Muscle Electrolytes (mmol/kg fresh muscle) | | | Intracellular Na⁺ (mmol/liter cell water) |
	Na⁺	K⁺	Cl⁻	
K^+-deficient	54.0	64.1	12.8	35.5
Control	17.1	110.9	11.5	4.5

*Young rats were fed K^+-deficient or K^+-adequate ("control") diets for 6–7 weeks; the muscle ion contents were then measured. The intracellular volume was calculated as the Cl^--free volume, assuming all Cl^- was extracellular. [From Heppel, 1939, Tables 1 and 2. Reprinted by permission of the American Physiological Society.]

What Heppel did do, however, was to show how rapidly Na^+ entered these muscle cells by using radioactive ^{24}Na as a tracer.[18] Radioactive isotopes were just coming into use, although they would not be commercially available in the United States for a decade, and Heppel took advantage of the cyclotron at the University of Rochester to produce ^{24}Na. This he injected into rats, noting that ^{24}Na equilibrated within 5 minutes with a volume equal to that of Cl^-. This was not surprising, since rapid exchange of Na^+ between blood and extracellular volume was a common conclusion. But he also showed that ^{24}Na reached more slowly a far larger volume of distribution than Cl^-: within an hour ^{24}Na was distributed between plasma and muscle in the same proportion as the total (nonradioactive) Na^+ (Table 3.5). This finding strongly supported his earlier interpretation.

Heppel thus demonstrated, convincingly, not only the presence on Na^+ within muscle cells but also its relatively easy access to cell interiors. These K^+-deficient cells clearly had mechanisms for regulating their ionic composition that could not be explained by the formulations of Mond and Netter, Fenn, or Conway and Boyle.

Other Studies with Radioactive Tracers

Until Ussing's notable papers of 1947–48, only five other reports on ^{24}Na and muscle appeared. Uncertainties about the relative magnitudes of extracellular–intracellular volumes probably contributed to the reluctance to undertake such experiments, for assigning ^{24}Na to intracellular compartments was then equivocal. Such difficulties are exemplified by experiments of George Hevesy and colleagues in Copenhagen:[19] after injecting ^{24}Na into rabbits they found 8.5% as much radioisotope in muscle as in plasma, so they concluded that ^{24}Na was "possibly not . . . extracellular."[20] But the next year they used the distribution of ^{24}Na to evaluate extracellular volumes.[21]

A third paper, from Oxford, was concerned chiefly with how radioactivity in blood changed with time after ^{24}Na was injected into rabbits. The authors noticed,

Table 3.5 Penetration of ^{24}Na into Muscles of K^+-deficient Rats*

Time after Injection (min)	^{24}Na (counts/min·g)		Muscle/Serum Ratio
	Muscle	Serum	
5	640	4700	0.14
10	770	4500	0.17
20	960	4200	0.23
31	350	1390	0.25
60	430	1400	0.31
260	1320	4000	0.33

*K^+-deficient rats were injected intraperitoneally with $^{24}NaCl$, and after the times indicated the animals were killed, the hind leg muscles removed and blood serum collected, and the radioactivity of the two samples measured. [From Heppel, 1940, Table 1. Reprinted by permission of the American Physiological Society.]

however, that heart differed from other organs in having higher concentrations of
^{24}Na than did plasma, and they attributed this to "absorption . . . into the cardiac
cells."[22] A fourth, from Berkeley, reported that after administering ^{24}Na to rats by
stomach tube, its volume of distribution in muscle was 16% in rats fed normal diets
but 19% in rats fed Na$^+$-deficient diets (no indication of the error of these data was
included so the significance of that difference is unknowable).[23]

In the fifth, Jeanne Manery and William Bale at Rochester concluded from
injections of ^{24}Na into animals that "the ^{24}Na distribution was identical to the Na
distribution, showing that there was no complete impermeability to injected sodium
in tissue phases which already contained sodium."[24] Thus, "the most reasonable
explanation . . . is that sodium has penetrated muscle fibre membranes hitherto
believed impermeable to it."[25] Manery and Bale also pointed out a crucial characteristic: adding ^{24}Na did not affect the measured content of Na$^+$, so radioisotopes did not
disturb cellular permeability.

Steinbach's Refutation of Conway's Model

Burr Steinbach (Fig. 3.5) was intermediate in age and career between Fenn and
Heppel. By 1940 he was an assistant professor at Columbia University, having
studied ion contents and electrical properties in various cells and organisms (and
having spent a year in Rochester). At this point Steinbach extended Heppel's studies
to experiments *in vitro*, akin to those of Conway and Boyle, and demonstrating that
increases in Na$^+$ content could be reversed *in vitro*.[26] He first soaked frog sartorius
muscles in solutions (nearly) free of K$^+$. He then measured the ion content of one of
the pair of muscles and soaked the other in a solution containing K$^+$. There was a
gain in K$^+$ and a loss of Na$^+$ during the second soaking, and no change in Cl$^-$ (Table
3.6). Steinbach concluded that "potassium and sodium [in muscle cells] are in some
sort of equilibrium with the ions of the external medium,"[27] but his nearest venture
to specifying a mechanism was in suggesting that intracellular ions might associate
with indiffusible organic molecules inside cells.

That same year Steinbach elaborated on these observations at a symposium,
noting that recent experiments by him and others demonstrated how Na$^+$ passed
into and out of cells "with considerable ease," a realization that "removes much of
the charm of the old selective permeability idea, since . . . the simplicity and clarity
of the scheme is destroyed," for "there must be a complete impermeability under all
reversible conditions or else there must be some mechanism present for pumping
out the sodium that wanders into the protoplasm."[28]

Steinbach's analysis was clear-cut. His commentary introduced the term
"pumping." But the mechanism he proposed was ill-fated: "the major mechanism
giving rise to the peculiar distribution ratios . . . are to be referred to the physical-
chemical balance between protoplasm and medium, with the permeability characteristics of the membrane playing only a subordinate role."[29] He considered it an
"unfortunate over-simplification to attribute much importance to permeability factors of a membrane [without considering] the rest of the protoplasm."[30] The focus

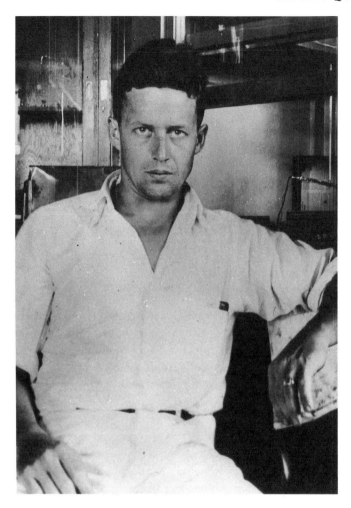

Fig. 3.5 H. Burr Steinbach. (Photograph courtesy of Joseph H. Steinbach.)

on permeabilty was in the narrow sense of Conway and Boyle's stipulation of permeable–impermeable. But the focus on protoplasm arose from Steinbach's attributing to protoplasm the generation of electrical responses and his commitment to ion binding.[31] Steinbach's reluctance to grant active roles to membranes undoubtedly reflected contemporary concepts of membrane structure, for Danielli's model depicted merely passive barriers.

But in 1941 Steinbach emphasized the central significance of steady-state systems, tying asymmetric distributions of ions to "active forces"; the explicit contrast was with a "diffusion equilibrium."[32] Still, a few years later he again referred to equilibria rather than to steady states,[33] perhaps because of his continuing advocacy of ion binding.

Table 3.6 Gain and Loss of Na⁺ by Frog Muscle in vitro*

| Ion measured | Ion concentrations (mmol/kg wet weight of muscle) | |
	Muscle after first soaking in K^+-free solution	Muscle after second soaking in 10 mM K^+
Na^+	5.04	4.04
K^+	3.44	4.99
Cl^-	3.79	4.10

*Pairs of frog sartorius muscles were soaked overnight in a K⁺-free solution. Ion contents were next measured in one of each pair; the other was then soaked in solutions containing 10 mM K⁺ for 6–8 hours and the ion concentrations then determined. [From Steinbach, 1940a, Table 1. Reprinted by permission of the American Society for Biochemistry and Molecular Biology.]

Dean's Arguments for Ion Pumps

Following Steinbach's survey at the 1940 symposium, the printed discussion began with a commentary by Robert Dean: "the . . . potassium saturation of frog muscle . . . can be explained if you assume a mechanism constantly excreting sodium."[34] And a year later he concluded that it was "difficult to see how . . . muscle can get rid of the sodium by any equilibrium process. . . . Therefore there must be some sort of a pump, probably located in the fiber membrane."[35] That assessment of the evidence prevailed, and his nomenclature endured.

Dean had studied chemistry at the University of California; for his doctoral training he moved to Cambridge, although the research, on transport across isolated frog skins, was performed partly in Copenhagen. After a further year in Cambridge he joined Fenn for a year in Rochester, where he studied radioactive isotope fluxes as well as the effects of metabolic inhibitors on the ion contents of muscle. But undoubtedly his major contribution was a paper written for a review volume that Fenn was editing.[36] Fenn had passed on to Dean the rather cryptic 1939 paper of Conway and Boyle, asking Dean to make it intelligible. This Dean did, generalizing the analysis as well. Most important, though, was Dean's contrast between Conway and Boyle's assumptions and data on the one hand, and the growing body of experimental results indicating permeability to Na⁺ on the other, with the conclusion in favor of a "pump" (quoted above). Dean also pointed out that "if the fiber is impermeable to anions, it makes little difference [theoretically] whether potassium or sodium is pumped [but] if chlorides were able to penetrate, a potassium pump would not work."[37] That analysis undoubtedly swayed opinion in favor of sodium pumps; the possibility of pumps transporting both Na⁺ *and* K⁺ was probably rejected on behalf of simplicity.

Although Dean nowhere spoke of evidence refuting hypotheses of membrane impermeability to Na⁺, he stated that Mond and Netter's formulation "fails to explain" and that Conway and Boyle's formulation "does not account for" the observations.[38] Dean was quite explicit, moreover, in placing certain principles beyond challenge: "In view of the fact that the second law [of thermodynamics] has been found to hold universally in chemistry, we are not justified in assuming that it is violated in biology."[39]

KROGH'S SUMMATION

August Krogh arrived at these concerns following a different route. His Ph.D. dissertation for the University of Copenhagen concerned respiration by frogs through their lungs and skin. His professor, Christian Bohr, believed that the exchange of oxygen and carbon dioxide was achieved by secretion; Krogh respectfully demonstrated that exchange was due to passive diffusion. A brilliant experimenter, Krogh extended these studies to blood flow through the capillaries, including hormonal and neural control of blood flow; for this work Krogh received the Nobel prize in 1920. In 1928, through the perceptive generosity of the Rockefeller Foundation, a Rockefeller Institute was established in Copenhagen.

Krogh's interests spanned the biological kingdoms, and in the 1930s he began studies on salt and water balance in freshwater and marine organisms, using both stable and radioactive isotopes as tracers. Toward the end of World War II, Krogh was invited by the Royal Society of London to deliver the Croonian Lecture for 1945 on this research. The published account was of heroic scope, encompassing transport epithelia in kidney, intestine, and capillary, as well as plasma membranes of plants, marine and freshwater organisms, and various vertebrates.[40] The introductory paragraph conveyed his central conclusion: "differences in concentration of individual [species of] ions between the interior of living cells and the fluids surrounding them [are] expressions of steady states, differing in principle from equilibria by the necessity of energy being supplied for their maintenance."[41]

Here only a few topics need be noted. Krogh agreed with Conway that a small fraction of the total muscle Cl^- was intracellular, but he complained that Conway "treat[s] the muscle fibres . . . as a physico-chemical system pure and simple," assuming permeabilities to only certain ions to "predict with surprising accuracy the mean change in potassium . . . and in cell volume."[42] Krogh summarized those experiments, noting, as Conway had admitted, that with physiological concentrations of K^+ the muscles paradoxically gained Na^+ and lost K^+. Next Krogh summarized Heppel's experiments, agreeing that there "can be no doubt about the fibre permeability for both [sodium and potassium] ions."[43] Krogh calculated the permeability to Na^+ of those cells, comparing it to permeabilities to K^+ he had measured. He also presented Steinbach's experiments with isolated frog muscle that showed gains and subsequent losses of Na^+ during sequential soakings; he remarked that Steinbach's report failed to state either the temperature of the experiments or the final weights of the muscles (to indicate swelling or shrinking). Krogh cautiously admitted that permeabilities to Na^+ in these experiments might be attributed to unphysiological exposures to K^+-free solutions in the first soaking, but that there could be no doubt that "the recovery process means an uptake of K and an elimination of Na . . . against the concentration gradient."[44] Krogh was evidently unaware of Dean's 1941 paper, but he reached the same conclusion: active transport of Na$^+$ could maintain the Na$^+$/K$^+$ asymmetries.

RESPONSES AND REBUTTALS

Fenn

"Potassium is of the soil and not the sea; it is of the cell but not the sap."[45] Fenn's review of 1940 surveyed contemporary knowledge, but he was unable to proceed mechanistically beyond that aphorism. He quoted Heppel's results (but not later papers by Steinbach and Dean), acknowledging that "intracellular Na . . . in muscles of rats raised on a low K diet readily exchange with injected radioactive Na [and this presents an] obvious difficulty" for Conway's theory.[46] Nevertheless, Fenn decided that "in spite of this demonstrated permeability to both K and Na, [Conway's] theory . . . seems to explain satisfactorily a good many features of the muscle electrolyte balance," and repeated that "in spite of this demonstrated permeability to K and Na the Donnan membrane theory [i.e., Conway's] appears to explain the electrolyte equilibrium . . . satisfactorily."[47] And in the summary Fenn referred to K$^+$ in muscle as being in "static equilibrium."[48]

At least part of Fenn's inability to anticipate reformulations by Steinbach, Dean, and Krogh is exemplified by his introductory essay to the volume containing Dean's 1941 paper: "The reader will undoubtedly be greatly dissatisfied with a theory which would explain the electrolyte equilibrium in muscle by postulating a 'pump' of unknown nature."[49] Nevertheless, he continued: "it is no small recommendation for a theory if it describes precisely what the 'pump' actually does."[50] By 1944 Fenn acknowledged the implication: "if Na is also regarded as a penetrating ion . . . some 'pump' theory like that of Dean . . . is required to explain the facts."[51]

Conway

Conway was more active in propounding theories of membrane impermeability and more vigorous in defending them. It is instructive to follow his responses over 3 years as they employ all the options available for rebutting scientific claims:

1. The data are incorrect for the experiments cannot be replicated.
2. The data are incorrect for contradictory principles or data exist.
3. The data may be correct but their interpretation is incorrect.
4. The data may be correct but the conclusion is unimportant.

Conway and Boyle did not refer to work by Heppel, Steinbach, or Dean in their 1941 paper; quite likely they had not seen these reports. Conway's 1945 review referred to Heppel's 1939 paper and Steinbach's 1940 papers only.[52] In response to

Heppel, Conway proposed that K^+-deficient diets *could* cause both permeability to Na^+ and impermeability to anions: then Na^+ could exchange for K^+ when plasma K^+ fell.[53] (Conway was thus reinterpreting Heppel's data, applying option 3 listed above, but by this mechanism the ratio of $[Na^+]_{in}$ to $[K^+]_{in}$ should equal the ratio of $[Na^+]_{out}$ to $[K^+]_{out}$, which was not the case.) In response to Steinbach, Conway complained that neither the compositions of the solutions nor the experimental temperatures were specified. More important, Conway stated that he had been unable to replicate Steinbach's results using a variety of experimental conditions. Conway noted that Steinbach had not mentioned whether muscle weights changed during the soakings, and he suggested that extracellular volumes had increased during Steinbach's experiments, which *could* account for the reported gains in muscle Na^+. Conway further suggested that "small increases of K content [during Steinbach's second soaking] are likewise easily explained both by a passive diffusion of the external K into the intercellular spaces and also into damaged fibres."[54] Conway not only applied option 1 but described how the results were misinterpreted, a viewpoint akin to option 3. He suggested as yet another possibility that additional extracellular compartments existed that could gain Na^+ during Steinbach's soaking of the muscles in K^+-free media, echoing Mond and Netter's placement of "extra" Na^+ on muscle surfaces and quoting Manery and Bale (erroneously) that muscle Na^+ is "essentially extracellular."[55]

In a brief review in 1946, Conway cited both Heppel's 1939 and 1940 papers, Krogh's Croonian lecture, and Steinbach's 1940 paper, but he still overlooked Dean's work.[56] Instead of repeating his previous reinterpretation of Heppel's results, Conway now pointed out that these papers failed to demonstrate any ability of the rats to lose the excess Na^+ and regain K^+, and he suggested that the change was irreversible—and thus irrelevant to physiological processes (option 4). He repeated his criticisms of Steinbach's report, stating that he had "consistently failed to obtain the reversibility [Steinbach] describes,"[57] and he attributed the initial gain in Na^+ to pathological losses of impermeabilty.

But in a major retreat, Conway now claimed that advocating a "complete exclusion of $[Na^+]$ has not been my intention."[58] His revised account featured "two regulatory mechanisms . . . (a) the 'standard' permeability [(meaning near impermeability to Na^+), and] (b) a very slow active extrusion of sodium."[59] Conway thus admitted that sodium pumps might exist, but if so they were functionally insignificant (they were denigrated as "very slow"); again Conway employed option 4. Still, he emphasized that "active extrusion of sodium ions . . . is a quite unproven hypothesis" and argued that "the reduction of the rate of entrance of sodium ions . . . would appear ideally the most advantageous, since it saves free energy."[60] Such an appeal to ideal design is attractive, for it would seem better to have tight membranes limiting Na^+ influx rather than energy-consuming pumps remedying leaky membranes.

Conway also attacked Krogh's claim that Na^+ influx was at least as great as K^+ influx, thus applying option 2. Since Conway did not have access to ^{42}K,[61] he estimated influx by measuring how fast muscles gained weight when transferred from media containing 2.5 mM K^+ to media containing 10 mM K^+.[62] To evaluate Krogh's claim, he set Na^+ influx equal to this calculated K^+ influx. Conway next calculated the rate of Na^+ *outflux* from muscle and its energy cost (the switch here from influx to outflux was justified by the two fluxes having to be equal, or a net gain or loss of Na^+ would result). Conway then reported that such a Na^+ outflux would cost 350 cal per kilogram of muscle/hour; that value he converted to oxygen consumption,[63] which he then stated was "about twice as much, or more, [than the] resting muscle metabolism of . . . frogs."[64] And, "realizing the very low efficiency of such excre- tion in general," it must follow that "sodium . . . enters only at a very minute fraction of the rate of potassium" and thus "the major, if not the entire cause of the exclusion of sodium from muscle tissue may be assigned to a membrane effect,"[65] i.e., to impermeability. Since the "hypothetical entrance of sodium . . . can only be a small fraction of the rate of entrance of potassium, the 'standard' permeability [must be] far more significant . . . than a hypothetical extrusion of sodium ions."[66] The chain of logic was correct, but the plethora of assumptions in his calculations sorely weakened the argument.

Conway's third defense of his proposal[67] cited Heppel's earliest paper showing some restoration of health when K^+-deprived rats were fed K^+. Conway then de- scribed his extension of Heppel's studies: rats kept on K^+-deficient diets lost K^+ and gained Na^+, but when they were then fed high K^+ diets they lost some of that excess Na^+, albeit slowly: the halftime was about 3 days.[68] He was now ready to admit that "a mechanism [exists] for the slow extrusion of Na," but he claimed that "the extrusion rate . . . is so slow that . . . the normal fibre membrane can be regarded as impermeable to the Na ion."[69] This conclusion fitted well his earlier argument, which no doubt accounts for the glaring error committed: equating *unidirectional outflux* of Na^+ from muscle cells (measurable by tracer studies) with *net outflux* (what Conway measured in rats losing excess Na^+), since net outflux is the differ- ence between unidirectional outflux and unidirectional influx.

Conway related once again his failure to confirm Steinbach's results. He now cited Dean's 1941 paper and acknowledged that certain cells might be engaged in rapid absorption and secretion of Na^+; in fact Conway had recently described active secretion by kidney cells.[70] But he maintained that, in general, the "standard" permeability bore the primary reponsibility. And although pumps might be required to make cells effectively impermeable to Na^+, all other ions were then distributed passively in response to that impermeability.

Steinbach

Steinbach's primary rebuttal to Conway's responses did not appear until 1951. In the interval, Steinbach found occasion to express disapproval of Conway's theory and to repeat his claim that the basic principle had been refuted: "there appears to be no justification for promoting a theory advocating complete sodium imperme- ability . . . in direct contradiction to experimental observations."[71] When Steinbach finally addressed the specific criticisms of his 1940 paper in 1951, he supplied information Conway and Krogh had requested: the composition of the media, the experimental temperature, and the final weights of the muscles.[72] Steinbach acknowledged that Conway had been unable to replicate his results, but stated that another investigator had.[73] And Steinbach then described similar and extended studies. The final weights of the muscles did not increase (agreeing with the new information about the old experiments), and the loss of Na^+ and gain of K^+ during the second soaking were nearly identical to those previously reported. Moreover, Steinbach presented a plausible explanation for Conway's failure to replicate his results, citing the necessity for accumulating sufficient Na^+ during the first soaking. Steinbach concluded that "cells do work to maintain their chemical identity," but he agreed with Conway that K^+ moved passively in response to Na^+ extrusion.[74]

Others

W. S. Wilde reported that the product of $[K^+]_{out} \times [Cl^-]_{out}$ did not always equal the product of $[K^+]_{in} \times [Cl^-]_{in}$, as required by Conway's formulation.[75] D. C. Darrow, in summarizing cell permeability in 1944, noted that Heppel's results had been replicated, and considered Na^+ and K^+ in muscle to be in a "dynamic state."[76] And three years later Walther Wilbrandt noted Conway's criticisms based on energy requirements, but he also deplored the failure of these calculations to include effects of electrical forces.[77]

USSING'S MEASUREMENT OF ^{24}Na OUTFLUX

Hans Ussing (Fig. 3.6) had served as a marine biologist on an expedition to Greenland, and his doctoral dissertation at the University of Copenhagen in 1934 dealt with plankton. After graduation Ussing consulted with Krogh, a former professor, about experimental techniques for classifying organisms; his career as a marine biologist ended when Krogh offered him a job. Krogh believed that isotopic tracers, which were just becoming available, would provide important new approaches to problems previously intractable. So Ussing's second career began: studying the incorporation *in vivo* of deuterium-labeled amino acids into proteins. But when Krogh retired in 1945, he persuaded Ussing to oversee his studies on ion and water movements. Thus began Ussing's third scientific career—which continues still.

Krogh had recognized that isotopes afforded unique opportunities for determining unidirectional fluxes. So Ussing's entry was an acknowledged variant of Heppel's determination of ^{24}Na uptake, but in this case he measured ^{24}Na outflux *in vitro*. Ussing published this study as a brief note and then, with Hilde Levi, as a full paper.[78] These studies can be described together.

Ussing first soaked frog sartorius muscles in media containing ^{24}Na until the tracer had equilibrated with all the Na^+ present. He then bathed the muscles in successive changes of unlabeled media and measured the amount of radioactivity

Fig. 3.6 Hans H. Ussing. (Photograph courtesy of H. H. Ussing.)

that flowed out of the muscle during these bathings. Plotting the logarithm of the radioactivity remaining in the muscle against time gave a curved pattern, and Ussing analyzed these data as two straight lines representing two simultaneous first-order outfluxes of ^{24}Na from the muscle (Fig. 3.7), which were attributable to extracellular (fast) and intracellular (slow) fluxes. The slopes of these lines reflected the rates of outflux; for the slower rate the halftime averaged 34 minutes. Ussing also measured the Na⁺ content of the muscle and calculated the intracellular concentration. With these data he then calculated the outflux from muscle cells as 15 mmol Na⁺ per kilogram of muscle per hour.

Two factors were required for calculating the energy needed to drive such outfluxes. The opposing concentration gradient Ussing calculated from intracellular and extracellular Na⁺ concentrations. He estimated from published reports that the opposing electrical potential was 60 mV, (Conway assumed that Na⁺Cl⁻ moved as a neutral complex, whereas Ussing assumed that Na⁺ moved as the cation, and was thus subject to transmembrane potentials). The calculated value for the energy required was then 59 cal/kg·hour. Ussing quoted Conway's value for the energy production of frog muscle, 175 cal/kg·hour. Thus, a third of the total energy produced was apparently consumed in pumping Na⁺ from cells.

This large fraction bothered Ussing[79] so much that he devised a mechanism that

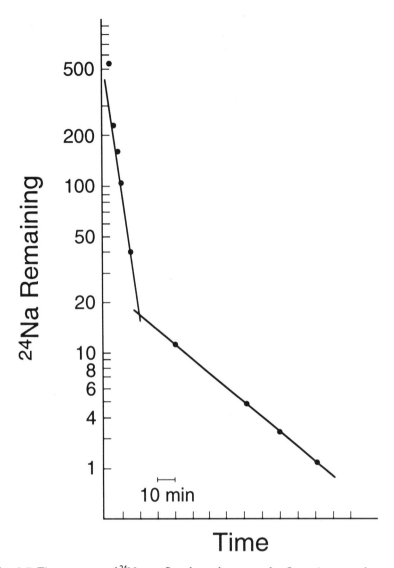

Fig. 3.7 Time course of ^{24}Na outflux from frog muscle. Sartorius muscles were first soaked in ^{24}Na-containing media to allow equilibration of isotope with all the muscle Na$^+$. Muscles were then placed in a bath whose contents were removed periodically to allow measurement of the ^{24}Na that had diffused from the muscle during that interval. The ^{24}Na remaining at the end of each successive bathing period was calculated by subtracting from the total radioactivity initially present, the cumulative ^{24}Na lost to that time. The logarithm of ^{24}Na remaining is plotted against the cumulative bathing time. The data are fitted by two straight lines (a first-order process gives a straight line in such plots). [Redrawn from Levi and Ussing (1948), Fig. 3. Reprinted by permission of the Scandinavian Physiological Society.]

explained why such fluxes need not represent energy-consuming processes. This mechanism he called "exchange diffusion": the tightly-coupled exchange of one intra-cellular Na^+ for one extracellular Na^+. Then ^{24}Na inside cells could exchange with unlabeled Na^+ outside—without a net change in Na^+ content and thus no energy consumption—while still resulting in measurable outfluxes of ^{24}Na. Ussing's mecha-nism foils all arguments against tracer fluxes based on energy requirements (although it is irrelevant to criticisms of energy requirements for *net* fluxes). The proposal is thermodynamically sound, and a membrane mechanism is easily envisaged.[80] The danger for interpretations of tracer fluxes cannot be ignored. But Levi and Ussing's data did not require such an exchange mechanism and they provided no direct evi-dence that such a mechanism actually existed.

A more extensive report by E. J. Harris at the University of London, which was evidently written before seeing Ussing's papers, described ^{24}Na and ^{42}K uptake and release by frog muscles *in vivo* and *in vitro*.[81] As in Ussing's study, a major concern was the analysis of isotope outflux. For ^{24}Na this was again interpreted as fluxes from extracellular and intracellular compartments; for the latter the halftime was 40–60 minutes. As an integral part of the analyses, Harris also estimated diffusion coefficients in the extracellular volume; tortuous pathways made calculations uncer-tain as well as extremely complex. For these and other reasons, the use of muscles to study ion transport fell from favor. But the permeability of muscle cells to Na^+ was no longer an issue.

Harris[82] subsequently calculated the energy required for Na^+ outflux per hour per gram of muscle: 0.045 cal, a value he compared with A. V. Hill's measurement of heat production by resting muscle, 0.24 cal. Harris showed that ^{24}Na outflux was no slower when K^+ replaced Na^+ in the bathing medium (then no Na^+_{out} would be available for exchange with $^{24}Na_{in}$), but when isosmotic sucrose replaced Na^+ and K^+ the outflux slowed: this implied that ^{24}Na left in exchange for an entering cation, thus maintaining electrical neutrality.

Steinbach also responded to Ussing's calculations.[83] He again loaded frog sar-torius muscles with Na^+ by soaking them in low-K^+ media and then measured net Na^+ outflux during a second incubation. From the halftime for Na^+ outflux and the calculated intracellular Na^+ concentration Steinbach reported a rate of 40 mmol/kg·hour, which he concluded was "very similar" to Ussing's value.[84] This net ouflux, unlike an exchange diffusion, required energy consumption, and this Steinbach estimated to be "around half the normal resting metabolism." But he too failed to measure the actual energy consumption.

COMMENTARY

Places and People.[85]

Obvious differences—in academic tradition, in availability of research funds, and in access to radioactive isotopes—distinguished Rochester from Dublin from Copenhagen. The consequences of personal leadership, too, were visible in Fenn's Rochester and Conway's Dublin and Krogh's Copenhagen. This account is neither an institutional history nor a biography. But it may be instructive, nonetheless, to note the course of several actors in this chapter. After the war Fenn concentrated more on administration than experimentation. His physiology department was judged among the best in the United States, and honors flowed: he was elected to the National Academy of Sciences and to the presidencies of the American Physiological Society and the Society for Experimental Biology and Medicine. Fenn died in 1971. Conway continued more actively in research, and we shall meet him again in later chapters; honors came to him as well, including the Boyle Medal of the Royal Dublin Society and memberships in the Royal Society of London and the Pontifical Academy of Sciences. Conway died in 1968. After medical school and a year of internship at Rochester, Heppel spent the war years at the National Institutes of Health, where he was involved with tedious studies on industrial toxicology. Subsequently, he had a distinguished career at the National Institutes of Health and Cornell University in several branches of biochemistry, but he did not continue with ion transport. He too was elected to the National Academy of Sciences. Steinbach, on the other hand, remained with membrane transport, moving from Columbia to Washington University, the University of Minnesota, and on to the University of Chicago, where he was chairman of the biology department. He too reaped rewards, including directorship of the Woods Hole Marine Biological Laboratory. Steinbach died in 1981. Dean left membrane transport and made even more moves: from Minnesota to California, where he took part in a synthetic rubber project for the War Production Board, followed by academic appointments in Hawaii and Oregon and service for the government and in industry. Krogh died in 1949, not long after his Croonian lecture, while Ussing developed in the latter 1940s an experimental model for studying transport of particular fruitfulness, which he and others continue to exploit (see Chapter 6).

Societies and Cultures

Although this account is not a social history either, certain contemporary events that marked the decade demand at least acknowledgment. The economic stringencies of the Great Depression compelled Heppel to become a physician (temporarily), Dean to travel on in quest of academic appointments, and Ussing to forego marine biology. The upheavals of World War II had quite different consequences in combatant United States (where military priorities and service took precedence), in neutral Ireland (where Conway continued to publish in British journals), and in occupied Denmark[86] (which Krogh fled). The postwar era brought transitions at different rates, affecting technologies, supplies, faculties, students, and communications—all crucial factors in directing the sequence of events and in selecting who first reached the common goals. It is a tenet of this account, however, that these factors did not dictate the ultimate conclusions about how ionic asymmetries are maintained.

The Currency of Scientific Argument

Scientific arguments are what this account addresses. And the units of these discourses, as sketched here, are the data of experimental tables and figures and the conclusions expressed as explanatory models. These are packaged in papers published and lectures presented. Rhetoric may embellish and awkwardness obfuscate, just as error may confound and carelessness corrupt. But the presented data are exposed to scrutiny. When rats fed low-K^+ diets gained Na^+ and lost K^+, no one long denied that Na^+ could enter muscle cells. Conway was a clever as well as a determined man, yet he could devise no alternative: for the rival camps shared communal standards of observation and measurement, as well as faith in the laws of chemistry. The underdetermination of theory by evidence is sharply constrained by such bounds: Hevesy could list in a few lines *all* the ways that the ionic asymmetry could be established,[87] and in 50 years no one has added to that roster.

With information incompletely available, however, different interpretations may endure and alternatives be pursued. Thus, Steinbach looked for Na^+ extrusion by muscle while Conway calculated energy demands that seemed to minimize such extrusion. Accommodation is surely slowed by ignorance (what was the *actual* energy consumed when frog sartorius muscles achieved net outfluxes of Na^+?).

The currency of experimental results—like the coinage of obscure realms— may be of uncertain value. But it is similarly available for assay when offered. Fenn's claim of impermeability to Cl^- foundered in part on Conway's more exacting examination; Conway's failure to replicate Steinbach's experiment was over- whelmed eventually by affirmations of its validity.[88] The public record of tables and figures is grist for the social construction of science, a process newly recognized by some commentators but realized and practiced by Fenn, Conway, Steinbach, Krogh, and Ussing. Of course, the crucial issue is how and from what the social construction is effected.

Scientific Summary

By the close of the 1940s, the permeability of muscle cells to Na^+ was inescapably established. The concept was accepted, but it was accepted by many in the form of Conway's revisionism: some Na^+ may leak into cells requiring pumps to extrude it, but the consequence is a *functional* impermeability to Na^+. The "double Donnan" distribution then accounts for the rest: K^+ and other ions are partitioned between cells and their environments in equilibrium distributions. Only tiny amounts of energy are needed to balance a wasteful leakiness to Na^+.

NOTES TO CHAPTER 3

1. Fenn (1936).
2. Although the ionic radius of K^+ is larger than that of Na^+ and Ca^{2+}, the hydrated radius is smaller. The assumption was that hydrated ions are the species that cross.

3. Fenn (1936), p. 462.

4. Fenn et al. (1934); Mond and Netter (1932).

5. Mond and Netter (1932). This possibility was explicitly examined by Manery et al. (1938) who concluded that proteins of the extracellular connective tissue could not represent the binding site for "excess" Na^+.

6. Fenn and Cobb (1936).

7. Ibid., p. 354.

8. Fenn (1936), p. 464.

9. Ibid., p. 451.

10. Ibid., p. 451. Earlier, Fenn et al. (1934) concluded that "it seems necessary to assume complete impermeability to either anions or cations in order to explain the retention of potassium" (p. 271); the possibility of secretion was not even raised.

11. Dean (1987) considered in retrospect that "pumps were unpopular because no one could imagine how they might work," and "no possible compound of sodium . . . was sufficiently different from the corresponding potassium compound" to provide the necessary selectivity; moreover, "physical chemists were confident that they could explain the results of all reactions by equilibrium theory" (pp. 451, 453, 454).

12. Conway and Boyle (1939); Boyle and Conway (1941). To eliminate uncertainty arising from alphabetic listings of authors in the *Journal of Physiology*, the paper specified that the "theoretical plan of the research is due to the senior author" (p. 62); in case someone might be uncertain about who was senior author, Conway later stated that "the theoretical part . . . was due to the present author" (1960, p. 26).

13. Conway et al. (1939); Manery and Hastings (1939); Heppel (1939); Boyle et al. (1941).

14. Boyle and Conway (1941), p. 44.

15. Ibid., p. 8.

16. Heppel (1939).

17. Heppel and Schmidt (1938). Heppel planned such experiments in Rochester, but, locked into the progression of medical school, he ran out of time.

18. Heppel (1940).

19. Hahn et al. (1939).

20. Ibid., p. 1556.

21. Hevesy and Rebbe (1940).

22. Griffiths and Maegraith (1939), p. 160. They did not mention muscle.

23. Greenberg et al. (1940).

24. Manery and Bale (1941), pp. 229–230.

25. Ibid., p. 224.

26. Steinbach (1940a).

27. Ibid., p. 700.

28. Steinbach (1940b), p. 244.

29. Ibid., p. 244.

30. Ibid., p. 251.

31. Ibid., p. 251.

32. Steinbach (1941), p. 63. He cited Burton's 1939 paper discussing steady states in biological systems, who in turn acknowledged A. V. Hill's seminal contributions.

33. Steinbach (1947), p. 870.

34. Dean, in Steinbach (1940b), p. 252.

35. Dean (1941), p. 333.

36. Dean (1941). This was in a relatively obscure journal, which, coupled with the disruptions caused by the war, contributed to Dean's paper being overlooked by many researchers.

37. Ibid., pp. 344–345.
38. Ibid., pp. 346–347.
39. Ibid., p. 347.
40. Krogh (1946).
41. Ibid., p. 140.
42. Ibid., p. 180.
43. Ibid., p. 183.
44. Ibid., p. 188.
45. Fenn (1940).
46. Ibid., p. 378.
47. Ibid., p. 378.
48. Ibid., p. 403.
49. Fenn (1941), p. 7.
50. Ibid., p. 7.
51. Fenn et al. (1944), p. 75.
52. Conway (1945).
53. This is Mond and Netter's proposal, coupled with permeability to Na$^+$.
54. Conway (1945), p. 70.
55. Ibid., p. 70. What Manery and Bale said was: "The most reasonable explanation at the moment is that sodium has penetrated muscle fibre membranes hitherto believed impermeable to it" (1941, p. 224).
56. Conway (1946).
57. Ibid., p. 717.
58. Ibid., p. 715.
59. Ibid., p. 715.
60. Ibid., p. 715.
61. Access to radioisotopes was limited not only by few institutions having cyclotrons to produce those isotopes but also by the short half lives of ^{42}K (12 hours) and ^{24}Na (15 hours), which made transport to distant locales impossible.
62. Particularly significant in this estimation are the number of assumptions: that weight gains parallel the K$^+$ gains; that soaking muscles in media containing only 2.5 mM K$^+$ has no deleterious effects (in contrast to Conway's earlier studies); that changes in bathing solutions do not affect permeability; that K$^+$ enters with Cl$^-$ as a neutral species (so that membrane potentials would not affect influx); etc.
63. Another assumption in his calculations is that all energy production is by oxidative means. Conway showed that adding cyanide (an inhibitor of oxidative metabolism) did not affect Na$^+$ movements (assessed as weight changes), but he did not cite the paper by Dean (1940) showing that both aerobic and anaerobic metabolism in frog muscle must be inhibited before changes in ion content occurred.
64. Conway (1946), p. 716.
65. Ibid., p. 716.
66. Ibid., p. 715.
67. Conway (1947a,b).
68. Conway and Hingerty (1948).
69. Conway (1947a), p. 598.
70. Conway et al. (1946).
71. Steinbach and Spiegelman (1943), p. 195; Steinbach (1944).
72. Steinbach (1951).
73. Keynes (1949b). Support for Steinbach's results is confined to a statement that Steinbach's "experiments . . . have recently been confirmed" (p. 169), with the citation of "Keynes, . . . unpublished observations."

74. Steinbach (1951), p. 287.

75. Wilde (1945).

76. Darrow (1944), p. 117. He cited four published replications.

77. Wilbrandt (1947).

78. Ussing (1947); Levi and Ussing (1948).

79. Peculiarly, Ussing (1980, p. 7, and 1994, p. 92) described these calculations as indicating that far more energy seemed to be required for pumping than frog muscles could generate.

80. Ussing (1947). He imagined a "particle" in the membrane that may face the interior of the cell and there bind a Na^+ in the cytoplasm. Then, by random thermal movement in the membrane, this particle may reorient to face the exterior of the cell where the transported Na^+ is released. A Na^+ from the extracellular medium may then bind for the reverse trip. If all Na^+ in the cytoplasm is labeled initially, that labeling will eventually be equally distributed by such a mechanism, even though there is no *net* movement of Na^+.

81. Harris and Burn (1949).

82. Harris (1950).

83. Steinbach (1952).

84. Ibid., p. 454.

85. Biographical information is from interviews, reminiscences (Ussing, 1980, 1994; Dean, 1987), and biographies (Maizels, 1969; Schmidt-Nielsen, 1995).

86. Krogh reported that experiments in Copenhagen using radioactive isotopes, begun in the winter of 1943, "had to be suspended during December and January owing to the seizure of the Institute for Theoretical Physics" (Holm-Jensen et al., 1944, p. 3).

87. Hahn and Hevesy (1941), pp. 60–61. They listed four possibilities: (1) K^+ cannot cross the membrane; (2) K^+ is constrained from leaving the cell by electrical forces; (3) K^+ is constrained from leaving by binding to constituents within the cell; and (4) the constant loss from the cell is balanced by an equivalent secretion into the cell.

88. Maizels (1969, p. 76) commented that "Steinbach's results are in fact easily reproducible [and] ultimately Conway accepted the reality of active and passive sodium fluxes."

chapter 4

ACCOUNTING FOR ASYMMETRIC DISTRIBUTIONS OF Na$^+$ AND K$^+$ IN RED BLOOD CELLS

C HAPTER 3 traced a reformulation of muscle cell properties from *absolute* impermeability to Na$^+$ to *functional* impermeability: active Na$^+$ extrusion balances small inward leaks. Investigators who differed over these properties at the beginning of the decade were in general agreement by the late 1940s. Although all experimental results and interpretations were not smoothly uniform, the qualitative principle was widely accepted. Contributing prominently to the creation of that consensus were parallel studies on red blood cells.

ESTABLISHING PERMEABILITY OF RED BLOOD CELLS

Simple explanations for how red blood cells maintain asymmetric distributions of Na$^+$ and K$^+$ between cells and surroundings invoked impermeability, but in this case, impermeability to *both* cations.[1] That view was derived from measuring volume changes when red blood cells were immersed in salt solutions of varying composition: water moved rapidly to achieve osmotic equilibrium, but without the secondary changes in volume expected if slowly penetrating ions eventually equilibrated.[2] And direct measurements of ion contents after bathing red blood cells in such media showed only small changes, attributable to damage during the experiments. Thus emerged the common interpretation enunciated by Hugh Davson and Danielli as "under physiological conditions the erythrocyte membrane is impermeable to cations";[3] by Peters as "human red blood cells under the conditions of these experiments are quite impermeable to . . . sodium [and] potassium";[4] and by Merkel Jacobs as "it is fairly certain that [experiments] do not indicate a permeability of the erythrocytes to cations under normal body conditions."[5] The conspicuous qualifications were to admit that red blood cells could be permeable—but only when injured. A prominent motivation for this insistence on impermeability is apparent in Jacobs's comment: "on those who wish to postulate a . . . long-continued permeability of the erythrocyte to . . . cations rests the burden of explaining how the destruction of such a delicate cell by swelling under the influence of the . . . Donnan equilibrium is avoided."[6]

An obvious reason why established theories are rarely falsified by single experiments is that such tests are generally brought to bear early, before a theory becomes established (control experiments are intended to rule out *all* explanations but the remaining correct *one*). The opportunity to disprove an accepted theory arrives late only when new critical tests are conceived through novel insights or techniques. The availability of radioactive tracers at the end of the 1930s permitted just such a critical test of the dominant theory. In 1939 Waldo and Elma Cohn, at the University of California where E. O. Lawrence's cyclotron provided radioactive isotopes, reported that experiment as an ancillary project tacked onto Waldo Cohn's doctoral research.[7] They injected ^{24}NaCl into dogs intravenously and, after waiting various time periods, removed samples of blood; they then waited additional time periods before separating blood cells from plasma by centrifugation; and finally they measured the radioactivity in these fractions (Table 4.1).

Over time, radioactivity in the red blood cells clearly approached that in plasma. Moreover, this uptake of ^{24}Na occurred similarly *in vivo* (between injection and sample removal) and *in vitro* (between sample removal and centrifugation). Comparisons between the two conditions—made economically in one experiment— effectively ruled out arguments that ^{24}Na uptake occurred artifactually during processing *in vitro*.

This impressive demonstration was clouded, however, by other reports that year. Hevesy and associates, using isotopes produced at the Niels Bohr Institute,

Table 4.1 Uptake of ^{24}Na by Dog Red Blood Cells in vivo and in vitro*

Time in vivo	Time in vitro	^{24}Na in Cells/^{24}Na in Plasma
5	143	0.32
7	3	0.06
27	4	0.10
397	3	0.44
1530	5	0.65

*^{24}NaCl was injected intravenously into dogs, and after specified times (in minutes, first column) blood was drawn and allowed to stand for specified times (in minutes, second column) before the red blood cells were separated by centrifugation. The radioactivity in the cells and the plasma was then measured (the ratio is in the third column). [From Cohn and Cohn, 1939, Table I. Reprinted by permission of the Society for Experimental Biology and Medicine.]

described troubling experiments with both ^{42}K and ^{24}Na.[8] After injecting ^{42}KCl into rabbits, they found that 24 hours later the red blood cells had only 66% of the radioactivity in an equal weight of plasma. Since the K⁺ concentration is roughly 20-fold higher in cells than in plasma, only 3% of the K⁺ in cells had been replaced by ^{42}K within 24 hours; this indicated an extremely low permeability. In corresponding experiments using ^{24}Na, they reported that red blood cells had only 6% of the radioactivity in an equal weight of plasma. In this case, however, they inexplicably failed to factor in the relative concentrations of cation in cells vs. plasma. This is about 6% as well, so equilibration had indeed occurred. Moreover, Anna Eisenman at Yale University reported similar results the next year, using isotopes prepared by the cyclotron there.[9] Through studying human red blood cells *in vitro*, she found that after 4 hours, ^{42}K had equilibrated to "a very limited extent if at all," but with ^{24}Na "equilibrium is established between sodium within cells and that in serum."[10]

In Rochester Dean and associates reexamined the permeability of red blood cells to ^{42}K, explaining away Hevesy's results as being due to contamination of his ^{42}K by ^{24}Na.[11] That argument was plausible, for, as they pointed out, it is quite difficult to separate K⁺ free from all traces of Na⁺, and small amounts of ^{24}Na would skew the results toward an apparent failure to equilibrate. Nevertheless, they could not demonstrate that such contamination had actually occurred in the Danish experiments. Instead, they purified their ^{42}K scrupulously, and with human red blood cells *in vitro* they then found an uptake corresponding to an exchange of 15% of the cell K⁺ in 10 hours. They noted that this was consistent with Eisenman's report of 4% in 4 hours. Acknowledging that added radioactivity might alter cell permeabilities, Dean stated the actual concentration of radioactivity used (7 μCi per liter) and cited a report that comparable amounts did not increase permeability of another cell type.[12]

Two further papers from Rochester in 1941 elaborated on these experiments and presented rate constants and permeability coefficients for ^{42}K fluxes in red blood cells from various species.[13] And in 1942 Hevesy revised his earlier conclusion: new experiments revealed equilibration of ^{42}K with dog red blood cells within 2 hours.[14] He also now calculated the ^{24}Na permeability in terms of Na⁺ concentrations in red blood cells and plasma, concluding that since "both sodium and potassium interchange between plasma and corpuscles, the great difference in the concentration ratio . . . cannot be due to a non-permeability of the . . . membrane."[15]

Was this conclusion inevitable from these papers, sufficiently to overthrow the previous consensus? Beyond the standard constraints of chemistry, the necessary assumptions were few: that measurements were accurate (always a haunting concern); that isotopes were pure (perhaps not the case in some experiments); that radioactivity did not affect the processes examined (specifically addressed in some but not all papers); and that radioactive isotopes behaved just as nonradioactive isotopes did (an accepted chemical principle[16]). Thus, common opinion had changed within a couple of years, as experimental results refuted established views. (Some uncertainty about the breadth of that generalization remained, nevertheless, for the most clear-cut demonstrations of permeability to Na⁺ were in dog red blood cells, and these differ from those of humans and most other species in having intracellular Na^+ high and K^+ low.)

DEMONSTRATING NET K^+ INFLUX AND Na^+ OUTFLUX

In Chapter 3 a mechanism was cited that allowed *tracer* fluxes to occur—without energetic cost—in the absence of *net* fluxes: Ussing's exchange diffusion. But Ussing did not propose exchange diffusion until nearly a decade later, whereas experiments on red blood cells showing *net* cation transport driven by metabolic energy were published almost simultaneously with the tracer studies described above.

John Harris

In the latter 1930s blood banks were being organized to facilitate transfusions. Among the significant problems to be resolved was how best to preserve the red blood cells. Included with such concerns was K^+ loss from red blood cells, since high concentrations of circulating K^+ are toxic: if K^+ leaked into the plasma during storage then administering such blood could cause severe reactions. (A further aspect was introduced in 1938 when G. Jeanneney reported that not only did red blood cells lose K^+ during storage but they also gained Na^+.[17])

John Harris (Fig. 4.1) published his first paper on this topic in 1939 in collaboration with two physicians, a study explicitly aimed at determining "the rate and mode of disintegration of human blood in various preservatives suitable for transfusion."[18] Harris received his Ph.D. in biochemistry from the State University of Iowa in 1940, and that year, as a research associate in the Department of Obstetrics and Gynecology, he published a second paper with the same physicians on factors affecting K^+ loss during cold storage (2°–5°).[19] Although the variables examined—Na^+, glucose, etc.—had no effect, the authors noted that K^+ loss was not simply due to hemolysis, to destruction of red blood cells; moreover, K^+ loss was "rapid during the first five days but bec[ame] gradually less."[20]

In an abstract for a meeting at the Woods Hole Marine Biological Laboratory, Harris extended these considerations without the physician collaborators.[21] Harris now interpreted K^+ outflux from stored red blood cells as a first-order reaction. This meant that the *rate* of outflux was the product of the K^+ concentration in the cell multiplied by a *rate constant*. Since outflux was a function of a rate constant from the very beginning, the mechanism by which K^+ escaped was operating from the very beginning: it was not due to some deterioration that developed during storage.[22] This was a salient interpretation, for it implied that K^+ loss was a general property rather than merely some consequence of degradation. Even more significant, Harris reported that K^+ loss during cold storage was halted by warming the blood to 37° (furthermore, a gain in K^+ "was usually seen");[23] on the other hand, adding fluoride, an inhibitor of glycolysis, caused K^+ loss even at 37°. Thus, "the

Fig. 4.1 John E. Harris (Photograph courtesy of John E. Harris).

potassium content of the erythrocytes of human . . . blood [appears to be] maintained in some way by the cellular metabolism."[24]

Those experiments were presented in more detail in 1941 in a full paper.[25] Now, however, the emphasis was on the general problem of maintaining ionic homeostasis. Harris cited Cohn and Cohn, Eisenman, and Dean on red blood cell permeability to radioactive cations, concluding that their results "suggest that there is a dynamic equilibrium between the cations in the erythrocyte and those in the surrounding medium."[26] He repeated his assertion that the first-order time course of K$^+$ outflux indicated that the factor responsible was immediately operative. And, to account for the net outflux in the cold, he proposed that at lower temperatures the metabolic rate—and thus the availability of energy to drive transport—dropped sharply. This hypothesis implied that warming the cells should halt net outfluxes. Harris found not just a halt in outflux but an actual reversal, to a net influx of K$^+$ against its concentration gradient. Moreover, adding glucose increased the net influx (Fig. 4.2A), with 0.8 mol of glucose metabolized per mol of net K$^+$ influx. Fluoride caused a marked loss of K$^+$ (Fig. 4.2C), but because that loss was delayed Harris concluded that glycolysis per se was not responsible for driving transport. He suggested that diphosphoglycerate might be involved: it was reported to be formed during glycolysis in red blood cells and to persist after glycolysis ceased. He did not mention ATP.

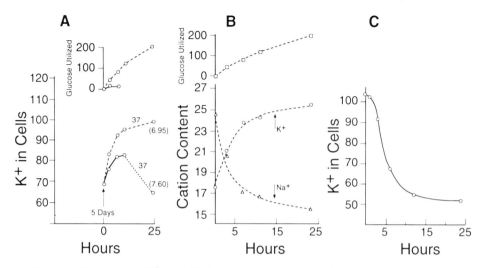

Fig. 4.2 Changes in K^+ and Na^+ content of cold-stored red blood cells after warming. In *A*, the K^+ content of cold-stored human cells is plotted against hours incubated at 37°. In the *inset* is plotted the amount of glucose consumed. Solid lines refer to experiments in which glucose was added to the media; numbers in parentheses indicate the final pH of the media. In *B*, changes in both K^+ and Na^+ contents are plotted against hours at 37°; the corresponding consumption of glucose is plotted above. In *C*, the K^+ content is plotted against hours at 37° during incubations with 0.04 M sodium fluoride. [Redrawn from Harris (1941), Figs. 3, 4, and 7. Reprinted by permission of the American Society for Biochemistry and Molecular Biology.]

Harris also showed for the first time that Na^+ was extruded against *its* concentration gradient when cold stored cells were warmed in media containing glucose (Fig. 4.2*B*). He concluded that "the view that the erythrocyte membrane is impermeable to cations . . . must be abandoned," for the cation "concentration [is] maintained by . . . metabolic function of the cells," making the membrane "functionally impermeable" to cations.[27]

(Harris did not consider electrical potentials across red blood cell membranes. If cell interiors were negative with respect to the exterior, then the equilibrium concentration of K^+ inside cells would be higher than that outside. Of course, Na^+ should then be distributed similarly. And if membrane potentials were primarily responsible for creating asymmetric cation distributions, the problem becomes one of accounting for those potentials.)

Danowski

Appearing a few months before Harris's paper was Thadeus Danowski's report in the same journal, which reached much the same conclusion.[28] This was taken from his thesis for the M.D. degree awarded by Yale University in 1940, which apparently antedated Harris's studies. Danowski, however, found only a fraction of what Harris described: a loss of K^+ to the plasma when red blood cells were stored at 7° (in his Fig. 1 this loss was drawn—from two points—with a constant rate rather than as a first-order process); a loss of K^+ from plasma when cells were then warmed to 37° ("evidently entering the cells"[29]), followed by a reversal of flow after 5 to 6 hours that coincided with exhaustion of the glucose present; a delay in the reversal when glucose was added; and inhibition of the flow from plasma to cells when fluoride was added. Danowski "inferred that the potassium in cells and in serum is not in diffusion equilibrium," and concluded that the study "suggests that definite quantities of . . . potassium . . . are moved across the boundary in connection with specific reactions."[30]

Danowski did his research in the Department of Medicine at Yale, where Peters's interests ranged from ionic imbalances associated with metabolic diseases to fundamental physiological processes. From that department had come Eisenman's report of ^{42}K fluxes (cited above) yet Danowski failed to measure ^{42}K fluxes even though he should have had access to radioisotopes (Harris in Iowa did not).

Six years later Danowski and Peters argued that "if energy is introduced into a system . . . the production of concentration gradients across membranes presents no insurmountable difficulties."[31] And they offered a critical insight:

> It seems to chemists an alien thought that a distinction should be drawn between simple inorganic elements, such as sodium or potassium, by a membrane or a protein. Nevertheless, such distinctions are regularly drawn by enzyme systems which have now been removed from the field of metaphysics.[32]

CRITICISMS AND CONTINUATIONS

Prior to the publication of Harris's and Danowski's studies, Wilbrandt described large losses of K^+ from red blood cells when they were incubated at 37° with fluoride or iodoacetate (another inhibitor of glycolysis).[33] In the published discussion following that paper, Danielli stated that in similar experiments he and Davson could "find no trace of salt leakage due to any enzyme poison, including the ones used by Dr. Wilbrandt."[34] In the full account of their experiments, Davson and Danielli described essentially no loss of K^+ from rabbit red blood cells after 1 hour with the inhibitors at 25°, but they acknowledged that Wilbrandt had found effects at a higher temperature.[35] They quoted Wilbrandt as being "satisfied that [the loss of K^+] is not due to [inhibition of] glycolysis *per se.*"[36] And they concluded that "it is quite clear that the high concentration of K^+ in the cells . . . is not maintained by a metabolic process"; "consequently the high concentration gradient must be considered as a static phenomenon."[37]

By contrast, Dean suggested that Davson's results might be due to cells excreting Na^+.[38] Davson compared their opposing viewpoints:

> Your picture of the erythrocyte as a cell with a slow but significant permeability to potassium, which maintains its high concentration difference . . . by a metabolic process . . . contrast[s] with my picture of a cell which is, in its normal environment, impermeable to potassium. . . . If your idea is correct, many of the results which I have just described . . . are susceptible of an explanation on the basis of a change in the metabolism of the cell rather than a change in membrane structure.[39]

And when Davson repeated his conclusion based on the poisons causing no inhibition, Dean responded that the "metabolic rate need by only about 3 p.c. of that observed to provide energy enough to replace potassium."[40]

The following year Davson admitted that fluoride and iodoacetate caused K^+ losses at 37°, just as Wilbrandt had reported earlier.[41] In the meantime, Wilbrandt concluded that such losses were due to toxic metabolites accumulating when inhibitors were added.[42] Davson agreed, arguing that inhibition "is definitely not simply [due to] immobilization of a potassium secreting mechanism."[43] One justification for that conclusion was the delay between adding inhibitors and the onset of K^+ outflux: an implicit assumption here was a tight coupling between glycolysis and cation transport, an assumption ignoring the possibility that glycolysis could build an energy store that in turn drove transport.[44]

Writing nearly 50 years later, Davson conceded that "we remained blind to the possibility of an active accumulation of K^+ by the normal erythrocyte until Harris, working on stored blood, showed that the slow leakage of K^+ that occurs during storage could be reversed by warming the cells to body temperature."[45] Indeed, the blindness cleared slowly and unevenly. In his 1945 book, Rudolf Höber interpreted Wilbrandt's results in terms of "normal glycolysis [maintaining] a structural situation, which retains cation."[46] Nevertheless, Höber admitted that "glycolysis may maintain ionic equilibria by releasing energy for the backtransport of K,"[47] citing Harris's 1941 paper. On the other hand, Krogh in his Croonian lecture cited Cohn and Cohn's paper on ^{24}Na penetration into dog red blood cells, Dean's paper on ^{42}K penetration, Harris's 1940 abstract (but not the 1941 paper), and Danowski's 1941 paper.[48] He also described his own unpublished experiments using rabbit red blood cells that replicated Danowski's results. Krogh summarized:

> Cations inside the erythrocyte exert normal osmotic pressure and are free to move in the whole of the corpuscle water [yet] the concentrations of the single ions differ more from the corresponding concentrations outside than can be explained by a Donnan effect. . . . [If] the erythrocyte membrane is definitely, even if slightly, permeable to these ions, the concentrations can be . . . *maintained* only by some active transport: we are face to face with a steady state and not with any equilibrium.[49]

Conway generally ignored studies on red blood cells. In his 1941 paper he cited without comment the experiments by Cohn and Cohn; in reference to other experiments using radioisotopes, however, he raised the possibility that radioactivity could alter membrane properties and concluded that this question was not "altogether settled."[50] He did turn to red blood cells in his 1947 lectures, citing Harris's 1940 abstract (not the 1941 paper) and Danowski's 1941 paper, but he denigrated the red blood cell and thus phenomena demonstrable with it: "the erythrocyte is essentially a moribund cell . . . that accounts for only 2 to 3 per cent of the bodyweight . . . affecting therefore very little the general ionic relations."[51] (Danowski, Peters, and associates took umbrage with this, complaining that "Conway has dismissed the red blood cell from consideration as an effete member of the biological system."[52]

PROGRESSING FURTHER

In London

Significant new experiments came from Montague Maizels. He had received his medical degree from Guy's Hospital and had been a clinical pathologist at University College Hospital, London, since 1931, where he studied red blood cell permeabilities. During World War II he turned to issues necessary for effective blood banking, and he published a paper in 1943 on the relationships between ion content and metabolism. Maizels acknowledged that cells must be "continuously tending to lose K and gain Na," but he considered that such changes could be remedied not only by the cells themselves but perhaps also by external means.[53] He cited Fritz Lipmann's magisterial review on ATP, but his discussion did not propose that ATP was an energy source for red blood cell work. Instead, he focused on diphosphoglycerate—as had Harris—although he could find no relationship between cellular concentrations of K^+ and those of organic phosphate compounds.

After the war Maizels completed the first real extension of Harris's studies. In an introductory paragraph, he remarked that attempted "repetition of Harris's experiments gave variable results."[54] Accordingly, he set about examining the pertinent variables, identifying the pH of the incubation media as the crucial factor. With that difficulty resolved, he endorsed Harris's results, agreeing that "expulsion of Na is an active process."[55]

Maizels addressed the question of whether K^+ was also transported actively in his succeeding paper.[56] In these experiments red blood cells were stored in the cold in media containing high concentrations of Li^+ but low concentrations of Na^+ and K^+: the cells then gained Li^+, although the intracellular concentrations of Na^+ and K^+ changed little. Next, these cells were incubated at 37° in media containing lower concentrations of Li^+ but higher concentrations of Na^+ and K^+: Na^+ now was extruded against its concentration gradient even though there was no net gain in K^+. Cells had thus achieved a net loss of Na^+ without a net gain of K^+. Maizels recognized that the absence of a net change could result from an active accumulation cancelled by a passive loss and noted that using radioactive tracers could be informative. But such experiments he did not do. He did report, however, that when red blood cells were incubated in media containing quite low concentrations of K^+, the rate of Na^+ outflux declined. Still, these studies furthered the concept of an active "sodium pump," a term that is still in vogue.

That interpretation was explicitly proclaimed in the introduction to Maizels's paper published a year later: whereas the "output of Na is active [the] uptake of K [is] secondary and passive."[57] Despite that bald assertion, Maizels returned to this issue, collaborating with E. J. Harris.[58] They first equilibrated human red blood cells with ^{24}Na, next washed the cells free of extracellular ^{24}Na, and then measured ^{24}Na outflux. From this they calculated permeability coefficients; when extracellular K^+ was reduced from 4.6 mM to 0–1 mM, the permeability coefficient for Na^+ fell by half. Moreover, the cellular Na^+ content actually increased when the media contained less than 1 mM K^+. This effect of extracellular K^+, they concluded, "may possibly be explained in terms of the energy required to transport a positively charged ion from the interior to the exterior of the cell."[59] They meant (I believe) that extracellular K^+ was necessary merely to maintain electroneutrality, replacing intracellular Na^+.

In a paper published the following year, Harris and Maizels prodded the issue further.[60] They recognized that if K^+ were merely distributed passively, following the membrane potential established by a pump transporting Na^+ outward, then the ratio of $[K^+]_{out}/[K^+]_{in}$ should be equivalent to that for any other ion distributed passively. A series of notable investigators had argued that Cl^- was distributed in this way, and Harris and Maizels, in reexamining the case, concurred. They found the ratio of $[Cl^-]_{in}/[Cl^-]_{out}$ to be 0.72 (at an extracellular pH of 7.42)—in the absence or presence of the metabolic fuel glucose. By contrast, the ratio of $[K^+]_{out}/[K^+]_{in}$ in freshly prepared cells at that same pH was only 0.03. They considered it unlikely that such a large disparity was solely due to differences in the thermodynamic activities of K^+_{in} and Cl^-_{in}. Consequently, they invoked "the possibility that inward transport of K is 'geared' to outward Na transport by the use of a common carrier."[61]

That "gearing," imagined as a shared mechanism alternately carrying Na^+ outward and K^+ inward, is quite a different solution from Eric Ponder's proposal in 1950 of a separate K^+ pump.[62] While a separate pump could establish distribution ratios for K^+ that were different from that for Cl^-, it would in itself lack the link to Na^+ transport that Harris and Maizels subsequently established.

In Boston and Oak Ridge

Meanwhile, experiments using radioactive isotopes were resumed in the United States following the interruption of World War II. From Harvard University J. W. Raker and associates reported in 1950 that all of the K^+ in human red blood cells was exchangeable with ^{42}K and at a single rate (implying no compartmentalization of K^+ within cells).[63] C. W. Sheppard and associates at Oak Ridge confirmed these observations in an accompanying paper, and the following year described ^{24}Na fluxes as well.[64]

In 1951 a lengthy and detailed account of ^{42}K and ^{24}Na fluxes in human red blood cells also appeared, written by a physical chemist at Harvard who had newly turned to transport studies, A. K. Solomon.[65] One issue Solomon addressed was the energy cost for transport. This he calculated to be 8.8 cal per hour per liter of red blood cells, based on the measured fluxes of ^{42}K and ^{24}Na and a value for the membrane potential,

-10 mV, evaluated from the ratio of $[Cl^-]_{in}/[Cl^-]_{out}$. By using the measured consumption of glucose and published values for the energy yield by glycolysis from glucose, he calculated that 110 cal was available per hour per liter of red blood cells. Consequently, active cation transport needed merely 8% of the energy available.

The other conclusion that Solomon reached was that "Na and K transport mechanisms are demonstrably independent."[66] This assertion he based chiefly on differences in calculated energies of activation for Na^+ influx and K^+ influx. Solomon carefully stated several assumptions—such as adequate mixing between radioactive isotopes and unlabeled species, and the absence of isotope effects and radiation damage—but several considerations he did not address explicitly: (1) that energies of activation for Na^+ *influx* and for K^+ *influx* ought to be identical if the same transport system were being used; (2) that Na^+ crossed the membrane by a single system, which could thus be compared to the single system that K^+ used; and (3) that calculated energies of activation reflected such a single system for each ion, rather than a composite from heterogeneous systems. These assumptions were by no means certain. In any case, his computed energies of activation for Na^+ *outflux* and K^+ *influx* were quite close.

Consensus

By the end of the decade orthodoxy had been transformed. No longer were red blood cells portrayed as impermeable to Na^+ and K^+, with the possibility of active transport being irrelevant. In his 1951 textbook Davson now acclaimed John Harris for making a "striking discovery" and acknowledged that the "return of K^+ to the cells obviously occurred against a concentration gradient [with] the energy of the metabolic processes . . . being utilized in the transfer."[67] And Charles Lovatt Evans's textbook published the following year attributed asymmetric distributions of Na^+ and K^+ "not to impermeability . . . but to active excretion of Na^+," concluding that losses of K^+ from red blood cells during storage occurred because "the 'sodium pump' ceases to act."[68]

COMMENTARY

Places and People[69]

To the Rochester of Fenn, Heppel, and Dean and the Copenhagen of Krogh, Ussing, and Hevesy several new sites were added. London was a prominent locale. Davson and Danielli received their doctoral degrees at the University of London before going their separate ways (Danielli to Princeton, London, and Cambridge before returning to London; Davson to Philadelphia, London, and Halifax before returning to London). E. J. Harris worked actively at University College, London after the war, while Maizels was at the University College Hospital throughout this era. Yale and Harvard Universities were also important places for research, together with the Universities of California and Iowa. As for the prominent actors, Hevesy received the Nobel prize in 1943 for his work with isotopic tracers, which stretched back to decades earlier than the episodes noted here (and included more successful studies). During the war years, Waldo Cohn went to Harvard, Chicago, and then Oak Ridge, where he remained; his research interests shifted to ion exchange chromatography and nucleic acid chemistry. His wife, Elma, died tragically young, not long after their joint paper. John Harris moved from Iowa to a postdoctoral position with Jacobs at the University of Pennsylvania just as the United States entered the war, so his research efforts were curtailed; after serving as an aviation physiologist in the Air Force, he entered medical school in Oregon and became an ophthalmologist. By 1958 he was professor and chairman of ophthalmology at the University of Minnesota. Danowski spent the war years at Yale, and in 1947 he moved to Pittsburgh as professor of internal medicine, where he too had a distinguished career in academic medicine. Maizels and E. J. Harris will return in Chapter 8.

Experimental Success and Reproducibility

In a widely praised book, Harry Collins asserted, from a handfull of selected cases, that "most experiments are delicate, and fail to work most of the time."[70] His interest lay in the insurmountable hindrances to resolving experimental conflicts, but his argument contrasts instructively with a scientist's view (leaving aside Collins's failure to document his quantitation). For example, Ponder wrote in 1950: "All . . . who have studied K accumulation by the red cell have had difficulty in reproducing their results, and all have concluded that quantitative work is impossible unless special care is taken to control a large number of variables."[71] Clearly, this is not a cry of despair but a call for action: he noted that confounding chemical impurities existed in some sources of lithium salts, and he prescribed remedies. In fact, the scientific culture inculcates a commonsense approach to achieving consensus in the face of conflict and contradiction. This was illustrated here in the search for uncontrolled variables (Miazels's attention to pH), for error (Dean's inference that ^{24}Na contaminated the ^{42}K used by the Copenhagen group), for significant differences in protocol (Wilbrandt's studies at 37° vs. Davson and Danielli's at 25°). These variables, errors, and differences were, once identified, easily appreciated, for they lay within a communal system that acknowledged the general concepts of pH, chemical separation, and temperature effects.

Change

Although the disruptions of World War II skewed the time course recorded here, it is nevertheless apparent that a new consensus formed by the mid 1940s. Studies with ^{24}Na and ^{42}K left little doubt that red blood cells were permeable to cations *in vivo* as well as *in vitro*. Studies with red blood cells warmed after cold storage left little doubt that such cells could extrude Na^+ and regain K^+. Experiments in Berkeley, Rochester, Copenhagen, Iowa City, New Haven, and London from the late 1930s through the 1940s provided results that changed the accepted opinion held previously by the authorities in the field. Those working in Fenn's department in Rochester and Peters's department in New Haven overthrew their leaders' opinions. Fenn's group took the lead in proclaiming what Fenn had earlier been loath to admit, pushing forward the new concepts. Danowski carried Peters's group into the commitment to permeability and transport. What else could Fenn and Peters do in the face of the experimental evidence coming from their own and other groups? Human frailties are well represented in the scientific community, but the repertoire of public behaviors is highly constrained. Lapses are newsworthy. The obvious conclusion— so obvious as to be unfashionable—is that experimental results are the salient determinant of scientific change.

Scientific Summary

With the introduction of radioactive isotopes, new experiments contradicted the reigning belief that red blood cells were normally impermeable to Na^+ and K^+. Such permeability implied that some means for transporting those cations was required in order that a steady state be maintained *in vivo*. That implication was met by demonstrations that red blood cells could regain their lost ionic asymmetries through processes requiring metabolic energy. The means for this active transport was called the sodium pump.

NOTES TO CHAPTER 4

1. If plasma membranes are impermeable to cations, how do cations get into red blood cells initially? A common explanation relied on the life history of these cells, which begin as stem cells with nuclei and the full complement of intracellular organelles. These stem cells differentiate, ultimately becoming non-nucleated structures without internal organelles, "mature" red blood cells. Thus, the particular cation content was established in earlier incarnations, becoming trapped when cell impermeability was attained. The special mystery of red blood cells was thus replaced by the general mystery of how all cells established their ionic composition.

Belief that red blood cells were impermeable to Na^+ and K^+ was also fostered by these cells indeed having far lower permeabilities than other cells (e.g., over two orders of magnitude lower than for muscle cells).

2. See Jacobs, 1962. For changes in cell volume with changes in concentration of permeant and impermeant solutes see Chapter 2.

3. Davson and Danielli (1936), p. 316.

4. Peters (1938), p. 302.

5. Jacobs (1935), p. 12.

6. Jacobs (1939), p. 9.

7. Cohn and Cohn (1939).

8. Hahn et al. (1939).

9. Eisenman et al. (1940).

10. Ibid., p. 170.

11. Dean et al. (1940).

12. Dean et al. cite a report that the effect of radiation would, if anything, *decrease* permeability. A more convincing demonstration is that measured ion contents did not change during experiments with radioisotopes. (Manery and Bale, 1941).

13. Noonan et al. (1941); Mullins et al. (1941).

14. Hahn and Hevesy (1942).

15. Ibid., p. 219.

16. Although some reports then described discrimination by cells between nonradio-active isotopes, subsequent studies failed to find appreciable distinctions among isotopes of the heavier elements such as Na^+ and K^+.

17. Jeanneney et al. (1938).

18. DeGowin et al. (1939), p. 126.

19. DeGowin et al. (1940).

20. Ibid., p. 857.

21. J. Harris (1940).

22. In a first-order reaction the rate of outflux would be proportional to the concentration of K^+ in the cells; thus, as the K^+ leaked out and the concentration fell, the rate of outflux would decline.

23. J. Harris (1940), p. 373. Although this is the first report of the restoration of ion content *in vitro*, there were by that time two reports of the restoration *in vivo*: Henriques and Ørskov (1936), and Maizels and Paterson (1940).

24. Harris (1940), p. 373.

25. J. E. Harris (1941).

26. Ibid., p. 579. The term "dynamic equilibrium" can be ambiguous. Currently it refers to a nonstatic equilibrium, but from the papers under discussion it could imply a steady state.

27. Ibid., pp. 591, 593.

28. Danowski (1941).

29. Ibid., p. 703.

30. Ibid., p. 704.

31. Hald et al. (1947), p. 348.

32. Ibid. For a discussion of enzymes discriminating between K^+ and Na^+ see Chapter 7.

33. Wilbrandt (1937).

34. Danielli, in Wilbrandt (1937), p. 964.

35. Davson and Danielli (1938).

36. Ibid., p. 993.

37. Ibid., p. 997.

38. Dean, in Davson (1940), p. 266.

39. Davson (1940), p. 266.

40. Dean, in Davson (1940), p. 267.

41. Davson (1941).

42. Wilbrandt (1940).

43. Davson (1941), p. 184.

44. J. E. Harris (1941) pointed out that the observed lag in inhibition could account for failures to observe effects of fluoride.

45. Davson (1989), p. 37.

46. Höber (1945), p. 251. On the preceding page Höber suggested that fluoride increased permeability by "loosening the linkage of colloidal aggregates."

47. Ibid., p. 251.

48. Krogh (1946).

49. Ibid., p. 173.

50. Boyle and Conway (1941), p. 49.

51. Conway (1947b), p. 665.

52. Hald et al. (1947), p. 348.

53. Maizels (1943), p. 175. Earlier, he suggested that "the spleen or some other organ reacts on the cell so as continually to reshuffle cation in favor of potassium" (Maizels, 1937, p. 964).

54. Maizels (1949), p. 247.

55. Ibid., p. 251.

56. Flynn and Maizels (1950).

57. Maizels (1951), p. 59.

58. Harris and Maizels (1951).

59. Ibid., p. 522.

60. Harris and Maizels (1952).

61. Ibid., p. 52.

62. Ponder (1950).

63. Raker et al. (1950).

64. Sheppard and Martin (1950); Sheppard et al. (1951).

65. Solomon (1952).

66. Ibid., p. 105.

67. Davson (1951), p. 248.

68. Lovatt Evans (1952), p. 516.

69. Biographical information is from interviews, telephone conversations with Waldo Cohn and John Harris, and from biographical notes (Stein, 1986).

70. Collins (1985), p. 41.

71. Ponder (1950), p. 745.

chapter 5

ION GRADIENTS AND MOVEMENTS IN EXCITABLE TISSUES

W HEN certain tissues, such as nerve and muscle, are perturbed in specific ways they change their properties in a characteristic manner. They are "excitable tissues," and when stimulated they respond. A nerve trunk stimulated electrically may cause the innervated muscle to shorten: something has traveled from the point of stimulation to and then along the muscle, inducing contraction.

BEGINNINGS[1]

By the mid-nineteenth century, Carlo Matteucci had shown that electric currents flowed between the muscle surface and its cut end; the associated voltage between the undamaged surface and a damaged region of nerve or muscle came to be called the "injury potential." Emil du Bois-Reymond had demonstrated not only that both nerve and muscle produced electricity but also that, after appropriate stimulation, they developed "action potentials," transient depolarizations from the injury potential measured at rest; Julius Bernstein found that the rapid decline and rise in voltage lasted only milliseconds. And Hermann von Helmholtz had determined that nerve impulses traveled at roughly 30 meters/second, which is much too slow for electricity passing along a conductor (such as wire), yet far too fast for substances flowing through tubes as thin as nerve.

A plausible explanation for this was provided by Bernstein's membrane theory, which was based on these observations and the new physical chemistry of Svante Arrhenius and Walther Nernst. In 1902 he proposed that plasma membranes served as insulating layers between cytoplasm and extracellular environment, and that when these membranes were at rest they were selectively permeable to K^+ but impermeable to other cations and to all anions. The injury potential then represented a preexisting membrane potential, a diffusion potential for K^+ that can be described by the Nernst equation:

$$E_m = -(RT/F) \ln ([K^+]_{in}/[K^+]_{out})$$

where E_m is the membrane potential, R the gas constant, T the absolute temperature, and F Faraday's constant. The injury potential represented this resting membrane potential because the damaged region completed a circuit from the external electrode on the undamaged surface, through the membrane, through the cytoplasm, and out through the leaky damaged region to the second external electrode. That loop was, however, shortcircuited by current passing through the bathing medium from one electrode to the other, so that exact measurements of resting potentials were unobtainable.

To account for action potentials, Bernstein proposed transient and reversible breakdowns of the membrane in which the selectivity to ions was transiently and reversibly abolished. With the consequent general permeability to all ions, the K^+ diffusion potential disappeared and the membrane was depolarized toward 0 volts. This action potential was then propagated along the axon by "local circuits" that stimulated further regions of the membrane ahead. Among the consequences of Bernstein's formulation were resting membrane potentials depending precisely on extracellular K^+ concentrations and action potentials equaling at most the resting potential in magnitude.

Appearing in the same volume that had Bernstein's 1902 study were two even longer papers by Overton. These included analyses of ionic requirements for muscle activity, from which he concluded that Na^+ was required in the bathing medium for excitability. On this basis Overton suggested that the transient flow of current during action potentials might involve an exchange of extracellular Na^+ for intracellular K^+. But that mechanism entailed a significant cost: "the greatest difficulty in entertaining such a hypothesis lies in explaining how . . . sodium ions should be extruded from the interior of the fibers [for] if some sodium ions enter and some potassium ions leave . . . then the differences between internal and external cation concentration would gradually be leveled out unless there is some mechanism at work which opposes this equilibrium."[2]

Three further characteristics of propagated action potentials were established early this century. (*1*) For excitation, a minimal magnitude of current had to pass in a given time: there was a "threshold" for excitation. (*2*) Stimulation either resulted in an action potential of a distinct magnitude or it did not; action potential magnitude was not a function of stimulus magnitude except as it was above or below threshold—the "all-or-none" characteristic. (*3*) For a brief time after the action potential, a second stimulus could not excite another action potential. This was called the "refractory period."

GIANT AXONS

In the 1930s came a discovery that changed experimental approaches decisively. J. Z. Young found very large nerve fibers while studying regeneration in octopods, and on behalf of comparative studies he pursued that discovery, describing "enormous" axons in decapods—squid (*Loligo*) and cuttlefish (*Sepia*).[3] In squid these giant axons were a half to a millimeter in diameter, which was 50 times the largest human axon. Young described the giant axons at a symposium in 1936. Neurophysiologists soon recognized the opportunity, although it came with a significant constraint: availability. Squid cannot be kept in captivity and the axons must be removed and used within hours; moreover, squid can be obtained at Woods Hole or Plymouth (the two prominent settings for this narrative) during only part of the year.

The first to exploit giant axons in their studies were H. J. Curtis and K. S. Cole, from Columbia University but working also at the Marine Biological Laboratory at Woods Hole. (Curtis had received his Ph.D. in physics from Yale University in 1932; Cole, already a stellar figure in electrophysiology, had received his Ph.D. from Cornell University in 1926, also in physics.) In 1938 they reported the transverse impedance, measured by passing an alternating current between electrodes on either side of axons, and calculated membrane capacitances of 1 microfarad/cm^2 at 1 kilohertz.[4] This value agreed with earlier measurements on red blood cell membranes. In 1939, again measuring transverse impedance, they calculated membrane resistance as action potentials passed:[5] it declined to approximately 25 ohm cm^2. Cole and Alan Hodgkin[6] then calculated resistance at rest: 1000 ohm cm^2. The drop to 25 ohm cm^2 thus represented a 400-fold decrease during the action potential. This change was, of course, the sort expected from Bernstein's membrane theory.

THE HODGKIN-HUXLEY MODEL

We come now to the collaboration of two heroic figures of twentieth century biology, Alan Hodgkin and Andrew Huxley (Fig. 5.1). Both were educated at Trinity College, Cambridge, and both began their research careers at the University's Physiological Laboratory. Hodgkin received his bachelor's degree from Cambridge in 1935, and proceeded directly to research on nerve conduction. In 1937 he travelled to the Rockefeller Institute for a fellowship year, and in New York he met Curtis and Cole, who invited him to Woods Hole for the summer of 1938. There, in addition to collaborating with Cole on measuring membrane resistance, Hodgkin and Curtis tried a few long-shot (and unsuccessful) attempts at pushing fine electrodes through the cut ends of giant axons. The following summer, back in England at the Laboratory of the Marine Biological Association in Plymouth, Hodgkin and Huxley succeeded spectacularly in that effort.

Huxley had recently received his bachelor's degree; he had entered Cambridge in 1935 to pursue studies in physical science or engineering, but he was enticed away by the lure of Cambridge physiology. (He went on to medical studies, then a

Fig. 5.1 Alan L. Hodgkin (left) and Andrew F. Huxley (right). (Photographs courtesy of the Cambridgeshire County Council and Andrew Huxley, respectively.)

common complement to academic physiology, but this course was interrupted by World War II. Like Hodgkin's doctoral degrees, Huxley's are all honorary.) In 1939 Huxley began by attempting to measure the viscosity of axon cytoplasm, but this turned out to be so great that injected mercury droplets did not budge. With that project a failure, Huxley suggested inserting an electrode into the axons for electrical measurements—and promptly succeeded. He pushed a thin glass tube (a "cannula") filled with seawater longitudinally down the axon, through a nick in its surface; a wire electrode ran from the cannula to measure the voltage between axon interior and an electrode in the bathing medium, thus providing a straightforward measure of membrane potential.

With this approach, Hodgkin and Huxley recorded resting membrane potentials of −50 mV (with the bath potential defined as 0 mV) and action potential magnitudes of 90 mV (these were measured from the resting potential and increased to +40 mV beyond the 0 mV of the bath: see Fig. 5.2).[7] Three weeks after their first measurement, war was declared with Germany. Hodgkin and Huxley reported their finding briefly in *Nature*[8] (immediately following Conway and Boyle's 1939 study), but the full account did not appear until 1945. One feature was readily apparent, even though they did not comment on it: the action potential was larger than the resting potential, with an "overshoot" of 40 mV. But by Bernstein's theory the action potential should at most equal the resting potential.

Curtis and Cole addressed this issue, with unfortunate consequences, in a paper

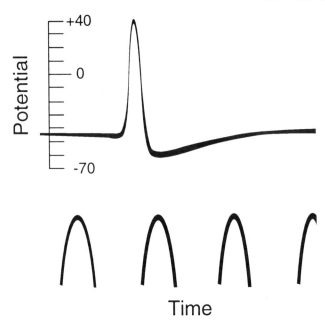

Fig. 5.2 Action potential of squid giant axon. Action potential and resting potential were recorded between a canula inside the giant axon and an electrode in the seawater bath. The vertical scale is in millivolts (bath at 0 mV) and the truncated sine waves at the bottom (500 Hz) indicate time. [From Hodgkin and Huxley (1939), Fig. 2. Reprinted by permission from *Nature*, ©1939 Macmillan Magazines Ltd.]

published in 1942.[9] They noted that "electrical characteristics of the microelectrodes and the amplifiers were seen to be unusually important for an accurate record of the action potential"; they admitted that "there is undoubtedly some error in the absolute values"; and they then reported that "the membrane potential is momentarily reversed in sign, so that the outside may be as much as 110 millivolts" positive.[10] The action potential illustrated had a total magnitude of 170 mV.

The magnitude of that overshoot, plus Curtis and Cole's report that replacing bathing media with isosmotic glucose had no appreciable effect on action potentials, undoubtedly[11] steered Hodgkin and Huxley away from one apparent interpretation of the overshoot: a transient increase in permeability to Na^+. If the membrane became selectively permeable to Na^+, then the action potential could, by the Nernst equation, rise to no more than +63 mV.[12] Moreover, substituting glucose for extracellular Na^+ should, with such a mechanism, abolish the depolarization.

In their 1945 paper, Hodgkin and Huxley made no mention of permeability to Na^+.[13] Instead, they proposed four alternative processes to account for the overshoot: (1) selective permeability to any anions present at higher concentrations in cytoplasm than in bathing media; (2) changes in the orientation of charged elements of the membrane; (3) participation of an electrical inductance in the membrane; or

(4) participation of an electrical capacitance that acquired a reversed voltage and that persisted when excitation occurred.

The possibility of Na^+ being involved was also ignored in two subsequent commentaries. In 1946 Höber proposed that "organic nonpolar-polar, hydrophobic-hydrophilic anions . . . present in the nerve membrane [were] liberated as the excitation wave travels along the fiber."[14] A year later Harry Grundfest summarized the views of Hodgkin and Huxley and of Höber, adding the possibility of potentials due to Donnan equilibria and of potentials generated by biochemical reactions.[15]

After the war years Hodgkin resumed his research in Cambridge. By immersing crab nerves in oil so that only a thin film of adhering seawater remained (thereby trapping in a small volume any ions leaving the axon), he and Huxley showed that when action potentials passed along the axon a K^+-like substance (in terms of its electrical effects on nerve) was released into the extracellular film.[16] Moreover, this "K^+" was then absorbed back into the nerve during subsequent rest periods. They calculated that with each action potential 1.7×10^{-12} mol of "K^+" flowed out through each square centimeter of axon surface.

In the introduction to this study, Hodgkin and Huxley noted that an "exchange of potassium and sodium is a more likely source of energy for nervous transmission than the leakage of potassium chloride,"[17] but they did not address the cause of the overshoot in that context. In the same volume, Bernard Katz described the slowing and eventual cessation of nerve conduction when sucrose was substituted for seawater,[18] but he made no claim that extracellular Na^+ was required.

In the summer of 1947 Hodgkin returned to Plymouth (the laboratory had been damaged in the war). Working again with squid giant axons, he—in collaboration first with Katz and later with Huxley—linked the action potential overshoot to a Na^+ influx: "dilution of sea water with isotonic [glucose] produces a . . . reversible decrease in the height of the action potential [while] the height . . . is increased by a hypertonic solution containing additional sodium chloride"; moreover, "to a first approximation the rate of rise [of the action potential] is directly proportional to the external concentration of sodium."[19] The action potential thus represented a depolarizing flow of Na^+ inward. Repolarization, on the other hand, represented a subsequent flow of K^+ outward.[20] This formulation required that resting membranes be more permeable to K^+ than to Na^+, with this selectivity reversed upon excitation.

Selectivity for Na^+ over K^+, however, clashed with contemporary models for ion pores. These pores were thought to select on the basis of hydrated ion radii, so pores that would pass larger hydrated Na^+ should—it would seem—also pass smaller hydrated K^+. To circumvent that problem, Hodgkin and Katz suggested that "sodium does not cross the membrane in ionic form, but enters into combination with a lipoid soluble carrier in the membrane which is only free to move when the membrane is depolarized"; on the other hand, "potassium ions cannot cross this membrane by this route because their affinity for the carrier is assumed to be small."[21] They published a diagrammatic model but did not specify how "lipoid soluble carriers" could discriminate between Na^+ and K^+.[22]

Hodgkin and Katz also recast David Goldman's[23] "constant field assumption" concerning the electrical potential within the membrane to give a nonequilibrium formulation for the transmembrane potential:[24]

$$E_m = -(RT/F)\ln\{(P_K[K^+]_{in} + P_{Na}[Na^+]_{in} + P_{Cl}[Cl^-]_{out})/(P_K[K^+]_{out} + P_{Na}[Na^+]_{out} + P_{Cl}[Cl^-]_{in})\}$$

where the P's represent relative permeabilities to the subscripted ions. The equation closely resembles the Nernst equation, which instead represents an equilibrium diffusion potential. It became widely used and cited under an eponymic composite, the "Goldman-Hodgkin-Katz equation."

In 1949 Hodgkin, Huxley, and Katz also reported their first experiments using a "voltage clamp."[25] This technique, which had been pioneered by Cole and his associate George Marmont,[26] involved holding the transmembrane voltage at a selected value, clamped by means of a feedback circuit that passed a current between an internal and an external electrode in response to the measured transmembrane potential from another pair of electrodes: the membrane potential could be set to a selected voltage and held there. The measurement of interest was the current flow occurring when the membrane potential was abruptly set at a new value. Moreover, this current flow could be identified with the flow of ions across the membrane, reflecting the time course of that permeability, and, in experiments varying ion concentration and content, the particular ionic species distinguished (Fig. 5.3).

In the summer of 1949, with an improved voltage clamp apparatus, Hodgkin, Huxley, and Katz completed experiments that resulted in five spectacular papers that underlay all thinking on action potentials in the succeeding decades.[27] The conclusions were: (1) resting membranes were freely permeable to K^+ but relatively poorly permeable to Na^+; (2) with depolarization to threshold values a nearly instantaneous outward flow of current followed from discharging the membrane capacitance; (3) this outward current flow was quickly followed by an inward flow, which was attributed to Na^+ influx down its electrochemical gradient; (4) Na^+ influx then ceased, the current "inactivated"; and (5) after a brief delay, an outward current flowed, which was attributed to K^+ outflux down its electrochemical gradient, although this current did not inactivate spontaneously. The action potential was thus characterized by a sequence of permeability increases: first to Na^+, causing depolarization and the overshoot; then to K^+, causing repolarization to the resting potential. Moreover, the increased Na^+ permeability turned itself off (was inactivated) in a time-dependent fashion, whereas the increased K^+ permeability did not. And Na^+ and K^+ crossed the membrane independently through separate and specific pathways.

Even more impressive was the presentation of these conclusions in the fifth and culminating paper. Here the electrical properties of the axon membrane were represented by an "equivalent circuit" (Fig. 5.4), a qualitative model that could be described mathematically. What then followed were solutions to the corresponding equation, with the parameters evaluated empirically to fit their data. This curve-fitting procedure, although not guaranteeing a unique solution, was highly constrained by the wealth of these data; for example, Hodgkin and Huxley were forced

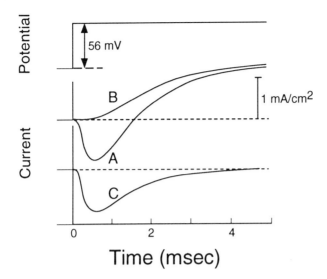

Fig. 5.3 Na^+ and K^+ currents. Voltage-clamped axons were abruptly depolarized by 56 mV (*upper trace*). Trace *A* shows the resultant current flow to hold the potential. Trace *B* shows the current flow when extracellular Na^+ was absent (replaced with choline$^+$), and thus represents K^+ outflux. Trace *C* shows the difference between *A* and *B*, representing Na^+ influx. I_{Na} thus increases rapidly and then inactivates. I_K increases more slowly and persists. [From Hodgkin and Huxley (1952a), Fig. 5, as redrawn in Hodgkin (1958), Fig. 9. Reprinted by permission of the Royal Society.]

to abandon their lipid carrier model for Na^+ entry since its kinetics were incompatible with measured current flows.

This accomplishment can be illustrated by their evaluation of the Na^+ current flowing during a stationary[28] action potential. The total current (I_{tot}) was assigned to a capacitative current ($C\,(dV/dt)$) plus variable ionic currents carried by Na^+, K^+, and all other species (a minor component termed the "leak" current):

$$I_{tot} = C\,(dV/dt) + I_{Na} + I_K + I_L$$

For the Na^+ current, I_{Na}:

$$I_{Na} = g_{Na}\,(V - V_{Na})$$

where g_{Na} is the Na^+ conductance of the membrane and $(V - V_{Na})$, the difference between membrane potential and equilibrium potential for Na^+ (the voltage that would maintain the given Na^+ gradient).

To account for the rapid rise and then inactivation of the Na^+ current, they expressed the Na^+ conductance as a function of an activating parameter m and an inactivating parameter h, where g'_{Na}, the maximal Na^+ conductance, is a constant:

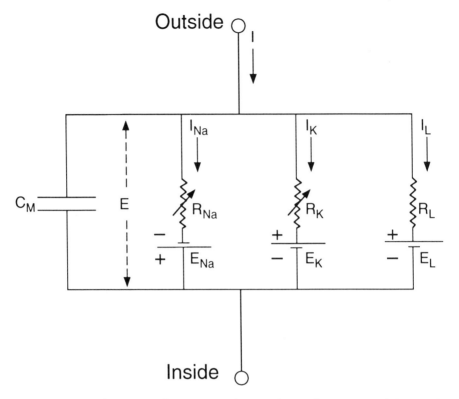

Fig. 5.4 Equivalent circuit for axon membrane. Electrical properties of the squid axon membrane are represented in terms of membrane capacitance (C_M), currents for Na^+, K^+, and "leak" (I_{Na}, I_K, and I_L), and the corresponding batteries (E) and resistances (R); resistances for Na^+ and K^+ are variable. [From Hodgkin and Huxley (1952d), Fig. 1. Reprinted by permission of the Physiological Society.]

$$g_{Na} = m^3 h\, g'_{Na}$$

(the third power of m reflects that rapid rise, and can be interpreted as [at least] three membrane elements participating in the conductance increase, as in the opening of a Na^+ pore).

The time dependences of m and h were specified:

$$dm/dt = \alpha_m(1 - m) - \beta_m m$$

$$dh/dt = \alpha_h(1 - h) - \beta_h h$$

where the αs and βs are voltage-dependent rate constants.

The delayed increase of the K^+ current was represented analogously.

The equation for I_{tot} with all these parameters evaluated was solved by Huxley

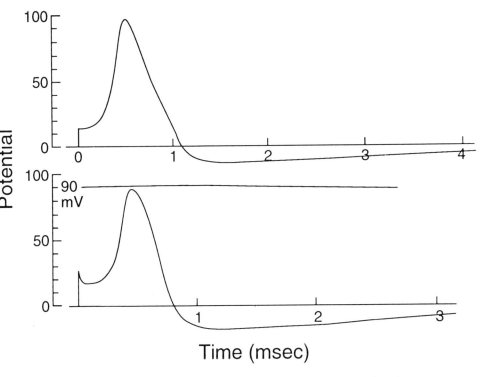

Fig. 5.5 Comparison of calculated and recorded action potentials. The upper curve is the calculated solution for a propagated action potential; the lower curve is a recorded potential trace. [From Hodgkin and Huxley (1952d), Fig. 14. Reprinted by permission of the Physiological Society.]

using a hand-cranked Brunsviga calculator to do the numerical integration—a truly herculean task involving more than a million values. The solutions also included that for propagated action potentials (Fig. 5.5). (Solving with a calculator was necessitated by the Cambridge electronic computer then being modified and hence unavailable. It is peculiarly gratifying that the first attempt to evaluate the Hodgkin-Huxley equations using an electronic computer was unsuccessful, attributed to programming errors.[29] Later attempts vindicated Huxley's calculations.)

Cautiously, Hodgkin and Huxley noted that "fairly simple permeability changes in response to alterations in membrane potential . . . are sufficient explanations of the wide range of phenomena that have been fitted by solutions of the equations."[30] That range included:

(a) The form, amplitude and threshold of an action potential under zero membrane current at two temperatures. (b) The form, amplitude and velocity of a propagated action potential. (c) The form and amplitude of the impedance changes associated with an action potential. (d) The total inward movement of sodium ions and the total outward movement of potassium ions associated with an impulse. (e) The threshold and response during the refractory period. . . .[31]

CONFLICTS AND CRITICISMS

Shortly after the squid giant axon was introduced, three papers reported magnitudes for the action potential that were less than the resting potential, greater by about 40 mV, and greater by 100 mV.[32] The smallest value, measured with external electrodes, could be explained away in terms of shortcircuiting. The largest value, which should have refuted hypotheses based on Na^+ influxes, disappeared equally fast; in a 1950 review Grundfest stated that the "report seems to have been of one observation, not seen again (personal communication from Curtis, spring, 1948) [and] Marmont uniformly obtains potentials of the same magnitude as those observed by Hodgkin and his colleagues (personal communication, summer, 1949)."[33] Cole later attributed that value to "an overcorrection for the electrode and amplifier input capacities."[34]

Equally catastrophic for Hodgkin and Huxley's Na^+ theory would be the occurence of action potentials in the absence of extracellular Na^+. This too Curtis and Cole had reported initially, but the phenomenon subsequently disappeared from their accounts. By contrast, the occurence of action potentials in the absence of Na^+ in the bathing medium was proclaimed vigorously by an unrelenting opponent of the Bernstein and Hodgkin-Huxley models, Rafael Lorente de Nó at the Rockefeller Institute. In 1944 he reported that frog nerve trunks continued to conduct impulses for 5 hours in the absence of Na^+ in the bathing medium.[35] And the eventual failure Lorente de Nó attributed to loss of Na^+ from *inside* the axons.

Lorente de Nó's interpretation was one facet of his comprehensive view of nerve potentials arising from cellular redox reactions. It is not necessary to summarize here his extensive studies, which appeared as a two-volume treatise in 1947;[36] the pertinent issue is whether Lorent de Nó's criticisms were insurmountable. Could alternative explanations account for the claim that removing Na^+ caused no loss of activity for hours? One obvious argument relied on the structure of frog nerve, which is composed of bundles of axons encased in supportive cells and surrounded by connective tissue sheaths. Such a structure would be expected to trap extracellular Na^+ around the axons, hindering its loss to the Na^+-free media. Indeed, in 1930 T. P. Feng and Ralph Girard had described "a striking difference in the behavior of nerve depending upon whether its connective tissue sheath was intact or split."[37] Lorente de Nó retorted that "it is utterly impossible to believe that the connective tissue sheath of frog . . . nerve could act as a diffusion barrier that could delay for considerable periods of time the penetration of solutes into the nerve."[38] So in 1949 Feng responded with a further study, using nerves from toads and from several species of frog whose sheaths he had removed:[39] desheathed nerves conducted action potentials at least as well as "intact" nerves and survived as long, but with intact nerves various ions penetrated far more slowly.

In a 1950 review Grundfest commented with obvious exasperation:

> Lorente de Nó claims that penetration . . . through . . . connective tissue sheaths of the
> bullfrog sciatic nerve is rapid;. . . . [Results] with frog nerve contrary to his own he

dismisses 'with assurance' as 'referable to a remarkable but fortuitous coincidence'. . . .
[He] rejects out of hand any contrary evidence based upon invertebrate material . . .
because those nerves are either 'essentially different from frog nerve' . . . or [because]
even if such nerves are comparable with frog nerve, the 'interpretation of the experimental observations was incorrect.'[40]

DIRECT MEASUREMENTS OF ION FLUXES

Although transmembrane electrical currents were readily interpretable as ion fluxes
through the membrane, especially when these currents were measured as a function
of ionic content, other processes were imaginable.[41] Consequently, more direct demonstrations were sought. But measuring net K^+ losses during stimulation was not
ideal for this purpose: such losses must represent a deterioration from the steady
state *in vivo* and might reflect merely exhaustion of the nerve from such stimulation. Optimally, ion fluxes should be measured while the nerve is able to maintain
its steady state; although no net changes in chemical content would then occur,
changes in radioactive isotope fluxes should be measurable.

This task Richard Keynes (Fig. 5.6) assumed. He had entered Trinity College in
1938 to study physiology; after the interruption of the war years, he returned for
his doctoral research with Hodgkin, which was completed in 1949. His preliminary
reports in 1948 and 1949 described: ^{42}K outflux from isolated crab nerve (first loaded
by soaking in ^{42}K), which increased during stimulation by 2.1×10^{-12} mol K^+/
cm^2·impulse; and ^{24}Na influx into isolated cuttlefish giant axons, which increased
during stimulation by 5×10^{-12} mol Na^+/cm^2·impulse.[42] Four papers, two of which
were in collaboration with the chemist P. R. Lewis, then presented this research in
detail.

Two of these dealt with crab nerve: (1) revealing that at least 97% of intracellular K^+ exchanged readily with ^{42}K and demonstrating that resting crab nerve
was permeable to K^+; (2) revising the influx per stimulus to $2.7-3.6 \times 10^{-12}$ mol K^+/
cm^2·impulse; and (3) calculating that this influx could easily repolarize the membrane from the peak of the action potential (i.e., the flux was large enough to do
what the Hodgkin-Huxley model required of it).[43] Control experiments indicated
that the dose of radiation did not affect the fluxes.

The other two papers dealt with cuttlefish giant axons.[44] Keynes and Lewis
found Na^+ influx and K^+ outflux to be essentially equal in unstimulated axons.
Moreover, the increments in these two fluxes remained equivalent during stimulation: 3.8×10^{-12} mol Na^+ and 3.6×10^{-12} mol K^+/cm^2·impulse. Because they
measured both fluxes and internal ion concentrations, their results "provide for the
first time direct proof that there is a net entry of sodium during nervous activity
which roughly balances the loss of potassium."[45]

At that same time Mortimer Rothenberg, working at Woods Hole, examined
^{42}K and ^{24}Na influxes into squid giant axons.[46] Although Rothenberg found that only
a tiny fraction of the ^{42}K equilibrated with axonal K^+ (which Keynes attributed

Fig. 5.6 Richard D. Keynes. (Photograph courtesy of Richard Keynes.)

plausibly to artifact[47]), he demonstrated an increased ^{24}Na accumulation during stimulation, calculating an influx of 4.5×10^{-12} mol $Na^+/cm^2 \cdot$impulse.

MAINTAINING IONIC ASYMMETRIES

Requiring an absolute impermeability to Na^+ seems particularly unrealistic for nerve. Even Bernstein's model invoked a loss of selective permeability during action potentials that would allow Na^+ entry, and Conway proposed a metabolic process to account for the asymmetry in nerve.[48] On the other hand, Hodgkin and Huxley remarked in their paper on K^+ release and reabsorption by crab nerve that while reabsorption "may be thought of as an active process of a secretory type . . . it can also be satisfactorily explained in terms of the type of Donnan equilibrium proposed by . . . Conway."[49] The calculations, however, were pertinent only to their experimental conditions: nerve immersed in oil and surrounded by only a thin film of seawater. The extracellular volume was then of the same magnitude as the intracellular volume, so losses of intracellular K^+ raised extracellular K^+ concentrations appreciably. Furthermore, the value for intracellular K^+, calculated to fit a Donnan distribution, was only 110 mM, a value they acknowledged to be "surprisingly low"[50] (Hodgkin[51] subsequently cited a measured value of 380 mM). And later in this study, Hodgkin and Huxley stated that the "preceding argument is put forward only because it is the simplest qualitative explanation of the facts," and that further considerations "suggest . . . that a secretory process is at work."[52] The problem became more acute when Na^+ influxes during action potentials were established, and Hodgkin and Huxley then pointed out that "some additional process must take place in a nerve in the living animal to maintain the ionic gradients."[53]

By 1951 Keynes firmly endorsed the need for active transport.[54] Moreover, he concluded that ample energy was available by comparing reported values for oxygen consumption by axons, equivalent to 0.8 cal/g·hour, to the calculated energy required to extrude the Na^+ influx, 0.05 cal/g·hour. He also showed that ^{24}Na outflux occurred even in the absence of extracellular Na^+, so the observed isotope flux was not merely a manifestation of Ussing's exchange diffusion that achieves no net flux.

GENERALIZATIONS

Myelinated Nerve

Axons of many vertebrate nerve cells are myelinated, surrounded by a thick coating of lipoidal myelin that is interrupted periodically at the "nodes of Ranvier," often a few millimeters apart. With a mechanistic model for the unmyelinated squid giant axon in sight, Huxley and Robert Stämpfli approached the myelinated axon. From analyses of current flow through nodes of Ranvier of frog myelinated axons, they concluded that "the action potential at each node excites the next node by current flowing forward in the axis cylinder and back in the fluid outside the myelin" coating:[55] conduction was saltatory in myelinated nerves, jumping successively from node to node, as R. S. Lillie had proposed in 1925. Then in two papers published in 1951, Huxley and Stämpfli linked that process convincingly to the mechanism characterized for squid giant axons.[56] They reported resting membrane potentials of −71 mV and action potential magnitudes of 116 mV (an overshoot of 45 mV), and pointed out the remarkable similarity of these values to those from invertebrate axons and vertebrate muscle (discussed below). As with squid giant axons, action potential magnitudes were a function of extracellular Na^+ concentration, whereas resting potentials changed according to the Goldman-Hodgkin-Katz equation (with limited permeabilities to Na^+ and Cl^- and a far greater permeability to K^+).

Huxley and Stämpfli also compared their results to those of Lorente de Nó and of Feng, commenting that the connective tissue sheath surrounding the nerve trunk contained an enveloping barrier of cells. And they cited Overton's inability to demonstrate that Na$^+$-free solutions blocked conduction in frog nerves, which Overton had attributed to extracellular Na$^+$ being trapped within that sheath.

Skeletal Muscle

Direct measurements of transmembrane potentials in muscle cells were obviously not possible with the intracellular electrodes used for giant axons. These electrodes were 100 μm in diameter, whereas skeletal muscle cells are often less than 70 μm across. To make electrodes small enough to penetrate such cells without undue damage (while forming an electrically tight seal between plasma membrane and electrode) was a formidable task. In 1946 Judith Graham and Gerard reported partial success, using electrodes pulled from glass capillaries to a tip diameter of 2 μm and filled with isotonic potassium chloride.[57] That success was qualified by the recorded variability: the measured range of resting potentials was −41 to −80 mV, which was attributed to damage by the electrodes. One apparent solution was to make the tip smaller.

Gilbert Ling had studied biology at the National Central University in Chungking, where he graduated in 1943. After winning a Boxer Indemnity Fellowship, he went to the University of Chicago to study for his doctoral degree with Gerard. From that association came the "Ling-Gerard microelectrode," which was pulled from glass capillaries to tip diameters of 0.5 μm or less. With it, Ling and Gerard found resting potentials for frog sartorius muscle cells averaging −98 ± 6 mV and varying with the extracellular K$^+$ concentration.[58] They did not describe muscle action potentials.

Hodgkin visited Ling and Gerard in Chicago, and upon his return to Cambridge he applied this technique in collaboration with W. L. Nastuk, who was visiting from Columbia University. They confirmed Ling and Gerard's value for the resting potential, finding a value of −88 mV.[59] They also measured the action potential magnitude, 119 mV, and demonstrated that it varied with the Na$^+$ concentration in the bathing medium.

Cardiac Muscle

In 1950 L. A. Woodbury and associates described resting potentials for frog heart cells of −62 mV and action potential magnitudes of 81 mV, using microelectrodes with tip diameters of 0.5 to 2 μm.[60] A more substantial report appeared the following year from two visitors to Cambridge, M. H. Draper and Silvio Weidmann.[61] By impaling dog Purkinje fibers (part of the electrical conduction system in heart) with microelectrodes having tip diameters of less than 0.5 μm, Draper and Weidmann found resting potentials of −90 mV and action potential magnitudes of 121 mV. However, the action potential shape was quite different from that in nerve and skeletal muscle; there was an initial rapid change in potential, a spike comparable to that in nerve and skeletal muscle, but this was followed by a drop in potential to a declining plateau (roughly two-thirds the magnitude of the spike). The plateau persisted for several hundred milliseconds before repolarization to the resting level. Still, as with nerve and skeletal muscle, removing Na^+ from the bathing medium did not affect the resting potential but abolished the action potential, and the magnitude of the action potential spike was proportional to the logarithm of the extracellular Na^+ concentration.

Although the explanation for this sustained plateau was not forthcoming for some time—and when it did it required a new ionic current—the parallels between all three excitable tissues were quite impressive. The membrane potential arose from asymmetric distributions of ions across the membrane, with a sign and magnitude representing specific and selective ionic permeabilities and alterable by changes in these permeabilities. Thus, in excitable tissues the asymmetric distribution of ions provided a membrane battery, one that was available for use in electrical signaling.

COMMENTARY

Places and People[62]

The marine biological laboratories at Plymouth and Woods Hole were, by the choice of experimental approach and the gathering of interested scientists, prominent locales, but the University of Cambridge dominated the landscape of this chapter. Nevertheless, persuasive arguments for the preeminence of people over place are obvious. There certainly was a Cambridge tradition of focus and thoroughness that came to fruition in Hodgkin and Huxley's construction of their apparatus, design of their experiments, analysis of their data, and synthesis of their model. But others, who were ahead in some aspects (for example, Marmont and Cole's design of the voltage clamp instrumentation), never approached the conclusive completion of their inquiries. Hodgkin remarked jocularly about assembling imposing racks of electronic apparatus for "cowing your scientific opponents or disuading your rivals from following in your footsteps."[63] But what daunted others were the persons and not the hardware; as Solomon recollected about choosing his own field of research, "it was not difficult to reach the conclusion that it would be the height of folly to enter a field in which Hodgkin and Huxley were already firmly ensconced."[64]

Hodgkin and Huxley shared the Nobel prize in 1963, but they did not collaborate after 1952. Hodgkin continued throughout the next decade to study axonal conduction and related problems. After collaborating with Stämpfli, Huxley turned to new fields. He concentrated on muscle contraction and, independent of but simultaneously with Hugh Huxley, formulated the sliding filament model, which he has continued to explore actively. In 1960 Huxley moved to University College, London; he returned to Cambridge in 1984, succeeding Hodgkin as Master of Trinity College.

Katz, who collaborated with Hodgkin and Huxley at Plymouth, had a most distinguished career at University College, London, where he focused on the process by which signals are transmitted from nerve to muscle. He was awarded the Nobel prize for these impressive studies in 1970. Keynes continued experiments on ion fluxes and ion transport as well as on axonal conduction. Subsequently, he was elected to the Royal Society and followed Hodgkin as head of the Physiological Laboratory at Cambridge. Cole was elected to the National Academy of Sciences and as a foreign member to the Royal Society, but his experimental contributions diminished after 1950. He died in 1984. Lorente de Nó remained opposed to the ionic mechanisms, to the Hodgkin-Huxley formulation, and to saltatory conduction, continuing to publish papers on frog nerve into the 1960s. He died in 1990. Ling developed an alternative view of ionic interactions in cells, but his adamant advocacy of that position while the field progressed rapidly in a different direction, left him increasingly isolated (see Chapter 13).

Communication

Some crucial conjunctures are recorded in the published papers: Curtis and Cole as well as Hodgkin and Huxley acknowledged Young's contribution; Hodgkin, Huxley, and Katz acknowledged Cole and Marmont's contributions, as did Nastuk and Hodgkin the work of Ling. Other indicators are the actual collaborations between investigators who came from different institutions. Hodgkin's memoir recorded the significance of his stay at the Rockefeller Institute in New York, his visit to Marmont and Cole in Chicago in 1948, and his attendance at the Paris colloquium in 1949.[65] Nevertheless, the broadest and most definitive mode of communication at that time was through published papers, and the quality of the exposition inevitably affected what was conveyed and to whom. For example, the Paris papers by Hodgkin, Huxley, and Katz depicted a strategic approach to defining the mechanism of nerve action potentials, whereas the first half of the paper by Cole was spent on the "linear characteristics" of the squid axon electrical properties (interpreting the phenomena chiefly in electrical terms), while the second half, which was on "non-linear characteristics," described electronic circuitry and only preliminary results using the voltage clamp technique.[66]

Technical Advances

Prominent in the successes of this decade and a half was the development and application of new techniques. Earlier, oscilloscopes and d. c. amplifiers with cathode followers had been introduced for recording electrical events, but the invention of voltage clamp circuitry enabled new measurements to be made that were essential for the ultimate analyses. Also crucial were intracellular recording techniques—first the "large" microelectrodes for squid giant axons and later the Ling-Gerard microelectrodes applicable to a wide range of cells. Availability of commercial flame photometers after World War II made measurements of cation contents less onerous, while the introduction of radioactive isotopes made possible, uniquely, the demonstration of minute ion fluxes under steady-state conditions.

Financial Support

To pay for those instruments, as well as for supplies and salaries, money was necessary. Before the war Hodgkin received a grant of £300 from the Rockefeller Foundation, which was crucial in paying for the electronic instruments. After the war the Rockefeller Foundation provided Lord Adrian, the head of the Physiological Laboratory, a grant of £3,000 a year for 5 years, and Hodgkin administered those funds. In the following 5 years the Rockefeller grant was reduced to £1,000 per year, but this support was supplemented by the Nuffield Foundation. In their papers Marmont and Cole and Gerard and Ling acknowledged financial support from the University of Chicago. But by the standards of largesse in later decades, money was tight and ingenuity necessary.

Models and Explanations

The equivalent electrical circuit reproduced in Fig. 5.4 is a stellar example of physiological explanation.[67] The phenomenon to be explained was how signals passed along nerves. The preliminary answer involved identifying the signal with an electrical impulse travelling along an axon, together with descriptions of that signal (including shape, magnitude, velocity, and all-or-none and threshold properties). The equivalent circuit not only reproduced those properties but accounted for them as electrical entities with plausible biological counterparts. The external and internal resistances were easily identified with those of extracellular fluid and cytoplasm. The lipid bilayer of plasma membrane models accounted for the transmembrane resistance and electrical capacitance. The Na^+ and K^+ batteries represented the Na^+ and K^+ concentration gradients. The variable resistors (more easily imagined as variable conductances) could not then be identified with known chemical or anatomical entities, but appropriate structures—pores that opened and closed to allow only certain ions to pass—were imaginable.[68] By contrast, Hodgkin and Huxley dismissed explanatory circuits containing inductances because it was "difficult to attach any physical significance to such a concept" in biological structures.[69] What remained, obviously, was identifying the biochemical bases of the variable resistors (see Chapter 14) as well as the biochemical bases for charging the membrane batteries (see Chapters 9, 11, and 12). That course represents the program of explanatory reductionism.[70]

Hodgkin and Huxley explicitly confined their analysis to their data from squid axons. Generalization—qualitatively and in many instances nearly quantitatively—to other systems demonstrated a fundamental commonality in biological functioning. Such generality is not guaranteed in biological research, but it is surely a gratifying culmination.

Scientific Summary

The Hodgkin-Huxley model portrayed electrical responses resulting from (1) an initial influx of Na$^+$ through a conductance pathway opened by depolarization, and then terminating; and (2) a delayed outflux of K$^+$ repolarizing the membrane potential; with (3) both fluxes driven by potential energy stored in the asymmetric distributions of Na$^+$ and K$^+$ across the membrane. Mechanisms for restoring those ionic asymmetries were obviously necessary. Studies on myelinated nerve and skeletal muscle revealed similar characteristics. But action potentials of mammalian heart cells displayed another component, a plateau of depolarization that extended for hundreds of milliseconds and that was not easily interpretable in terms of Na$^+$ influx and K$^+$ outflux alone.

NOTES TO CHAPTER 5

1. For relevant historical accounts see Brazier (1959); Hille (1992); Hodgkin (1992); and Tasaki (1959).
2. Translations quoted are from Weidmann (1955), p. 116, and Glynn (1989a), p. 39.
3. Young (1936).
4. Curtis and Cole (1938). They acknowledged Young's introducing them to squid giant axons.
5. Cole and Curtis (1939). R. L. Post (personal communication, 1995) pointed out that Cole and Curtis could have derived the membrane currents from their measurements, applying their analysis of cable properties, by taking derivatives of the action potentials. He quoted Curtis as not having had sufficient confidence in those measurements to make such analyses then.
6. Cole and Hodgkin (1939).
7. The resting membrane potential is expressed relative to the potential of the bathing medium defined as 0 mV; it has then a negative sign. The action potential *magnitude* is expressed as an absolute number, relative to the membrane potential defined now as 0 mV. Algebraic addition of the resting potential and the action potential magnitude gives the overshoot magnitude.
8. Hodgkin and Huxley (1939). They too thanked Young for his contribution.
9. Curtis and Cole (1942).
10. Ibid., pp. 136, 141, 142.
11. Hodgkin (1992), p. 252.
12. This calculation is based on intracellular concentrations subsequently reported by Steinbach and Spiegelman (1943); with values available previously (Bear and Schmitt, 1939) the maximal value would be +14 mV.
13. Hodgkin and Huxley (1945).
14. Höber (1946), p. 388.
15. Grundfest (1947).
16. Hodgkin and Huxley (1947).
17. Ibid., pp. 341–342.
18. Katz (1947).
19. Hodgkin and Katz (1949), pp. 73–74; Hodgkin et al. (1949).
20. Hodgkin et al. (1949).
21. Hodgkin and Katz (1949), p. 65.

22. Hodgkin et al. (1949).

23. Goldman (1943).

24. Hodgkin and Katz (1949).

25. Hodgkin et al. (1949).

26. Cole (1949); Marmont (1949).

27. Hodgkin et al. (1952); Hodgkin and Huxley (1952a,b,c,d).

28. Stationary action potentials were obtained with electrodes clamping potentials over appreciable lengths of axons (a "space clamp").

29. Cole (1954) and Cole et al. (1955) reported the failure to replicate Hodgkin and Huxley's solution; Cole (1958) retracted those reports.

30. Hodgkin and Huxley (1952d), p. 541.

31. Ibid., pp. 543–544.

32. Webb and Young (1940); Hodgkin and Huxley (1939); Curtis and Cole (1942).

33. Grundfest (1950).

34. Cole (1968), p. 145.

35. Lorente de Nó (1944).

36. Lorente de Nó (1947).

37. Feng and Gerard (1930). The quotation is from Feng and Liu (1949), p. 1.

38. Lorente de Nó (1947), as quoted in Feng and Liu (1949), p. 1.

39. Feng and Liu (1949).

40. Grundfest (1950), pp. 17–22.

41. For example, Lorente de Nó (1947) proposed that "K^+ and Na^+ can play a specific role only in so far as they directly or indirectly participate in enzymatic reactions" (p. 107), with the membrane potential then "referable to the oxidation-reduction systems of the chain of oxidative enzymes" (p. 106).

42. Keynes (1948, 1949a). Keynes used cuttlefish axons because cuttlefish, unlike squid, could be kept in Cambridge.

43. Keynes and Lewis (1951a); Keynes (1951a). The incremental flux, due to the action potential, was calculated to be greater than 2.3×10^{-7} coulombs per impulse, whereas repolarizing the membrane required only 1.4×10^{-7} coulombs.

44. Keynes (1951b); Keynes and Lewis (1951b).

45. Keynes and Lewis (1951b), p. 179.

46. Rothenberg and Feld (1948); Rothenberg (1950).

47. Keynes and Lewis (1951a).

48. Boyle and Conway (1941).

49. Hodgkin and Huxley (1946), p. 377.

50. Hodgkin and Huxley (1947), p. 364.

51. Hodgkin (1951).

52. Hodgkin and Huxley (1947), p. 365.

53. Hodgkin and Huxley (1952d), p. 541.

54. Keynes (1951a,b).

55. Huxley and Stämpfli (1949), p. 339.

56. Huxley and Stämpfli (1951a,b).

57. Graham and Gerard (1946).

58. Ling and Gerard (1949a).

59. Nastuk and Hodgkin (1950). They thanked Ling for "demonstrating his experimental technique" (p. 72).

60. Woodbury et al. (1950).

61. Draper and Weidman (1951).

62. Biographical information is from interviews, reminiscences (Hodgkin, 1992; Keynes, 1989), and biographies (Huxley, 1992).

63. Hodgkin (1976), p. 11.

64. Solomon (1989), p. 129.

65. Hodgkin (1992).

66. Hodgkin (1949); Hodgkin et al. (1949); Cole (1949). Cole's paper is also notable for its contrasting focus on electronic currents to the neglect of ionic currents.

67. Expressing nerve activity in terms of equivalent circuits was not a novel idea (see, for example, Rushton, 1937).

68. Some ideas about these structures were, however, often far from today's proposed structures; see Cole (1968), pp. 537–539, for conceptions of that period.

69. Hodgkin and Huxley (1945), p. 192.

70. See, for example, Robinson (1986a, 1992).

chapter 6

EPITHELIAL TRANSPORT BY FROG SKIN

IN multicellular organisms not only must substances pass into and out of individual cells, those substances must pass across layers of cells to reach yet other cells beyond. Among the means for satisfying such basic needs is epithelial transport. An epithelium is a layer of closely juxtaposed cells covering a surface of animals or organs. Like plasma membranes of cells, epithelia are both barriers and entryways. Indeed, the central physiological role of an epithelium in many cases is absorption and excretion, as in transport epithelia lining the intestinal lumen, where foodstuffs are absorbed, or those lining the kidney tubules, where wastes are excreted and nutrients salvaged.

Epithelia consist of layers of cells that are often atop a supporting "basement membrane" (an extracellular sheet providing mechanical strength without impeding the passage of small molecules). Beneath the basement membrane is characteristically a connective tissue layer containing blood vessels. Fig. 6.1A shows diagrammatically the arrangement of these layers in frog skin, the focus of this chapter. Fig. 6.1B shows—even more diagrammatically—a current view of cells forming an epithelial layer abutting a basement membrane, as in kidney tubules. Two surfaces are distinguishable topographically and functionally: one facing the outside (called "apical" or "luminal," since it faces tubular lumens), and one facing the interior (called "serosal," since it faces the blood supply, or "basolateral," since tight junctions near the apical ends of cells separate the apical surface from the remainder).

To minimize confusion with processes involving single cells, I will refer to apical exposures as "out," serosal exposures as "serosal," and intracellular exposures as "in." Correspondingly, I will refer to fluxes from out to serosal surfaces as "influx$_{epi}$" and the opposite fluxes as "outfluxes$_{epi}$". Influx$_{epi}$ thus involves influx across the apical plasma membrane, diffusion through the cytoplasm, and outflux through the serosal plasma membrane.

Another prerequisite for this discussion is defining the concept of active trans-

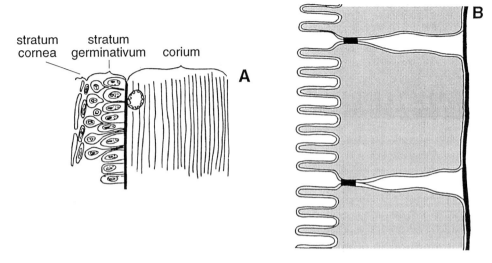

Fig. 6.1 Epithelia and epithelial cells. Panel *A* shows the frog skin in cross section, with the stratum cornea cells on the outside surface and the stratum germinativum cells against the basement membrane; the corium (containing a blood vessel) lies on the inner side of the basement membrane. [Redrawn from Koefoed-Johnsen and Ussing (1958), Fig. 6. Reprinted by permission of the Scandinavian Physiological Society.] Panel *B* shows kidney tubule cells with their brush border membranes (facing the tubular lumen) separated from their basolateral membranes by tight junctions between adjacent cells; the cells lie against the basement membrane. [Drawing courtesy of Peter Holohan.]

port. This term was used casually in preceding chapters and by many whose work was cited there. In 1948 Thomas Rosenberg provided a precise, explicit definition of the process: the transport of a solute against its electrochemical gradient.[1] However, it is also necessary to keep in mind the distinction between the *process* of active transport and a *system* capable of effecting active transport, for an active transport system need not always be working against the electrochemical gradient, even though it has the capacity to do so. The absence of active transport, the absence of the process, does not indicate that an active transport system is not participating.

EARLY STUDIES ON FROG SKIN

In the mid-nineteenth century, du Bois-Reymond described electrical potentials across frog skins, that were positive on the serosal side, and by the turn of the century, E. W. Reid had shown that isolated frog skins caused water to flow from outside to serosal side. Frog skin thus became a favorite experimental preparation for students of both electrophysiology and permeability. These two phenomena were often explored separately, however, and different interests led to different approaches.

For those focusing on electrical potentials, a central concern was how frog skins generated the 50–100 mV observed. By the mid 1930s there were two schools of thought.[2] One concentrated on redox reactions, noting the correlations between oxidative metabolism and voltage. Thus, E. J. Lund observed that the metabolic inhibitor cyanide markedly reduced frog skin potentials, the rate of oxygen consumption correlating with voltage.[3] In electrochemical cells, redox reactions (which involve gain and loss of electrons) induce electron currents through metallic conductors. How redox potentials were to be converted into currents across a frog skin remained unspecified.

The other school also linked cellular respiration to voltage, but through formation of ionic diffusion potentials. In this formulation, metabolism provided energy for separating ions; their asymmetric distribution then produced diffusion potentials. Depicting frog skin potentials as ionic diffusion potentials also provided a link between electrical processes and flows of salt and water; the latter could be interpreted as osmotic fluxes secondary to solute transport. But these models left unanswered fundamental questions of how metabolic processes drove ion transport.

Early this century, G. Galeotti showed that Na^+ was required for current flow, and he proposed that frog skin favored Na^+ diffusing more readily from outside to serosal side. In 1933 Steinbach reported that reducing the external Na^+ concentration or substituting K^+ for Na^+ decreased electrical potentials across frog skin.[4] Subsequently, Steinbach concluded that K^+ moved from serosal side to outside; he did not measure Na^+.[5] In the mid-1930s Ernst Huf[6] argued that water and salt moved from outside to serosal side; he measured Cl^- but not Na^+. In addition, Huf demonstrated that metabolic substrates energized this flux, whereas inhibitors blocked it.

The most clear-cut experiments before World War II, however, were by Leonard Katzin working at Woods Hole.[7] He reported that with chemically "similar" solutions on both sides of the skin, ^{24}Na influx$_{epi}$ exceeded outflux$_{epi}$ by 12–20×10^{-8} mol/ $cm^2 \cdot$ hour, whereas ^{42}K outflux$_{epi}$ exceeded influx$_{epi}$ by 2–6×10^{-8} mol/$cm^2 \cdot$ hour. The net ion movements were in opposite directions and in different amounts. And he found that the ratio of ^{24}Na influx$_{epi}$ to outflux$_{epi}$ increased dramatically when a mixture of NaCl and KCl replaced pure NaCl in the bathing media.

USSING'S FROG SKIN MODEL

In the 1930s Krogh noticed that frogs could absorb Na^+ from quite dilute solutions.[8] When Ussing obliged Krogh by assuming leadership of the transport group in Copenhagen, he inherited that phenomenon as a process to be explained. Ussing's first approach, published in 1947,[9] involved studies *in vivo* on another amphibian, the axolotl, a choice abandoned thereafter because the wrinkled skin complicated experiments and calculations. Nevertheless, those experiments led to an associate's investigating how epinephrine affected transport across isolated frog skin,[10] as well as to Ussing's first paper on the subject, published 2 years later in 1949.[11]

In these and succeeding studies, Ussing mounted pieces of frog abdominal skin between two chambers; he then measured radioactive tracer fluxes by adding iso-topes to either bath. At the same time, he recorded electrical voltage (or current) between electrodes in each. From the ^{24}Na fluxes$_{epi}$ in both directions, Ussing calcu-lated that influx$_{epi}$ far exceeded outflux$_{epi}$ (except when Na$^+$ concentrations outside were extremely low): this confirmed and extended Katzin's experiments. With Na$^+$ concentrations in *both* chambers comparable to that in the frog's extracellular fluid, he found influx$_{epi}$ to be 10-fold greater than outflux$_{epi}$, even though the serosal side was electrically positive. Na$^+$ thus moved against its electrochemical gradient, dem-onstrating by Rosenberg's definition active transport. Furthermore, adding cyanide reduced influx$_{epi}$ drastically without affecting outflux$_{epi}$.

On the other hand, the influx$_{epi}$ of ^{38}Cl was less than that of ^{24}Na, and Ussing calculated that the net Cl$^-$ movement could be attributed to a passive distribution, following the membrane potential: Cl$^-$ influx$_{epi}$ did not require active transport. To account for steady-state movements of more Na$^+$ than Cl$^-$ inward across the skin, Ussing noted that either some cation was lost at the outside surface or it was exchanged for Na$^+$ at the serosal surface. Ussing included in his model an exchange of Na$^+$ for H$^+$ from the serosal medium.

Ussing also argued that the site of active transport must be at the serosal surfaces of the deepest layer of epithelial cells (the stratum germinativum, the cells abutting the basement membrane), whereas the passage of Na$^+$ through the outer (apical) plasma membranes was passive. This cellular polarization was thus the basis of epithelial transport:

> The difference between epithelial cells and other cells . . . is not that they extrude Na$^+$, but that they do so only on the side turning inwards. . . . [A] symmetrical extru-sion . . . would lead to the rapid loss of all the NaCl in the animal.[12]

Ussing's qualitative demonstration of active Na$^+$ transport across the frog skin was straightforward and convincing (to most). There remained, however, a funda-mental problem in deciding which ions are actively transported. Such a distinction necessitated his quantitating the electrochemical driving forces. The difficulty lay in the unknown characteristics within the barrier. The problem is identical to that noted in Chapter 5, where I cited Goldman's and Hodgkin and Katz's simplifying assumption for axons: a linear gradient across the plasma membrane. Across hetero-geneous and more complex barriers, however, that constant field assumption cannot be justified. Ussing's clarifying insight was recognizing that the unknown potential profile experienced by an ion moving from outside to serosal side is the inverse of that experienced in moving from serosal side to outside. The unknown factors cancel when the two routes are compared.[13]

Consequently, Ussing measured the fluxes$_{epi}$ simultaneously[14] in each direction and compared their ratios to the chemical activities in the two chambers (which are usually identical) times a function of the measured electrical potential across the skin:

$$\text{influx}_{epi}/\text{outflux}_{epi} = [(A)_{out}/(A)_{ser}]\ e^{z\,F/\,\Delta\Psi)}$$

where (A) represents the chemical activity of ion A in a chamber, z the valence of A, and $\Delta\Psi$ the electrical potential across the skin.[15] The criterion for active transport follows from Rosenberg's definition: if the flux ratio (on the left) exceeds the passive driving force (on the right), then some further process must be acting. This was clearly the case for Na^+, where the flux ratio was 100-fold greater than could be accounted for by passive diffusion.[16]

Conversely, Ussing showed in analogous experiments that Cl^- fluxes did not require active transport, nor did K^+ outflux$_{epi}$.[17] But, as pointed out above, such calculations do not guarantee that K^+ was not passing through an active transport *system*, and other studies published at this time implied further roles for this cation. These studies were by Huf, who had emigrated from Germany to Richmond, Virginia and resumed studies on isolated frog skins. In 1951 he reported that removing K^+ from the bathing media sharply reduced net Na^+ influx$_{epi}$.[18] Why active Na^+ influx$_{epi}$ was so dependent on K^+ he attributed to K^+ being either "one important element in the mosaic of cellular structure and thereby regulat[ing] permeability" or "an activator of enzymes upon which depend the energy liberating reactions which are geared with active salt uptake."[19] The possibility of a linked Na^+/K^+ transport, then being advocated by Harris and Maizels, was not mentioned.

Ussing accomplished yet another experimental feat at this time. In association with Karl Zerahn he developed a method for comparing the number of Na^+ moving through the skin with the total current flowing across it.[20] To do this they added another pair of electrodes to the chambers on either side of the frog skin (Fig. 6.2), and through these they passed sufficient current to make the voltage across the skin zero: they shortcircuited the skin. The amount of current required to bring the voltage to zero, the "shortcircuit current," then equaled the net current of all ions passing through the skin, the algebraic sum of all ionic fluxes. When Ussing and Zerahn compared the measured ^{24}Na flux$_{epi}$ with the measured shortcircuit current, they found, with identical solutions in each chamber, that ^{24}Na influx$_{epi}$ averaged 105% of the shortcircuit current and outflux$_{epi}$, 5%. Thus, the active Na^+ flux accounted for all the net current flow.

With this technique a visitor from Stanford University, Frederick Fuhrman, explored the dependence of transport on metabolism.[21] Among the inhibitors he tried was dinitrophenol, which had recently been shown to decrease ATP formation without decreasing oxygen consumption (it "uncoupled" oxidative phosphorylation). So Fuhrman suggested that "high-energy phosphate compounds are involved in active sodium transport in the frog skin."[22] By contrast, Huf concluded that ATP was not involved, since adding ATP to the bathing media did not increase Na^+ transport.[23]

Fuhrman also collaborated with Ussing in examining how extracts of posterior pituitary[24] increased Na^+ transport across the skin:[25] the extracts increased both ^{24}Na influx$_{epi}$ and ^{24}Na outflux$_{epi}$, although influx$_{epi}$ remained greater. The extracts also increased electrical potentials across the skin and the flow of water. Ussing and Zerahn considered three possible mechanisms: (1) an increased rate of the transport mechanism, (2) a decreased rate of Na^+ diffusing back into the skin from the serosal

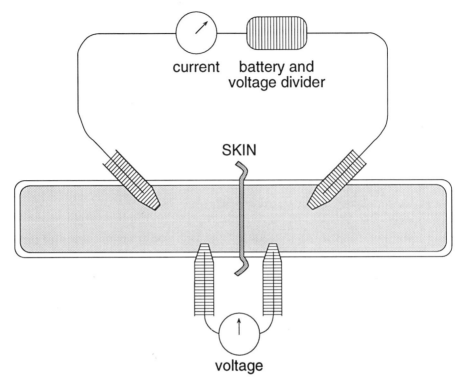

Fig. 6.2 Measuring shortcircuit current. The piece of skin lies between two baths (representing the out side and the serosal side). Voltage is monitored by one set of electrodes and is kept at zero by current passing through a second set of electrodes, with this shortcircuit current also measured. [Redrawn from Ussing and Zerahn (1951), Fig. 1. Reprinted by permission of the Scandinavian Physiological Society.]

side, or (3) an increased rate of Na^+ passing through some rate-limiting diffusion barrier.[26] This last possibility they considered most likely, for the shortcircuit current in the presence of these extracts still equalled ^{24}Na influx$_{epi}$.

Finally, Ussing and associates[27] returned to the peculiar action of epinephrine noted earlier. They now documented a marked increase in Na^+ outflux$_{epi}$ together with a comparable Cl^- outflux$_{epi}$ when epinephrine was added to the serosal bathing medium. The voltage across the skin remained positive on the serosal side. From flux-ratio analyses they concluded that Cl^- was transported actively to the outside. And from measurements of the shortcircuit current they identified only two sources of ionic current: Cl^- outflux$_{epi}$ plus the usual Na^+ influx$_{epi}$. They assigned the active Cl^- transport system not to the stratum germinativum cells, however, but to mucus glands in the skin.[28] Thus, their explanation acknowledged more complexity in the frog skin as well as in the transport processes.

COMMENTARY

Places and People

Why did the definitive explanation come from a handful of investigators in Denmark when before World War II a host of scientists from Germany, Italy, England, Scotland, and the United States had been exploring the processes? (1) One factor surely was resources. The Copenhagen group had access to radioactive isotopes, and the use of these isotopes was essential. Katzin was well on his way to defining the basic fluxes with radioactive tracers when his studies were interrupted by the war (during which his interests moved to other topics). And, as noted in Chapter 4, the use of radioactive isotopes for transport studies was delayed elsewhere for several years after the war; such studies were not resumed in the United States until after Ussing's first results were announced. (2) Once Ussing's reports began to appear, the remaining experimental issues were rapidly addressed by him as well. Thus, there were few loose ends and few opportunities for others to exploit. The Copenhagen group focused on the necessary issues and resolved them expeditiously. (3) The Copenhagen group may also have profitted from being in a small nation where their success was even more visible, attracting not only the best of collaborators[29] but the necessary financial support as well. (Those dispensing the funds were perceptive of the individual if not the science: Ussing confided that, after his first demonstration of active transport across the frog skin, his grant from the Carlsberg Foundation was awarded for 3 years, with the unofficial comment that by then Ussing will have discovered where his error lay.[30]) (4) Ussing was also not shy about presenting his results at meetings and publishing them in journals. He pushed hard to complete the crucial shortcircuit current experiments[31] in the 10 days before the International Physiological Congress held in Copenhagen in 1950. (5) But from the perspective of practicing scientists, the outstanding determinant was Ussing. He exploited his resources and focused his attention, he proceeded expeditiously and sought collaborators and research funds effectively. He also communicated widely and was sharper than his potential competitors—experimentally and theoretically. The scientific world is filled with entrepreneurs who are more successful in such domains than Ussing but who leave no mark in intellectual history. What distinguishes Ussing is his insight and effort.

Progress

Reports before World War II reflected a broad interest but yielded a fragmentary, varied, and inconclusive picture, one from which little sense of what underlay the electrical and fluid currents could be divined. Then in just a few years after the war, Ussing dramatically reformulated the issues, clarified the phenomena, and established a solid experimental and theoretical basis that survives as a cornerstone of ion transport studies. Foremost among these achievements was the unequivocal demonstration of active transport across biological membranes. Na^+ was moved from one bulk phase to another where the chemical potentials were the same, and it was moved against a demonstrable electrical gradient. Such epithelial transport left no reason to doubt that active transport occurred across the plasma membranes of cells (even though the activity coefficients of ions in the cytoplasm were not directly measurable and thus remained a point of argument for some committed skeptics). In addition, his demonstration was achieved by notable experimental approaches and theoretical analyses. Ussing and associates developed techniques to distinguish active transport from passive diffusion (the flux-ratio method) and for comparing and quantitating ion fluxes and currents (the shortcircuit current method), techniques that were founded on sound theoretical bases.

These results were, moreover, a significant element in reformulations of cellular and tissue permeability. Where cells had been regarded as impermeable to Na^+, such permeability was now a common aspect of cell functioning. Where Na^+ had been seen to enter passively through general pores, that entry was now a site of regulated, hormonal control. Where Cl^- had been considered to be diffusing passively, it too was transported actively. Where transport had been depicted isotropically, polarized cells now effected different transport functions at different surfaces.

Indeed, this chapter exemplifies an aspect of biological, physiological, and biochemical explanation worth belaboring. The frog skin is a complex structure and an intrinsic part of the whole animal. Yet in Ussing's earlier studies he not only used the isolated skin but viewed it as a simplified structure: a layer of cells endowed with passive diffusion to Na^+, K^+, and Cl^- on one surface, and on the opposite surface active transport of Na^+ together with passive diffusion of the other ions. This model explained well the experimental fluxes of Na^+, K^+, and Cl^- measurable under particular controlled conditions. Then to this simplified model were added, one by one as the experimental conditions were varied further, new capabilities and properties. Administering posterior pituitary extracts required adding a hormone-dependent Na^+-influx to the plasma membrane. Administering epinephrine required adding mucus glands among the epithelial cells that actively transported Cl^-. Why such emendations were not merely *ad hoc* lay in the independent evidence for them: as demonstrable entities as well as observable processes. Adding more and more detail to increasingly complex models is the most salient characteristic of how biological explanation progresses.

Scientific Summary

Measurements of electrical current and ion fluxes across isolated frog skin provided unequivocal evidence for active transport of cations: movement against their electrochemical gradients. These studies also led to complex models of transcellular as well as transmembrane fluxes. In this endeavor, Ussing provided essential theoretical analyses and experimental approaches, which included flux-ratio comparisons and shortcircuit current measurements.

NOTES TO CHAPTER 6

1. Rosenberg (1948).
2. See Ponder and Macleod (1937).
3. Lund (1926, 1928); Lund and Moorman (1931).
4. Steinbach (1933).
5. Steinbach (1937).
6. Huf (1935, 1936a,b).
7. Katzin (1939, 1940).
8. Krogh (1937).
9. Jorgensen et al. (1947).
10. Jorgensen (1947).
11. Ussing (1949b).
12. Ussing (1948), p. 198.
13. Ussing (1949a), p. 55.
14. Because the properties of frog skins could change during experiments, simultaneous measurements in each direction were essential for comparisons to be valid.
15. An ancestral formulation is Behn's equation (1897), cited in Ussing (1952). Ussing's argument is far more complex than presented here, and addresses explicit assumptions and limitations.
16. Levi and Ussing (1949); Ussing (1952).
17. Ussing (1949a); Levi and Ussing (1949); Koefoed-Johnsen et al. (1952a).
18. Huf and Wills (1951).
19. Ibid., p. 259.
20. Ussing and Zerahn (1951).
21. Fuhrman (1952).
22. Ibid., p. 277.
23. Huf and Parrish (1951). The assumption was that ATP acted on the extracellular surface, for it was known not to cross plasma membranes.
24. Posterior pituitary gland extracts were known to affect urinary output; the active substance was subsequently identified as the hormone vasopressin.
25. Fuhrman and Ussing (1951).
26. Ussing and Zerahn (1951).
27. Koefoed Johnsen et al. (1952b).
28. The presence of mucus glands in the skin and the contribution of glands to electrical activity had been explored earlier (e.g., by W. M. Bayliss and J. R. Bradford in 1886).
29. For example, Ussing remarked that Zerahn was a friend who was "good with wires" (interview, 1993).
30. Ibid.
31. Ibid.

chapter 7

CONTEMPORARY EVENTS: 1939 – 1952

A wealth of new concepts, discoveries, and techniques emerged in this period. Here I will note briefly some that were (or soon became) relevant to further investigations of membrane transport and cellular energetics.

STRUCTURE

The first complete amino acid content of a protein was published in 1945 and the first amino acid sequence (Frederick Sanger's description of insulin) in 1951.[1] That year Pauling and Robert Corey proposed α-helix and β-sheet conformations within proteins, based on measurements on model compounds.[2]

Metabolic studies using isotope tracer techniques by then had demonstrated the "dynamic" state of proteins:[3] many proteins were constantly being synthesized and then degraded in cells, some turning over within hours. Oswald Avery and associates showed in 1944 that genetic information in bacteria was carried in deoxyribonucleic acid (DNA);[4] earlier, George Beadle and Edward Tatum's researches on yeast led to the dictum linking one gene to one enzyme.[5]

On the other hand, little new information about membrane structure was forthcoming. Danielli, however, modified his membrane model in an influential book published in 1943, depicting more intimate interactions between membrane proteins and the lipid bilayer they lay atop: amino acid side chains now penetrated the bilayer, in hydrophobic association with the fatty acid tails of membrane lipids.[6]

METABOLISM

In one of the most influential review articles of this century, Lipmann in 1941 introduced the concept of "energy-rich" phosphate bonds, designated $\sim$P, which served as the proximate energy source for a range of cellular functions.[7] He identified four chemical types of "energy-rich" linkages, estimating the free energy liberated by hydrolysis as roughly 10 kcal, and contrasted these linkages with ester phosphates (as in glucose-6-phosphate) having free energies of hydrolysis of only 2–4 kcal. Lipmann concluded that "phosphate is introduced into compounds [during intermediary metabolism] as a prospective carrier of energy," with "glycolysis represent[ing] reactions where the energy, derivable from . . . glucose, is converted quantitatively into phosphate bond energy."[8]

Curiously, Lipmann's review did not mention the ATPase activity of muscle myosin, which had been reported 2 years previously,[9] although that would seem to represent an exemplary transformation of chemical to mechanical energy. In the following decade this ATPase activity was examined vigorously, but in complex and changing contexts and without a satisfying resolution. In 1942 F. B. Straub identified the protein actin,[10] a component of previous preparations of "myosin," and Albert Szent-Györgyi suggested a new nomenclature: actin, myosin, and, for the complex, actomyosin.[11] Moreover, the ATPase activity of (purified) myosin was altered by adding actin, although the characteristics were not yet untangled.[12] In any case, myosin ATPase then represented the preeminent instance of ATPase activity, hydrolyzing ATP to ADP plus inorganic phosphate,[13] coupled to cellular work. But even beginning to understand the process of muscle contraction would take another decade.

Moreover, the uniqueness of this myosin ATPase was challenged by reports of two other ATPases. In 1948 Wayne Kielley and Otto Meyerhof described a labile ATPase reaction in muscle not associated with actin or myosin.[14] Whereas myosin ATPase was activated by Ca^{2+}, this ATPase was activated by Mg^{2+} but inhibited by Ca^{2+}. By contrast, Benjamin Libet found an ATPase reaction activated by Ca^{2+} in the sheaths of squid giant axons.[15]

During this period, accumulating detail was consolidated into the central formulations of the glycolytic pathway and of the Krebs cycle. The oxidations associated with these two sequences—achieved through extractions of hydrogen atoms—were linked to the reduction of nicotinamide and flavin coenzymes. Also recognized was the ultimate role of cytochrome oxidase in catalyzing the union of oxygen with hydrogen, forming water. Functioning between the extraction of hydrogens and the synthesis of water were several protein cytochromes with distinctive spectra, constituting the respiratory chain of redox reactions.

Severo Ochoa[16] demonstrated that a variety of glycolytic and Krebs cycle intermediates were metabolic fuels for producing ATP, with P/O ratios near 2. He thus confirmed Belitser and Tsybakova's calculation of P/O ratios greater than 1 (see Chapter 2). Ochoa also endorsed their conclusion that such ratios indicated more than one point of ATP synthesis along the redox chain from oxidizing substrate to reducing oxygen. And in 1952 Henry Lardy argued persuasively that the P/O ratio was 3 with most substrates (measured values were less, but general assumptions required integral ratios, and excuses for finding nonintegral values, such as concomitant ATP hydrolysis, were obvious).[17] Furthermore, the rate of redox reactions through the respiratory chain seemed to be regulated by the availability of phosphate, thus establishing the notions of "respiratory control" and "tight coupling": respiration was so linked to ATP formation that respiration ceased if phosphorylation were blocked.

Formation of several ATP molecules for each atom of oxygen consumed (represented by the P/O ratios) was imaginable as one ATP being formed at each of several steps along the respiratory chain. Since individual components of that chain differed in their redox potentials, a step-wise process was anticipated, but defining the precise order and identifying the actual coupling sites turned out to be an arduous task, one that extended well beyond the next decade. Moreover, it was not until 1951 that Albert Lehninger demonstrated ATP formation driven by added NADH in lieu of oxidizable substrate.[18]

In 1949 Lehninger and Eugene Kennedy showed that mitochondria could accomplish oxidative phosphorylation.[19] Analogous investigations revealed that mitochondria contained the enzymes for the Krebs cycle and for fatty acid oxidation but not for glycolysis. And in 1952 George Palade published the first high-resolution electron micrographs of mitochondria, showing an organized system of membranous septae within, termed "cristae." Palade suggested that the enzymes were "arranged in the proper order in linear series . . . built . . . in the solid framework."[20]

Among metabolic poisons affecting oxidative phosphorylation, a prominent reagent was dinitrophenol, and in 1948 Lipmann and F. W. Loomis reported that it blocked phosphorylation while allowing oxidation to proceed, "reversibly uncoupl[ing] phosphorylation from oxidation."[21] Earlier, Lardy had suggested that dinitrophenol could act either by "allowing oxidations to occur without phosphorylation or actually cause dephosphorylation of high energy phosphate."[22] Indeed, he reported that dinitrophenol increased ATP hydrolysis by minced muscle.

Finally, two papers bearing indirectly but significantly on the central theme of this book appeared during this period and they deserve celebration.[23] In 1942 Lardy, Paul Boyer, and Paul Phillips reported that added K^+ dramatically stimulated—by more than 10-fold—the formation by muscle extracts of creatine phosphate from creatine, AMP, and 3-phosphoglycerate. The following year they narrowed the site of stimulation to the generation of ATP from phosphoeneolpyruvate and ADP, a reaction now attributed to the enzyme pyruvate kinase. Moreover, they showed that Na^+ not only failed to stimulate enzymatic activity but actually inhibited stimulation by K^+. (It required a half century more to reveal the structure of the K^+-binding site on pyruvate kinase, as defined by X-ray crystallography.[24])

The titles of both papers contained the phrase "the role of potassium" and should have attracted the notice of those interested in how K^+ affects biological systems. Certainly the conclusion that an enzyme can discriminate between Na^+ and K^+ should have piqued the curiosity of those in the transport field at a time when mechanisms for discriminating between those cations were uncertain and when lipid carriers were being proposed to achieve that feat. Presumably, these papers were the work to which Pauline Hald and colleagues referred in 1947 (without citing a reference) when they noted that distinctions between Na^+ and K^+ were "regularly drawn by enzyme systems."[25] By contrast, Leonor Michaelis, commenting the next year on cation binding during transport, invoked the image of carboxyl groups in artificial collodion membranes instead of carboxyl groups in proteins.[26]

METHODS

Undoubtedly the major technological advance for biochemistry was the development of chromatographic methods for separating different species of molecules: column chromatography, paper chromatography, and gas-liquid chromatography.[27] These techniques, which were pioneered by A. J. P. Martin, R. L. M. Synge, and associates in the early 1940s, made separations easy that heretofore were difficult or impossible.

As emphasized previously, the introduction of radioactive isotopes created vast new opportunities for research, and after World War II these endeavors were expedited by commercially available radioactivity counters.

Also of great importance was the development of cell fractionation techniques, which involved an initial mechanical dispersion of tissues ("homogenization") followed by centrifugation at successively higher forces and for longer times to isolate lighter and lighter particles. These procedures, begun successfully by Albert Claude, were extended and standardized by George Hogeboom, Walter Schneider, and their colleagues into a four-fraction protocol.[28] Tissues were homogenized in isotonic sucrose and centrifuged briefly at low speed (that pellet contained "nuclei and cellular debris"). The unsedimented material was then centrifuged at intermediate speed (that pellet contained "mitochondria"). Finally, the unsedimented material was centrifuged at high speed for a long time (that pellet contained "microsomes";[29] the unsedimented fraction from this step was termed the "supernatant"). This approach was essential for allocating specific biochemical functions to cellular organelles as well as for preparing organelles for further characterization. The various "particulate" fractions (those sedimenting) represented cell membranes and membrane-bounded vesicles, and these became central preparations for future studies on membranes and transport.

In addition to the major advances in electronic instrumentation noted previously, three other developments facilitated research on membranes and bioenergetics during this period. (*1*) High-speed preparative centrifuges made possible the isolation of membranes and vesicles for experimental study (initially, the "multispeed attachment" to the International centrifuge was used for cell fractionation, but the Spinco model L became the standard tool after its introduction in 1950). (*2*) Photoelectric colorimeters (such as the Klett photometer, which used optical filters to select ranges of wavelengths) offered greater ease as well as accuracy for colorimetric analyses than did the visual comparisons previously relied on. Still better were photoelectric spectrophotometers (such as the Beckman model DU, first marketed in 1941); these permitted measurements from the ultraviolet through the visible range for determining characteristic absorption spectra as well as for colorimetric assays. (*3*) Flame photometers, introduced at the end of the decade, allowed convenient measurements of Na^+ and K^+ (as their emission bands when injected into a flame), replacing the onerous methods previously employed. Fenn reminisced that a "flame photometer calls to mind the years wasted [with the older methods,] precipitating and titrating those hundreds of sodium and potassium samples."[30]

NOTES TO CHAPTER 7

1. Brand et al. (1945); Sanger and Tuppy (1951a,b). The latter described one of insulin's two chains; the sequence of the other was published in 1953.
2. Pauling et al. (1951) and seven subsequent papers that year.
3. Schoenheimer (1942).
4. Avery et al. (1944).
5. Beadle and Tatum (1941).
6. Davson and Danielli (1943), Fig. 16b, p. 65. However, none of the amino acids shown corresponds to an ordinary one (all the standard amino acids of proteins were known by the time the book was published), and the bond angles in some lipids were distorted. Without these liberties, the amino acid side chains could not penetrate the hydrophobic region of the bilayer.
7. Lipmann (1941). In the text he used the symbol ~ph rather than ~P, presumably to distinguish between the phosphate group and the element phosphorus.
8. Ibid., pp. 100, 137.
9. Engelhardt and Ljubimowa (1939).
10. Cited in Szent-Györgyi (1947).
11. Szent-Györgyi (1947).
12. Straub (1950); Dubuisson (1952).
13. Straub (1950). He claimed that "no other enzyme known . . . splits off only one inorganic phosphate residue from ATP except myosin" (p. 377).
14. Kielley and Meyerhof (1948).
15. Libet (1948).
16. Ochoa (1941, 1943).
17. Lardy and Wellman (1952); Copenhaver and Lardy (1952). Notable exceptions were P/O ratios near 2 with succinate and 4 with α-ketoglutarate.
18. Lehninger (1951). The technical problem was thought to lie in making the mitochondria leaky so added NADH could penetrate to the mitochondrial interior, but without destroying catalytic activity. This study also demonstrated that phosphorylation occurred at steps beyond substrate oxidation, an important distinction for discriminating between substrate-level phosphorylation and oxidative phosphorylation (cf. Chapter 16).
19. Kennedy and Lehninger (1949).
20. Palade (1952), p. 439.
21. Loomis and Lipmann (1948), p. 808.
22. Lardy and Elvehjem (1945), p. 16.
23. Boyer et al. (1942, 1943).
24. Larsen et al. (1994).
25. Hald et al. (1947), p. 348.
26. In discussion to Ussing (1948), p. 200.
27. Martin and Synge (1941), and a series of papers thereafter.
28. Hogeboom et al. (1948); Schneider (1948).
29. Microsomes were identified with fragments of the endoplasmic reticulum. The location of plasma membranes within this scheme was then uncertain.
30. Fenn (1962), p. 1.

chapter 8

CHARACTERIZING THE Na$^+$/K$^+$ PUMP

B Y the early 1940s, new kinds of experiments had revealed that some cations were not distributed passively across cell membranes and that membranes were not absolutely impermeable to them. Extensive studies of muscle, red blood cells, axons, and epithelia supported proposals for membrane pumps actively extruding Na$^+$. Through the 1950s—the period considered here—the implications of these proposals were developed, with details added and capabilities revised. The focus here is on how new experiments were devised to clarify issues and to examine mechanistic models, but I will begin with a serendipitous discovery that provided a crucial experimental tool for further studies.

CARDIOTONIC STEROIDS INHIBIT THE PUMP

H. J. Schatzmann (Fig. 8.1), a recent medical school graduate working in the early 1950s with Wilbrandt in Bern, happened on a major discovery from examining, at his mentor's suggestion, how mineralocorticoid hormones affect cation transport in red blood cells.[1] The rationale was: (1) mineralocorticoids such as aldosterone increase Na$^+$ reabsorption by the kidney; (2) mineralocorticoids are lipids (steroids) containing polar groups (hydroxyls and carbonyls) that could form complexes with cations; and (3) mineralocorticoids might promote Na$^+$ reabsorption as components of the Na$^+$ pump, serving as cation carriers (the notion of lipoidal carriers being then in vogue). But Schatzmann soon showed that mineralocorticoids had no effect on red blood cell cation transport.

Another group of steroids that were structurally similar[2] also had prominent pharmacological effects—the cardiotonic steroids. These could stimulate the heart; indeed, they represented a major therapy for such ills as congestive heart failure. Schatzmann realized that Na$^+$ and K$^+$ affected the force of cardiac contraction, and he

Fig. 8.1 Hans J. Schatzmann. (Photograph courtesy of H. J. Schatzmann.)

considered that cardiotonic steroids might act in the way proposed for mineralo-corticoids. When he tested these on red blood cells, he found, however, not stimula-tion of transport but inhibition.[3] (Schatzmann measured K^+ uptake and Na^+ loss at 37°, using cold-stored human red blood cells: he added cardiotonic steroids to John Harris's experiment.)

Various metabolic inhibitors were then known to decrease active transport, so finding another inhibitor might have seemed trivial. But when Schatzmann mea-sured red blood cell glycolysis, he found no inhibition by cardiotonic steroids. His conclusion, which was by no means trivial, was that "the transport mechanism is [affected] directly."[4]

These experiments and his interpretation, published in 1953, had an enormous impact, for Schatzmann had discovered a highly selective inhibitor. As work through

Fig. 8.2 Ian M. Glynn. (Photograph courtesy of Ian Glynn).

succeeding decades revealed, that specificity not only established cardiotonic steroids as preeminent experimental tools, it provided a defining criterion: cardiotonic steroids inhibited the pump, and whatever cardiotonic steroids inhibited reflected an aspect of the pump.

Attaining the status of a defining criterion began soon after, with the replication of Schatzmann's results in 1954 and the extension of those studies to muscle in 1954, frog skin in 1957, and squid axon in 1959.[5] And three further characteristics were soon reported from Cambridge. Ian Glynn (Fig. 8.2) used cardiotonic steroid inhibition to estimate the number of pump sites per red blood cell, "not . . . greater than about 1000 per cell."[6] Glynn also noted that extracellular K$^+$ diminished inhibition by cardiotonic steroids.[7] And Keynes and Peter Caldwell reported that the cardiotonic steroid ouabain inhibited transport when it was added to the bathing medium but that it had no effect even at far higher concentrations when injected into squid axon cytoplasm:[8] ouabain acted at one surface of the pump.

None of these papers, however, included measurements of metabolism. The nearest approach was Glynn's study of red blood cell fluxes: he identified "downhill" fluxes through the pump that were inhibited by the cardiotonic steroid digoxin, and he noted that such an effect on passive processes "suggests that the inhibitors act on the transport mechanism itself rather than on the energy supply to the pump."[9]

Others, however, addressed this issue explicitly. In 1957 H. A. Kunz and F. Sulser in Bern reported that cardiotonic steroids did not alter ATP levels in red blood cells.[10] The following year, Ronald Whittam in Cambridge showed that (1) cellular contents of both K[+] and ATP fell together when glucose was removed; (2) adding digoxin reduced K[+] content without affecting either ATP levels in cells at 37° or rises in ATP levels when cold-stored cells were warmed to 37° in the presence of glucose; and (3) digoxin did not affect K[+] content when glucose was absent, but in the presence of glucose digoxin produced the same K[+] loss as did the removal of glucose (Table 8.1).[11] Whittam concluded that "the effect of digoxin . . . was not due to a depletion of intracellular ATP [but] may be due to the inactivation of a membrane carrier directly."[12]

When Schatzmann's conclusion about cellular metabolism was eventually reexamined, however, the result was quite different, more reasonable, and far-reaching in significance. In 1961, with the link between ATP consumption and transport firmly established (see below), Whittam pointed out that "the rate of metabolism might be regulated by the rate of active transport [since metabolism] is governed by the concentration of . . . ADP . . . which is determined in turn by the activity of the various mechanisms causing hydrolysis of ATP."[13] He then showed, using slices of brain and kidney tissue, that ouabain reduced K[+] contents by 70% and 64% and oxygen consumptions by 50% and 45%, respectively. Whittam concluded that "active transport may act as a pace-maker [for metabolism, accounting] for about half the respiration of brain and kidney."[14] Equating a decrease in respiration with the share of cellular energy devoted to transport clearly required further definition, but the magnitude of the apparent share was nonetheless startling. Instead of Conway's characterization of remedial pumps balancing an insignificant leakiness, Whittam's

Table 8.1 Effect of Glucose and of Digoxin on Cellular Content of ATP and K[+]*

	Cellular Content (mmol/L cells)	
Conditions	ATP	K[+]
Before incubation	0.95	104
After incubation:		
with glucose	0.85	103
with glucose plus digoxin	0.90	94
without glucose	0.60	95
without glucose plus digoxin	0.60	94

*Fresh human red blood cells were incubated for 5 hours at 37° in the presence or absence of 11 mM glucose or of 0.01 mM digoxin, and the ATP and K[+] contents then measured. [From Whittam (1958), Table 5. Reprinted by permission of the Physiological Society.]

analysis implied that pumps demanded half the cell's energy output. At first glance, this seemed a waste of horrendous magnitude.

Three years later, Whittam pointed out that even in red blood cells the cardiotonic steroids decreased metabolism.[15] But these cells, whose permeability to Na$^+$ and K$^+$ is far lower than that of kidney or brain cells, allocated a far smaller fraction of metabolic energy to the pump: ouabain reduced glycolysis by only 16%. This small decrease was missed by Schatzmann's analytical method. Indeed, that was another stroke of good fortune, for if Schatzmann had first studied another cell type he would have found obvious parallels between inhibition of fluxes and that of metabolism, and he might well have judged cardiotonic steroids to be merely another metabolic inhibitor. Just that conclusion had been drawn earlier. In 1938 W. L. Francis and O. Gatty in Cambridge reported that a drop in frog skin potential "occurs with the respiratory inhibitors arsenite, Ouabaine [*sic*], fluoride, and iodoacetate."[16] They did point out, however, that compared with iodoacetate the "action of Ouabaine is . . . relatively more lethal to the potential and less so to the oxygen uptake,"[17] i.e., for a given reduction in potential, ouabain decreased oxygen consumption less: ouabain did not act precisely as iodoacetate. Francis and Gatty even suggested that ouabain "might affect the potential because it changed the properties of the cell surface in . . . a different way,"[18] but they did not pursue this issue.

COUPLING Na$^+$ AND K$^+$ TRANSPORT: A Na$^+$/K$^+$ PUMP

Is there Coupling?

As described in Chapter 4, Maizels and associates demonstrated that extracellular K$^+$ was required for net Na$^+$ outflux and that K$^+$ distributions could not be attributed solely to passive equilibration. In 1953, at a notable meeting in Bangor, Wales, Maizels concluded that there was "no evidence" that transport of Na$^+$ and K$^+$ "are independent [and it is] more likely [to be] closely integrated"; he suggested, moreover, that there were "common carriers for Na and K" with Na$^+$ liberated by a specific enzyme at the exterior face of the membrane and K$^+$ by a specific enzyme at the interior.[19] E. J. Harris argued similarly for "a single mechanism [for] active Na extrusion and active K accumulation."[20]

In 1953 Hodgkin and Keynes reported that omitting K$^+$ from media bathing giant axons markedly reduced ^{24}Na outflux, and they suggested that "there may be a coupling between K influx and Na outflux."[21] Although Hodgkin and associates had shown that the asymmetric distribution of K$^+$ across squid axon membranes was close to that calculated for a passive equilibration, Hodgkin and Keynes now envisioned an active transport of K$^+$ coupled to that for Na$^+$. At the Bangor symposium they cited parallels with Harris's and Maizels's results and concluded that "there may be some more specific form of linkage between potassium influx and sodium efflux" with "the most reasonable explanation [being that] potassium ions may . . . be drawn into the cells by a metabolic process coupled to one which simultaneously extrudes sodium."[22] By 1955 Hodgkin and Keynes proposed a "coupled system which ejects sodium from the axon on one limb of a cycle, and absorbs potassium on the other."[23] Keynes reported in 1954 that Na^+ outflux from muscle was reduced in K^+-free media, and he suggested "that there may be, as in cephalopod axons, some form of coupling between sodium efflux and potassium influx."[24]

Ussing had meticulously calculated that active K^+ transport need not occur in frog skin, but in 1954 he reported that omitting K^+ from the serosal bathing medium reduced Na^+ influx$_{epi}$.[25] Ussing wondered "whether there is always a compulsory transport of K in the direction opposite to that of the Na transport," but he warned that there was "still no evidence for such a process in . . . frog skin."[26] By 1958, however, Valborg Koefoed-Johnsen and Ussing had reanalyzed cation movements, and they now depicted a complex model that included a coupled pump (a crucial experimental factor was their blocking anion fluxes so potentials could be analyzed solely in terms of the cations present).[27] Voltage varied with the K^+ concentration in the serosal medium, according to the Nernst equation, but not with the Na^+ concentration; i.e., the serosal barrier was passively permeable to K^+ but only minimally to Na^+. Conversely, voltage varied with the Na^+ concentration in the medium bathing the outside surface, also according to the Nernst equation, but not with the K^+ concentration; i.e., the outer barrier was passively permeable to Na^+ but minimally to K^+. These characteristics also agreed with electrical potentials measured by penetrating microelectrodes, which changed in two steps across the cell: interpretable as the potential from outside medium to cytoplasm, and from cytoplasm to serosal medium.[28] Koefoed-Johnson and Ussing's model (Fig. 8.3), which established the precedent for such diagrams, showed a downhill (passive) flow of Na^+ from outside to cytoplasm and an uphill (active) flow from cytoplasm to serosal medium; this active outflux was coupled to an active influx of K^+ at the serosal surface, which in turn balanced a passive outflux across that same membrane. This model thus portrayed a net flux of Na^+ across the cell without a net flux of K^+. Ussing concluded that there was "a carrier-linked and possibly obligatory exchange of sodium . . . against potassium" and that the "simplest mechanism thus seems to be a coupled Na/K pump."[29]

Is there a Fixed Coupling Ratio?

Early studies on ^{42}K influx into human red blood cells had revealed *no* (detectable) change in rate with change in extracellular K^+ concentration,[30] but these experiments suffered from a wide scatter in the values. In 1954 David Streeten and A. K. Solomon found an increasing rate over a low range of K^+ concentrations; they fitted these data to a hyperbolic, "saturable" model (equivalent to substrate-velocity relationships of enzyme kinetics).[31] That same year Daniel Tosteson, recognizing that influx rates continued to increase at higher concentrations of extracellular K^+, proposed a composite scheme—a saturable influx plus a linear influx—but he did not examine fluxes in the lower concentration range that would have characterized the

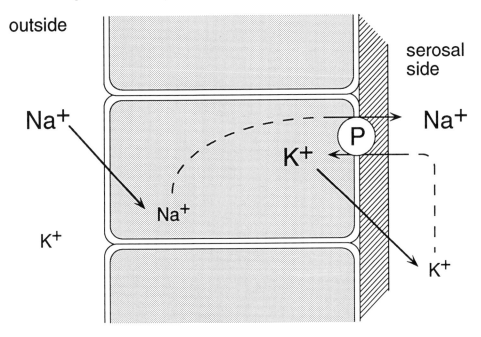

Fig. 8.3 Cation fluxes in epithelial cells of frog skin. Na$^+$ diffuses passively into the cell from the out side and is extruded to the serosal side by a coupled pump (P) that transports K$^+$ in from the serosal side; this K$^+$, however, then diffuses back out on the serosal side. [Redrawn from Koefoed-Johnsen and Ussing (1958), Fig. 6. Reprinted by permission of the Scandinavian Physiological Society.]

saturable component (Fig. 8.4; Tosteson presented data from previous reports as well, illustrating the difficulties in discerning just what the relationship actually was).[32]

Glynn now fitted his admirably precise ^{42}K influx data (Fig. 8.5) to such a two-component scheme, identifying an active transport pathway (saturable component) and a diffusion pathway (linear component).[33] (Trevor Shaw, who received his Ph.D. in 1954 at Cambridge, had just published similar analyses from studies on horse red blood cells.[34]) Glynn also reported that ^{24}Na outflux slowed when extracellular K$^+$ was reduced, and that, most strikingly, the concentration of extracellular K$^+$ was the same for both (1) the half-maximal activation of the saturable component of ^{42}K influx and (2) the half-maximal activation by K$^+$ of Na$^+$ outflux. This strongly tied K$^+$ influx (1) to Na$^+$ outflux (2). He also judged the active fluxes of Na$^+$ and K$^+$ to be "roughly the same magnitude," which "suggest[ed] a one-to-one exchange of sodium for potassium."[35] This stoichiometry satisfied a simple model for coupled transport (Fig. 8.6), although it contradicted Maizels's calculation of 2 Na$^+$/K$^+$ and Harris's 1.76 Na$^+$/K$^+$.[36]

Glynn's 1:1 stoichiometry was challenged within a year in a 1957 paper by Robert Post (Fig. 8.7) at Vanderbilt University.[37] Post measured net fluxes (avoiding

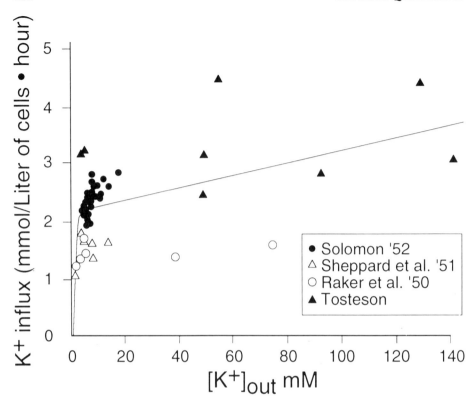

Fig. 8.4 K^+ influx as a function of extracellular K^+ concentration. The rate of ^{42}K accumulation by human red blood cells is plotted against the K^+ concentration in the medium. Included also are results of similar studies by Solomon, Sheppard et al., and Raker et al. [Redrawn from Tosteson (1955), Fig. 1. Reprinted by permission of the American Physiological Society.]

tracer fluxes, which are potentially confounded with exchange processes), using red blood cells that had been stored in the cold to alter their cation composition. When warmed to 37°, the cells lost Na^+ and gained K^+ (Fig. 8.8A)—again, a repetition of John Harris's experiment. The measured ratio of net fluxes, for a range of extracellular K^+ concentrations, was a spectacularly constant 3 Na^+ for 2 K^+ (Fig. 8.8B). A similar ratio resulted from experiments with a range of intracellular Na^+ concentrations. Furthermore, when active transport was blocked by cardiotonic steroids or induced by adding metabolic substrates, the ratio of active fluxes was still 3:2.

But others, studying different systems, did not find a fixed coupling ratio. For example, H. G. Hempling, after measuring Na^+ and K^+ transport by isolated tumor cells and observing different sensitivities of the individual fluxes to temperature and to metabolic substrates, concluded that "there are separate mechanisms for sodium and potassium."[38] And Abraham Shanes reported an uptake of K^+ by frog nerve in the absence of Na^+ outflux, which he thought "point[ed] to a mechanism capable of potassium transport independent of sodium."[39]

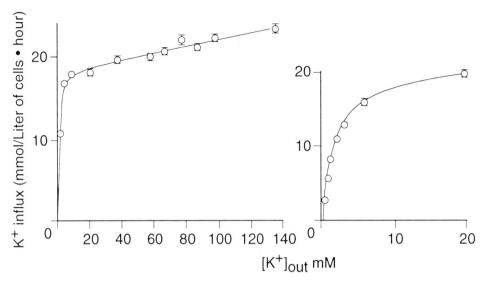

Fig. 8.5 K$^+$ influx as a function of extracellular K$^+$ concentration. The rate of ^{42}K accumulation by human red blood cells is plotted against the K$^+$ concentration in the medium. Data points are fitted with a curve representing an active hyperbolic (saturable) component and a passive linear component. The two experiments (with different lots of cells) cover different concentration ranges. [From Glynn (1956), Figs. 1 and 2. Reprinted by permission of the Physiological Society.]

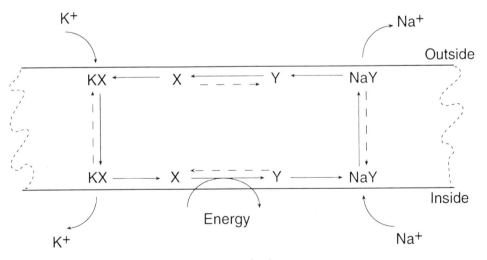

Fig. 8.6 A carrier model for coupled Na$^+$/K$^+$ transport. A carrier Y transports Na$^+$ across the membrane from inside to out. There it discharges Na$^+$, assumes a different character (becomes X), and accepts K$^+$, which it transports from outside to in. Metabolic energy drives the cycle, as in converting X to Y. [From Glynn (1956), Fig. 15; redrawn from Shaw's 1954 Ph.D. thesis. Reprinted by permission of the Physiological Society.]

Fig. 8.7 Robert L. Post (Photograph courtesy of Robert Post.)

MAINTAINING THE STEADY STATE: MATCHING PUMP FLUX TO LEAK FLUX

If Na^+ and K^+ can leak across cell membranes, active transport fluxes must be matched to these leak fluxes to ensure a steady state. That necessity underlay early discussions of cation pumps, but Tosteson and Joseph Hoffman presented an explicit and quantitative analysis in 1960, based on their examination of contrasting properties of red blood cells from two types of sheep.[40] "HK" sheep have red blood cells with high K^+ contents (like their other cells), whereas "LK" sheep have red blood cells with low K^+ contents (unlike their other cells), with those properties reflecting a heritable difference due to a single gene locus. From measuring ion fluxes in these two varieties of red blood cells, Tosteson and Hoffman constructed a specific model in terms of a coupled Na^+/K^+-pump, passive leaks for the cations, and exchange diffusion. That model accounted economically for the steady state in each cell type, despite rates for pump, leak, and exchange fluxes that differed widely between them.

(Exchange diffusion they equated to the fraction of ^{24}Na outflux abolished by replacing extracellular Na^+ with nontransported cations such as choline$^+$.[41] That operational definition agreed with Ussing's original formulation: 1:1 exchange demonstrable as tracer fluxes but not as net fluxes. Ussing had shown that exchange

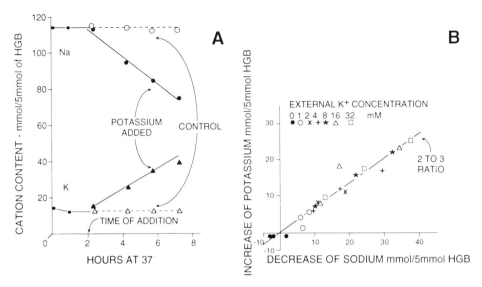

Fig. 8.8 Na⁺ loss and K⁺ gain by red blood cells. In *A*, cold-stored human red blood cells were warmed to 37° and after 2 hours, K⁺ was added to the medium (but not in control experiments). The measured cation content is plotted against incubation time: Na⁺ content decreased and K⁺ content increased (expressed relative to hemoglobin content, HGB). In *B* the results of such experiments, with a range of extracellular K⁺ concentrations, are plotted to show the ratio of Na⁺ loss to K⁺ gain. [From Post and Jolly (1957), Figs. 1 and 3. Reprinted with kind permission of Elsevier-NL, Amsterdam, The Netherlands.]

diffusion did not contribute appreciably to ^{24}Na fluxes across frog skin, but Keynes found with frog muscle that after replacing extracellular Na⁺ with choline⁺, the ^{24}Na outflux dropped sharply.[42])

ENERGY SOURCES FOR ACTIVE TRANSPORT

Redox Pump Hypotheses

Redox reactions are attractive candidates for cation transport mechanisms because they involve charge movement and their energy yield is huge. As noted in Chapter 6, Lund proposed in 1928 that redox processes underlay the electrical currents across frog skin. A decade later, Robert Stiehler argued similarly for redox reactions participating in transport epithelia of brain (choroid plexus) and eye (ciliary body), pointing out that electrical potentials across a barrier could drive cations in one direction and anions in the other.[43] Then, beginning in the late 1930s, H. Lundegårdh developed models for H⁺ transport in plant roots that was driven by redox reactions, including portrayals of rotating elements in the membrane that carried reactants and products from one surface to the other.[44] After World War II, R. N. Robertson

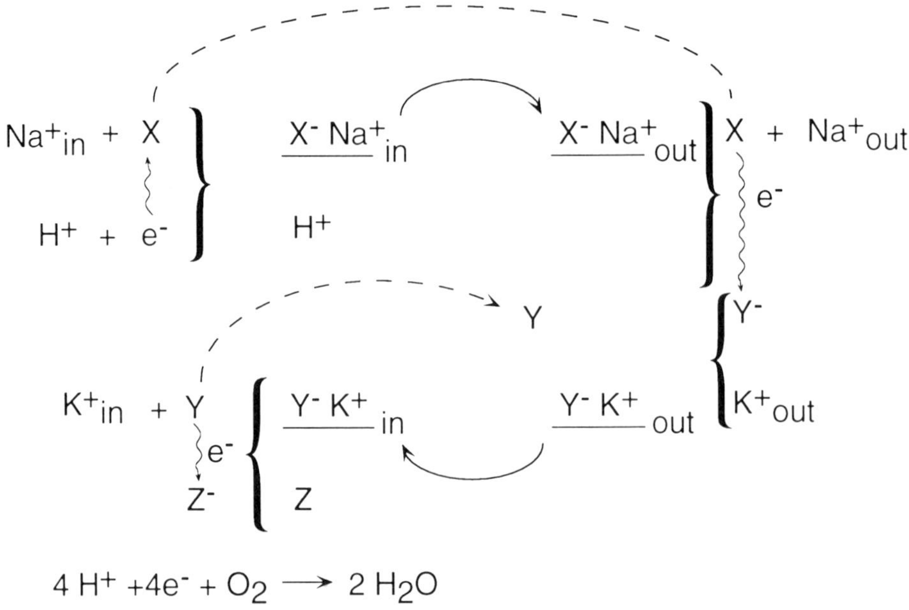

$$4 H^+ + 4e^- + O_2 \longrightarrow 2 H_2O$$

Fig. 8.9 Redox Na^+/K^+ pump. A redox carrier X accepts an electron from a hydrogen atom to become X^-, and then accepts Na^+ at the cytoplasmic face of the membrane. This complex moves to the extracellular face where Na^+ is released and X^- is oxidized back to X by donating an electron to Y. Y^- accepts K^+ at the extracellular surface and this complex moves to the cytoplasmic surface. There K^+ is released and Y^- donates its electron to Z. Z^- then donates its electron to the final acceptor, oxygen, which picks up H^+ to form water. For each O_2 reduced, two molecules of water are formed and four cations transported in each direction. [Redrawn from Conway (1953), Scheme III. Reprinted by permission of Academic Press.]

applied similar mechanisms to ion transport in plants, and Conway and R. E. Davies applied them in separate studies to acid secretion by the stomach.[45]

Here I shall concentrate on Conway's studies, for he vigorously promoted redox pumps as the *common* mechanism for ion transport, developing arguments for redox Na^+ pumps in muscle, nerve, and frog skin (as well as in yeast, which he also studied enthusiastically). A 1953 version of a pump effecting Na^+/K^+ exchange (Fig. 8.9) shows the transport of four Na^+ for each O_2 reduced.[46] This is a significant characteristic to which we shall return.

First, however, let us consider arguments advanced in favor of the redox pump. One was simplicity: "the most attractive feature of the 'redox pump' is the fact that the active carrier and the energy source are one and the same system."[47] Another was generality: "The systems are essentially the same for yeast and frog skin."[48] Beyond these, Conway listed five lines of experimental evidence.[49] Net extrusion of Na^+ in frog muscle was entirely inhibited by (1) cyanide, which was known to block cellular respiration (the sequence of redox reactions terminating in the reduction of oxygen); (2) by anoxia, which also indicated a dependence on cellular respiration; and (3) by reducing the temperature to 0°, which demonstrated a reliance on processes having strong temperature sensitivities, such as enzymatic redox reactions. (4) Net extrusion, however, was "relatively unaffected" by dinitrophenol, implying that oxidative phosphorylation was not involved since dinitrophenol uncouples oxidation from phosphorylation, thus prevent ATP formation. (5) Net extrusion of Na$^+$ was coupled to oxygen consumption with a ratio of 2 to 4 Na$^+$ transported for each O$_2$ consumed (which was in accord with Conway's scheme allowing a maximal ratio of 4 Na$^+$/O$_2$).

Two misgivings should be noted before questioning how robustly Conway's five characteristics argue against a specific, contemporary alternative. First is the issue of generality: notably absent from his list were red blood cells. Their omission was due to the simple fact that they cannot metabolize oxidatively; they lack the necessary enzymes. Conway would have been forced to grant them a different sort of pump if he had mentioned red blood cells, which he did not in his reviews.[50] Second is the problem of metabolic efficiency intrinsic to the stoichiometry of 4:1. The free energy required to move one Na$^+$ against a 10-fold concentration gradient and a 70 mV potential is 3 kcal; thus, the energy needed to move four Na$^+$ is 12 kcal. By contrast, the same amount of oxygen consumed by Conway's proposed pump to transport four Na$^+$ could instead generate five ATP, with their free energy of hydrolysis totaling roughly 50 kcal. On the other hand, Conway believed (as noted in Chapter 3) that transport was an inefficient process, so this concern probably bothered him little.

As for the five lines of evidence, the first three are, of course, equally consistent with ATP being the required energy source, for the bulk of the ATP generated in most animal cells is by oxidative phosphorylation. The stoichiometry of four Na$^+$ per O$_2$ is also compatible with ATP being the proximate energy source, but with ATP the energy yield would in principle allow a considerably higher ratio. Thus, only the fourth characteristic is incompatible with a role for ATP. Keynes, too, had reported that dinitrophenol failed to inhibit Na$^+$ outflux from frog muscle, but he found that cyanide failed also.[51] He argued that, since dinitrophenol did not block muscle contraction either, frog muscles evidently retained some energy reserves through even prolonged periods of metabolic inhibition.

In 1946 Conway had stated that cyanide failed to inhibit cation movements, and again in 1959 he admitted that in some instances it, as well as dinitrophenol, failed to inhibit.[52] He now reported ATP contents of frog muscle treated with dinitrophenol: ATP fell 23%, to 1.5 mmol/kg. Still unanswered, though, was the crucial question of whether that change was functionally significant. On the other hand, by 1955 dinitrophenol had clearly been shown to block Na$^+$ transport in frog skin and in squid axons.[53] Conway did not cite in his 1957 review these studies that surely undercut his claim to generality.

The point on which redox pump proposals was challenged most vigorously, however, was the issue of stoichiometry. In 1956 Zerahn (who had moved temporarily to Aarhus, a place prominent in Chapter 9) compared shortcircuit currents across frog skin, equated to net Na^+ transport, with oxygen consumption by the skin.[54] From the "associated" consumption (the difference between consumption when Na^+ was absent from media bathing the outer surface of the skin, so no transport could occur, and that when Na^+ was present and shortcircuit currents generated) Zerahn calculated that 16–20 Na^+ were transported for each O_2 consumed (Table 8.2). This was far beyond Conway's maximum of four. One obvious loophole Zerahn addressed: "It is possible that the oxygen consumed by the skin during the control periods [when Na^+ was absent from the medium] may be used for sodium transport when sodium becomes available."[55] But even if *all* the oxygen consumption were exclusively coupled to Na^+ transport, the ratio still averaged to more than four (Table 8.2). Zerahn noted that with a Na^+/O ratio of 9 and a P/O ratio of 3, then the "Na/ATP ratio is approx. 3."[56] Independently, Alexander Leaf and A. Renshaw, using slightly different approaches, found the same ratio: 18.2 ± 2.3 Na^+ transported per O_2 consumed.[57]

Before Zerahn's data were published, Ussing presented them in 1955 at a meeting in Madison, Wisconsin that was attended also by Conway.[58] Conway's printed rebuttal was "other electron acceptors, besides oxygen, may be present in relatively high concentrations [and so an] efficiency over 100 per cent . . . does not outrule the application of the redox-pump theory."[59] What these other acceptors might be he did not state, nor did he address why they were inoperable when oxygen removal halted transport. In his 1957 review, which quoted papers published after that meeting was held, he did not cite Zerahn's work; indeed, he did not refer to frog skin experiments at all, nor did he in a 1960 review that still proclaimed the redox pump.[60]

Table 8.2 Correspondence between Oxygen Consumption by Frog Skin and Na^+ Transport across it*

		Rate (μmol/hr)	
Date	Na^+ Transport	Total O_2 Consumption	Net O_2 Consumption
April 5	5.87	1.790	0.355
19	10.00	1.618	0.628
22	5.67	1.308	0.203
27	11.9	1.655	0.595
August 23	8.2	1.35	0.418
23	9.0	1.40	0.443
Mean values	8.44	1.52	0.44

*Na^+ transport was measured by the shortcircuit method. Net oxygen consumption is the difference between the total consumption and the consumption when no transport was occurring, i.e., when distilled water was in contact with the outer surface of the skin. [From Zerahn (1956b), Table 1. Reprinted by permission of the Scandinavian Physiological Society.]

ATP Provides the Energy for Transport

By the mid-1950s, demonstrations that dinitrophenol inhibited transport in several cell types provided a strong indication that "energy-rich" phosphate compounds— most likely ATP—drove the transport machinery. From his experiments with cardiotonic steroids (Table 8.1), Whittam had concluded that "ATP is at least one of the links in the coupling between glycolysis and active transport."[61] A year earlier E. T. Dunham, at the National Institutes of Health, described analogous experiments: as red blood cells were kept in glucose-free media, the decline in ^{42}K influx paralleled the decline in cellular ATP.[62] But a second abstract that year went significantly further.[63] Red blood cells were kept in glucose-free media until all intermediates of the glycolytic pathway were exhausted and the only (known) energy reserves were nucleotide phosphates (notably ATP and ADP) and 2,3-*bis*phosphoglycerate.[64] Added iodoacetate then caused no change in ^{42}K influx. This was explained by the absence of glycolysis that iodoacetate could inhibit. But adding glucose plus iodoacetate caused ^{42}K influx to decline. Glucose then served as a sink for the ATP present (consumed to form glucose-6-phosphate) without glucose being able to generate new ATP (since iodoacetate blocked metabolism of glucose-6-phosphate through the glycolytic pathway). And when ATP concentrations fell, ADP concentrations rose. Dunham identified "ATP as the principal [nucleotide] from which . . . energy is removed for transport work."[65]

Still, demonstrating declines after removing substrates is less satisfying than recording gains after adding them. The necessary ploy was adding ATP intracellularly. Such an experiment had been accomplished before Whittam's and Dunham's experiments, but it was published in 1954 in an obscure journal, *Acta Physiologica Academiae Scientiarum Hungaricae.*[66]

Gyorgy Gárdos used a technique described in 1953 in the same journal (the 1953 study also went unnoticed): F. B. Straub had developed a method for "resealing" red blood cell ghosts, making their membranes once again poorly permeable after the leakiness was induced during hemolysis.[67] Since the leakiness allowed the interior of the ghost to equilibrate with external media, resealing ghosts could trap within them new contents of the experimenter's choice.[68] Gárdos used this procedure not only to adjust cation contents within ghosts but also to incorporate ATP.[69] When cells containing more Na$^+$ than K$^+$ were enriched with ATP, they then accumulated K$^+$ from the medium, against the gradient. No glucose was required, and arsenate (which inhibits ATP production from glycolysis) did not block transport by ATP-enriched ghosts. Moreover, ATP levels fell in the presence of arsenate even though lactate production continued, and K$^+$ was then not accumulated. But adding ATP to arsenate-containing ghosts drove K$^+$ uptake.

Initially unaware of Gárdos's experiments, Hoffman had begun similar studies. While pursuing the recalcitrant problem of just what happens during hemolysis, he developed independently a procedure for resealing ghosts.[70] In an abstract published in 1960 and a subsequent full paper, Hoffman described ^{24}Na outflux from resealed ghosts:[71] outflux followed saturation kinetics, required extracellular K$^+$, was sensitive to the cardiotonic steroid strophanthidin, and depended on metabolic energy. Added ATP was most effective, although glycolytic substrates that promoted ATP generation (such as phosphoenolpyruvate) also drove outflux if ADP were present in "catalytic" amounts.[72] Higher concentrations of ADP alone were also effective; this was attributed to ATP being formed from ADP by adenylate kinase.[73]

Hodgkin and Keynes undertook analogous experiments using squid axons. In 1956 they reported injecting ATP into axons previously treated with metabolic inhibitors (to deplete endogenous stores of ATP), but they found no "really dramatic" recovery of ^{22}Na outflux (they noted, however, that "there was . . . doubt about the purity of the ATP sample").[74] The next year, Keynes and Caldwell showed that injecting ATP indeed restored ^{24}Na outflux from cyanide-poisoned squid axons.[75] Injecting arginine phosphate, however, had similar effects (arginine phosphate serves in squid and other invertebrates, as a high-energy source in lieu of creatine phosphate used by mammals), and the resolution of these results in favor of ATP required further examination.[76]

MECHANISMS DEPICTED FOR ACTIVE TRANSPORT

At the 1953 Bangor meeting, Danielli described five transport models (Fig. 8.10A–D).[77] The first model was a mobile carrier that diffused within the membrane, bearing sites to which transported solutes bind specifically. This possibility had been endorsed by such investigators as Davson, Hodgkin, Solomon, and Ussing;[78] moreover, attempts to isolate and identify carriers were being reported in the 1950s.[79] To achieve active transport, the carrier was coupled to systems that altered the affinity for the solute at one or both faces of the membrane. Transit across the thickness of the membrane relied on passive diffusion (on thermal energy).

The second model was a propelled carrier, which was linked to mechanisms directly moving the carrier plus bound solute across the membrane. Danielli suggested a contractile protein for the motor. Such mechanisms would merely facilitate equilibration, so further modifications were needed to alter affinity at one face or the other. R. J. Goldacre had envisioned contractile proteins that upon unfolding exposed sites, permitting binding; upon refolding these sites were lost, forcing release.[80]

The third and fourth models were rotating carriers or membrane segments. These models used rotating rather than reciprocating action, which had a certain mechanical appeal and fitted neatly with cyclical schemes such as those in Fig. 8.6. Although Danielli considered the energy required for the torque, he did not address the formidable problem of moving segments from the aqueous environment at either face of the membrane through the hydrophobic interior.

The fifth model depicted expanded lattices, which consisted of polypeptide chains whose relative spacing changed (driven by metabolic energy) so as to bind within their interstices certain solutes—and then release them. With asymmetric orientations and time courses, such expansions and contractions could propel solutes preferentially in one direction across the membrane. Danielli did not draw this model, but it is similar to three proposals that appeared soon after.

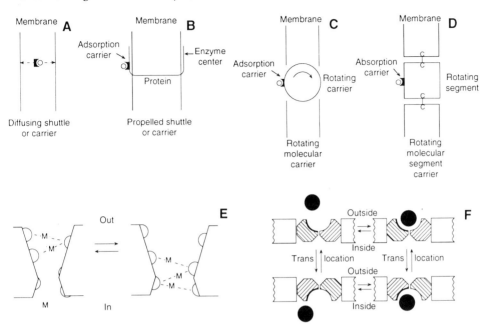

Fig. 8.10 Transporter models. [*A* through *D* are redrawn from Danielli, 1954, Fig. 1. Reprinted by permission of the Society for Experimental Biology. *E* is redrawn from Burgen, 1957, Fig. 3. Reprinted by permission of the National Research Council of Canada. *F* is redrawn from Mitchell (1957), Fig. 1. Reprinted by permission from *Nature*, ©1957, Macmillan Magazines Ltd.]

In 1956 G. S. Adair imagined labile cross-links between peptide strands that thus formed transient binding sites within proteins spanning the membrane. Propulsion could then be achieved sequentially across the membrane by creating and destroying these binding sites.[81] The next year, Arnold Burgen drew a conical, oscillating pore across the membrane (Fig. 8.10E). Binding sites would trap the ion at its narrow end (where binding was tightest) until dilatation of the pore (an oscillation that increased the diameter and thereby lessened the binding) released the ion preferentially to the nearer surface.[82] Also that year, Peter Mitchell, a former student of Danielli and just a few years from a still more salient formulation (see Chapter 17), proposed a rocking mechanism driven by metabolic energy, one that opened first to one face and then to the other (Fig. 8.10F).[83] These later models are direct ancestors of current "moving barrier" or "canal lock" models.

COMMENTARY

Places and People

To Dublin, Cambridge, London, and Copenhagen several new sites have been added in this chapter, but none with such broad significance as the National Institutes of

Health in Bethesda. The research on campus, in the intramural program, grew rapidly after World War II. This was accompanied by an influx of able scientists accommodated in a proliferation of new laboratories (Dunham, Hoffman, and Tosteson worked at NIH). Financial support of research off campus, in the extramural program, expanded even more prodigiously (see Chapter 10).

This chapter also has introduced a new generation of scientists, including Caldwell, Glynn, Shaw, and Whittam in Cambridge, and Dunham, Hoffman, Post, and Tosteson in the United States. (Of these, Glynn, Shaw, and Whittam received doctoral degrees from Cambridge, and Caldwell, Hoffman, and Tosteson spent postdoctoral years there.) But Dunham was soon incapacitated by illness, Shaw died quite young and Caldwell in his prime, and Tosteson's talent for administration diverted him to a department chair at Duke University and then deanships at the University of Chicago and Harvard University. Glynn, Hoffman, Post, and Whittam will reappear prominently in later chapters.

Communication

As recovery from the war proceeded, large international congresses attracted scientists, as did national meetings sponsored by individual disciplines. Probably more productive for those involved with membrane transport, however, were the smaller meetings focusing on that field, such as the meetings in Bangor in 1953 and in Madison in 1955. And in 1958 Hoffman and Tosteson founded the Red Cell Club, which continues to provide a salutary forum for personal and professional exchanges.

The fruits of those communications are evident in the spread of similar experiments as well as in the responses to explicit challenges. The use of cardiotonic steroids jumped rapidly from studies in red blood cells to those in muscle, squid axon, and frog skin, as did experiments omitting extracellular K^+ and adding metabolic inhibitors. Conversely, those outside this common circle risked oblivion, their work invisible; Straub and Gárdos published in German in an obscure journal, and their valuable work went unnoticed for years.

Accuracy

Since some commentators are so confident that accuracy can assume no responsibility for choice of theory, even a single instance to the contrary may be therapeutic. Compare Fig. 8.4 with Fig. 8.5. Fig. 8.4 provides little sense of how K^+ influx responds to the concentration of extracellular K^+. Indeed, several creators of these data asserted that influx did not change with concentration. Fig. 8.5, however, allows two mechanistically plausible processes to be fitted readily. In the years following the publication of Fig. 8.5, that explanation flourished as data similarly accurate became commonplace. The conceptual transition—to recognizing systems of simple diffusion in parallel with active transport—thus accompanied more accurate measurements. And that greater accuracy was due to identifiable changes in experimental approach.

1. The gain of ^{42}K by cells was measured rather than the loss of ^{42}K from the media.
2. Washed red blood cells were incubated in artificial media rather than whole blood, and washed cells were vastly diluted so that K^+ leaking from them, or K^+ removed by them, would not alter extracellular K^+ concentrations appreciably.
3. At the end of the incubation, cells were carefully washed free of incubation media.
4. The ^{42}K was expressed relative to hemoglobin contents rather than cell volumes, representing changes in *amount* of ^{42}K rather than *concentration*.

Conflict Resolution

Three episodes here represent a range of circumstances in which experiments and interpretations clashed, with rival camps responding variously.

The most clear-cut resolution was Zerahn's demonstration that Conway's redox pump hypothesis could not apply to frog skin. Zerahn's values for the Na^+ to O_2 ratio were soon confirmed, and after an initial attempt to salvage the redox pump for that site, Conway tacitly acquiessed and no longer advocated his model for frog skin. It would require considerable imagination to recast the resolution of that conflict without acknowledging a primary role for experimental results. H. M. Collins and G. Cox opined that "if writers wish to soothe themselves . . . by endorsing an 'objective' world, they should make clear exactly where they think it must intrude in accounts of scientific practice."[84] How about when reproducible measurements showed that 18 Na^+ moved per O_2 consumed?[85]

Less clear-cut and possibly more typical was the debate over the redox pump in frog muscle. Here experimental results were at odds and far from conclusive, requiring the participants to judge among disparate lines of evidence. Conway repeatedly asserted that dinitrophenol's inability to block Na^+ outflux argued against ATP being the proximate energy source, whereas cyanide's ability to block Na^+ outflux favored the redox pump hypothesis. Keynes challenged the latter observation, however, and Conway later acknowledged that inhibition was variable. The former observation, on the other hand, was agreed on by both. But what did it mean? Ideal science *obligates* the experimenter to demonstrate that the inhibitor is actually doing what is claimed. Conway eventually reported the ATP levels after treatment with dinitrophenol: they fell from 1.97 to 1.51 mmol/kg.[86] But still unanswered was whether that drop made a functional difference to muscle cells. One easy approach— one that Conway did not take—was to compare the drop in ATP due to cyanide (which [sometimes] inhibited outflux and which blocks ATP production by inhibiting the respiratory chain) with the drop in ATP due to dinitrophenol. If dinitrophenol reduced the ATP level at least equally but only cyanide blocked Na^+ outflux, then Conway's argument would move forward.

In designing new experiments, a scientist in an unsettled field often must draw, even if tentatively, from one among rival hypotheses. In this case, most pursued an ATP-driven pump. My reading of the papers written through the 1950s endorses their choice as one substantiated by the evidence then available. Events in the decades thereafter confirm these scientists' assessment.

Yet another course is exemplified by the reception accorded to Post's report that the red blood cell pump moved three Na^+ out for every two K^+ moved in. No other studies covered so broad an experimental range or fitted so consistently a single ratio.[87] Still, 3:2 was an awkward ratio, and the inability to formulate a tidy mechanism to achieve that stoichiometry may have motivated the public response to that paper, rather than some established authority advocating an alternative view. In any case, the response was to ignore Post's paper, an inattention due perhaps to skeptical dismissal or agnostic restraint. In the 3 years following its publication, only eight papers cited it, and of these only three mentioned the 3:2 ratio. One study contrasted it with the authors' finding a 2:1 ratio in muscle; another contrasted it with their finding no fixed ratio in nerve; and the third contrasted it with their calculation of a 1:1 ratio in red blood cells (evaluated from radioactive tracer fluxes).[88] No one published an attempt to replicate the study.

The 3:2 pump ratio also implied an electrogenic pump, since more charge is moved in one direction than the other (assuming no other cations were transported in or some anion out). But no one explored that possibility either, despite experiments published that same year: J. M. Ritchie and R. W. Straub interpreted the hyperpolarization of nerve following repetitive stimulation as a consequence of increased pump activity (they showed that ouabain blocked hyperpolarization).[89] They suggested two alternative mechanisms: an electrogenic pump that directly hyperpolarized the membrane, or a pump that exhausted the extracellular K^+ outside the membrane by pumping K^+ into the axon faster than it could diffuse from the bath, so that the concentration gradient for K^+ across the membrane was then increased. They favored the latter explanation, but not on the basis of any experimental evidence.

Scientific Summary

Cardiotonic steroids, such as ouabain, inhibited active Na^+ outflux in muscle, red blood cells, squid giant axon, and frog skin, directly and specifically. For these cell types, experimental evidence increasingly supported the notion of a coupling between active Na^+ outflux and active K^+ influx. Instead of a Na^+ pump, these cells thus seemed to use a Na^+/K^+ pump. Studies on red blood cells were most extensive and most convincing, with demonstrations that the K^+ sensitivity for K^+ influx was identical to the K^+ sensitivity for Na^+ outflux and that the coupling ratio was three Na^+ pumped out for every two K^+ pumped in. With other cell types, the evidence was less comprehensive and less compelling, but Na^+ outflux generally was seen to decrease when extracellular K^+ was absent. The proximate energy source for the pump, according to a number of diverse studies, was ATP.

NOTES TO CHAPTER 8

1. For autobiographical notes see *Current Contents* (1984), number 45, p. 23, and Schatzmann (1995).

2. Both mineralocorticoids and cardiotonic steroids contain a "cyclopentanoperhydrophenanthrene" nucleus, but they differ significantly in the folding of their rings.

3. Schatzmann (1953). Of the large family of cardiotonic steroids, two glycosides are important here: digoxin, which is widely used clinically, and ouabain, which is favored in laboratory studies because of its solubility in water (it can also be purified by recrystalization, an important advantage during this period). Ouabain minus its sugar, the aglycone, is strophanthidin; this highly lipid-soluble compound readily diffuses across cell membranes, unlike ouabain.

4. Ibid., p. 354.

5. Joyce and Weatherall (1954); Matchett and Johnson (1954); Koefoed-Johnsen (1957); Caldwell and Keynes (1959).

6. Glynn (1957a), p. 172. Current estimates are on the order of a hundred sites.

7. Solomon et al. (1956a) had also reported antagonism between cardiotonic steroids and extracellular K$^+$.

8. Caldwell and Keynes (1959).

9. Glynn (1957a), p. 172.

10. Kuntz and Sulser (1957).

11. Whittam (1958).

12. Ibid., p. 494.

13. Whittam (1961), p. 603.

14. Ibid., p. 604.

15. Whittam et al. (1964).

16. Francis and Gatty (1938), p. 136.

17. Ibid., p. 136.

18. Ibid., p. 137.

19. Maizels (1954), pp. 211, 224.

20. Harris (1954a), p. 236.

21. Hodgkin and Keynes (1953a), p. 47P.

22. Hodgkin and Keynes (1954), p. 434.

23. Hodgkin and Keynes (1955a), p. 55.

24. Keynes (1954), p. 359.

25. Ussing (1954); he cited the earlier publication by Huf and Wills (1951) reporting similar observations.

26. Ibid., p. 20.

27. Koefoed-Johnsen and Ussing (1958). This model should be contrasted with a quite different view described in 1954 by Linderholm.

28. Engbaek and Hoshiko (1957).

29. Koefoed-Johnsen and Ussing (1958), p. 305; Ussing (1960a), p. 146.

30. Raker et al., (1950); Sheppard et al. (1951); Solomon (1952).

31. Streeten and Solomon (1954).

32. Tosteson (1955); the publication date was a year later than the meeting. Tosteson pointed out that although the line fitting the higher concentrations had a slope greater than zero, the slope was not statistically different from zero.

33. Glynn (1956).

34. Shaw (1955).

35. Glynn (1956), p. 302.

36. Maizels (1954); Harris (1954).

37. Post and Jolly (1957).

38. Hempling (1958), p. 580.
39. Shanes and Berman (1955), p. 188.
40. Tosteson and Hoffman (1960).
41. Coupled transport of Na^+ with K^+ was impossible since extracellular K^+ was also absent. In addition, the exchange flux was insensitive to cardiotonic steroids.
42. Ussing and Zerahn (1951); Keynes and Swan (1959).
43. Stiehler and Flexner (1938); Friedenwald and Stiehler (1938).
44. See, for example, Lundegårdh (1939, 1940).
45. Robertson and Wilkins (1948); Conway and Brady (1948); Crane et al. (1948).
46. Conway (1953).
47. Conway (1951), p. 273.
48. Conway (1955), p. 387.
49. Conway (1957a).
50. Conway (1953, 1954, 1955, 1957a, 1960).
51. Keynes and Maisel (1954); Frazier and Keynes (1959).
52. Conway (1946); Carey et al. (1959).
53. Fuhrman (1952); Hodgkin and Keynes (1953b, 1955a).
54. Zerahn (1956a,b).
55. Zerahn (1956a), p. 938.
56. Zerahn (1956b), pp. 312–313. At that time, oxidative metabolism was thought to provide P/O rations of three.
57. Leaf and Renshaw (1957).
58. Ussing (1957).
59. Conway (1957b), p. 194.
60. Conway (1957a, 1960). Earlier, he had cited repeatedly (Conway 1953, 1954, 1955) Linderholm's paper (1952) advocating redox pumps in frog skin. However, in 1958 Conway cited studies on frog skin and then suggested that ATP might react with the system to increase the stoichiometry (Conway, 1959).
61. Whittam (1958), p. 495.
62. Dunham (1957a).
63. Dunham (1957b).
64. 2,3-bisphosphoglycerate can serve as a fuel through conversion to phosphoeneol-pyruvate; phosphoenolpyruvate is then converted to pyruvate in a reaction coupled to ATP formation.
65. Dunham (1957b).
66. Gárdos (1954).
67. Straub (1953).
68. There were limits to such substitutions, however, and introducing polyvalent ions such as ATP was particularly tricky.
69. Later assessment of those experiments raised the possibility that alternative processes were involved; see Hoffman (1962a, 1980).
70. Hoffman et al. (1960).
71. Hoffman (1960, 1962a). The submission date of the latter paper was September, 1960.
72. As long as tiny amounts of ADP were present, ATP could be formed from it. When that ATP was used to drive transport the ADP then formed would be available for conversion back to ATP. Thus, ADP was recycled: like a catalyst, it was not consumed.
73. Adenylate kinase catalyzes the conversion of two ADP to one ATP and one AMP; since the equilibrium constant for the reaction is near 1.0, ADP can thus maintain an ATP concentration near its own.
74. Hodgkin and Keynes (1956), p. 611.

75. Caldwell and Keynes (1957).

76. Caldwell et al. (1960a,b). Na^+ outflux had an absolute requirement for extracellular K^+ with arginine phosphate, as in unpoisoned axons, but not with ATP. This suggested that arginine phosphate played the physiological role. Caldwell and associates argued, however, that the coupling of Na^+ outflux to K^+ influx was sensitive to the concentration of ADP. Adding ATP increased ADP levels in axons poisoned with cyanide (ATP was hydrolyzed but could not be resynthesized in poisoned axons), whereas adding arginine phosphate led to a recycling of ADP back to ATP over uninhibited pathways. Subsequent studies confirmed this hypothesis.

77. Danielli (1954).

78. See, for example, Davson and Reiner (1942); Hodgkin and Katz (1949); Ussing (1949c); and Solomon (1952).

79. See, for example, Solomon et al. (1956b); Kirschner (1958); and Sanui and Pace (1959).

80. Goldacre (1952).

81. Adair (1956).

82. Burgen (1957).

83. Mitchell (1957a). Jardetzky (1966) reinterpreted this as an "allosteric" pump.

84. Collins and Cox (1976), p. 438.

85. That determination was, of course, filtered through layers of enabling theories of electricity, electrochemistry, chemistry, and gas laws. But that theory was shared by Conway and Zerahn.

86. Carey et al. (1959).

87. See, for example, Maizels (1954); Harris (1954); Glynn (1956); and Hodgkin and Keynes (1955a).

88. Fuhrman (1959); Shanes (1958); Gill and Solomon (1959).

89. Ritchie and Straub (1957).

chapter 9

IDENTIFYING THE
Na⁺/K⁺-ATPase

C HARACTERIZATION of the Na^+/K^+ pump was by no means completed in the 1950s. Essential concerns—from assessing conflicting reports to generalizing isolated claims—awaited resolution. But beyond clarifications and elaborations lay a basic question: What is the Na^+/K^+ pump? The reductionistic program of biological explanation translates that query as: What is the chemical nature of the pump? A related question is: How does this chemical nature endow it with the functional properties observed?

In a thorough review of red blood cells published in 1957, Glynn pointed in a likely direction: "energy [for the pump] seems to be made available as ATP [so] the role of the ATP-ase on the outer surface of the red cell . . . might repay investigation."[1] But how might one link an ATPase to the pump and reveal its transport capabilities? Glynn discussed enzymes that were stimulated and inhibited by Na^+ and K^+ and contemplated transport mechanisms being "more active when binding one [cation] than the other."[2] Accordingly, he considered mechanisms by which "energy is fed into the transport system through reactions of a K^+ activated enzyme situated on the cell surface," and referred to "sodium or [emphasis added] potassium activated enzymes . . . responsible for transport."[3]

Studies published by the time Glynn wrote his review had recorded only meager progress toward such goals, however. In 1952 Maizels described ATPase activity localized to the outer surface of red blood cell membranes.[4] Maizels contemplated a transport system associated with phosphorylation at the inner surface (linked to K^+ release and Na^+ uptake) and hydrolysis at the outer (linked to Na^+ release and K^+ uptake), but he did not explore that possibility. In 1956 Edward Herbert in Philadelphia described ATP hydrolysis by red blood cells as well as ghosts; however, he found that "the presence of Na ions up to a concentration of about 0.4% has practically no effect on the activity."[5] In his second abstract of 1957, Dunham reported attempts to relate membrane ATPases to transport systems

Fig. 9.1 Jens Christian Skou. (Photograph courtesy of J. C. Skou.)

through their similar sensitivities to pH changes.[6] That feeble[7] attempt by a perceptive investigator underscores the inability of those then active in this research to design experiments linking an ATPase to Na^+/K^+ transport. Nevertheless, such experiments had actually been completed by the time Glynn's review and Dunham's abstract appeared, completed by someone outside the transport field geographically as well as conceptually.

SKOU DISCOVERS THE Na^+/K^+ ATPase

What Skou Reported

Early in 1957 appeared an extraordinary paper by a physiologist at the University of Aarhus who had been studying local anesthetics, Jens Christian Skou (Fig. 9.1).[8] Skou began by pointing out that Na^+ extrusion is required for nerve to function but that "the mechanism of this transport is not known"; he also cited arguments for energy-rich phosphate esters being utilized and for squid axon membranes possessing ATPase activity.[9] Skou next described his preparation of a membrane fraction

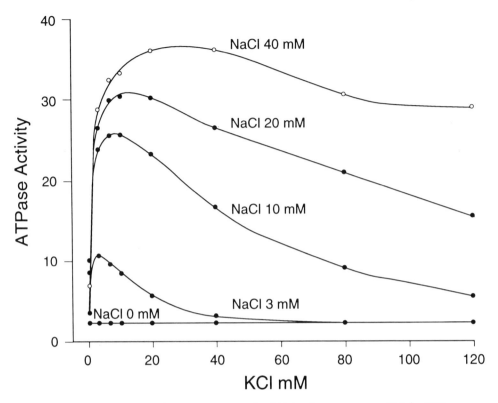

Fig. 9.2 Stimulation of ATPase activity by K^+ in the presence of Na^+. ATPase activity (μg of phosphate released from ATP during 30-minute incubations at 36°) of a crab nerve membrane fraction is plotted against the K^+ concentration of the medium, in the presence of the Na^+ concentrations indicated. [From Skou (1957), Fig. 5. Reprinted with kind permission of Elsevier-NL, Amsterdam, The Netherlands.]

from crab leg nerves (because squid axons were unavailable). He also specified how ATP was neutralized without adding the usual inorganic cations[10]—a novel approach (as well as a detail generally omitted in previous accounts, where the cations involved were rarely identified). Skou then described the ATPase activity of that membrane fraction. (1) ATP was hydrolyzed to ADP and phosphate, whereas ADP was not hydrolyzed. (2) Mg^{2+} was required for ATPase activity, whereas Ca^{2+} inhibited activity when it was added with Mg^{2+}. (3) Na^+ added in the presence of Mg^{2+}, however, increased activity fourfold, whereas K^+ added with Mg^{2+} had no effect. And (4) K^+ added in the presence of Mg^{2+} plus Na^+ increased activity even more, to 10-fold that with Mg^{2+} alone (Fig. 9.2).

Skou had thus discovered a $(Na^+ + K^+)$–stimulated ATPase activity[11] whose function he explicitly proposed: "the crab nerve ATPase . . . seems to fulfil a number of conditions that must be imposed on an enzyme . . . involved in the active extrusion of sodium ions."[12]

How Skou Found the Na^+/K^+-ATPase[13]

Skou received his medical degree from the University of Copenhagen in 1944 and proceeded to hospital training in Hjørring for a career in surgery. There, from the rigors of such training and of that era, he developed rheumatic fever, spent a month in bed, and decided to find a less strenuous job. Accordingly, he moved in 1946 to an orthopedic hospital in Aarhus; some months later when a position opened in the Department of Physiology at the University of Aarhus, he took it.

Research theses were necessary for careers in academic medicine, and Skou set about studying the mechanism by which local anesthetics block nerve conduction, extending interests developed during his surgical practice. He chose as a model for nerve membranes a monolayer of lipid—extracted from nerve—spread atop an aqueous solution. From measuring changes in the monolayer surface pressure when he added local anesthetics, Skou developed a plausible hypothesis: local anesthetics affected the membrane proteins governing nerve activity by increasing the surface pressure in the surrounding lipid phase. To explore that proposal, Skou needed protein in his lipid monolayer model, specifically one whose function he could monitor. Acetylcholinesterase, an enzyme with high intrinsic activity (and particular neural significance) seemed a good candidate, and Skou received a 2-month fellowship to visit David Nachmansohn at Columbia University, the foremost authority on acetylcholinesterase, to learn how to prepare that enzyme. Skou arrived in July 1953, followed Nachmansohn to Woods Hole, and then returned to New York in August to prepare acetylcholinesterase, which he took back to Aarhus.

While at Woods Hole, Skou was offered a book (edited by Nachmansohn) that cited Libet's description[14] of Ca^{2+}-activated ATPase activity in squid axon sheath, which Libet thought might be involved in nerve impulse conduction. In 1954 Leo Abood and Gerard reported that the axonal ATPase activity was instead greater with Mg^{2+} than Ca^{2+} (they also described a centrifugation scheme that Skou adapted to crab nerve).[15]

With that background, Skou began preliminary experiments in November and December 1954, but he found the measured ATPase activity of crab nerve membranes to be inexplicably erratic—high in some experiments, far lower in others. Skou stopped at this point. His interests centered on local anesthetics, and he had published a series of papers that were attracting considerable attention[16] and had secured his promotion to associate professor in 1954. Still, at the beginning of June 1955 he resumed his studies on crab nerve ATPase, and by the end of that month he had traced the variability to different cations added when ATP was neutralized: sometimes he had used Na^+, sometimes K^+.[17] When he returned from vacation in August, Skou compared Na^+ and K^+ salts of ATP and soon established the cation dependence of the ATPase activity. These results he submitted for publication in June 1956.[18]

Merely working out the unanticipated dependence on Na^+ and K^+ was a significant achievement, one that no previous investigator had undertaken. But recognizing the physiological significance of that activation was an even greater achievement. Skou's interest in local anesthetics had familiarized him with studies on nerve conduction so he, like Libet, first thought in terms of the Na^+ and K^+ fluxes of action potentials. Skou quickly realized, however, that the ATPase was far more likely to participate in restoring the ion gradients, although he was generally unfamiliar with the membrane transport field (his report cited no papers on muscle, red blood cells, or frog skin). And fearing that the term "pump" in the title would be "too provocative," he settled for "The Influence of some Cations on an Adenosine Triphosphatase from Peripheral Nerves."

PRECURSORS

In addition to those who had measured ATPase activity in red blood cells and squid axons cited above, others had been measuring ATPase activity of brain for still different reasons. One interest was exploring why Na^+ decreased glycolytic activity *in vitro*. In 1950 Merton Utter in Cleveland argued that Na^+ stimulated the destruction of ATP.[19] ATP hydrolysis by brain homogenates increased with the Na^+ concentration added, from 3 to 80 mM (Fig. 9.3A); 24 or 48 mM potassium bicarbonate buffer was present. Thus, he found an ATPase stimulated by Na^+ in the presence of K^+. Moreover, Utter noted that the activity of the homogenate could be sedimented by high-speed centrifugation: it was "particulate" (membrane-bound).

Utter was a first-rate biochemist—he subsequently received the Paul Lewis award for enzymology and was elected to the National Academy of Sciences—but he did not explore this phenomenon further. He had accurately located a point at which Na^+ inhibited glycolysis *in vitro* and evidently was uninterested in what function Na^+-stimulated ATPases might have *in vivo*. In 1954 Steinbach reported "personal communications" with Utter about cation effects on enzymes,[20] but that interaction inspired neither of them to study these effects. Glynn cited Utter's report in his 1956 paper as well as in his reviews[21]—it was the only example of a Na^+-stimulated enzyme—but he made no comment on K^+ being also present in the reaction media.

Through 1960, 46 papers cited Utter's study, but only four dealt with monovalent cation effects on ATPases.[22] One was Skou's 1957 paper. Another was by J. A. Muntz and J. Hurwitz in Utter's department.[23] They noted that in the presence of 25 mM sodium bicarbonate adding 10 mM K^+ increased ATP hydrolysis by a brain extract, more than did adding 10 mM Na^+. But their interest was the antagonistic effects of Na^+ vs. K^+ on metabolism. The third, by Alex Novikoff and associates in Burlington, Vermont, described their failure to find such responses with liver homogenates: "neither sodium nor potassium ions had any significant effect on [ATPase] activity" at concentrations from 3 to 30 mM, "in a system in which all the other salts were potassium."[24]

The fourth paper, published in 1951, was by Marion Gore in London,[25] whose

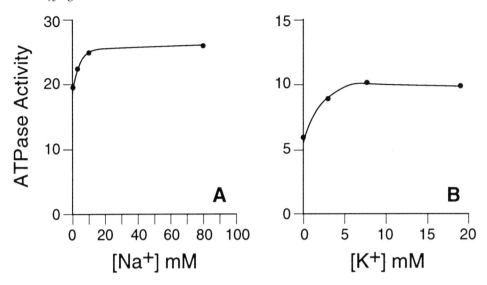

Fig. 9.3 Utter's Na$^+$-stimulated ATPase and Gore's K$^+$-stimulated ATPase. *A:* ATPase activity of a particulate brain fraction is plotted against the concentration of Na$^+$ added to the medium; 24 or 48 mM potassium bicarbonate was also present (Utter did not state which). [Redrawn from Utter (1950) Fig. 5. Reprinted by permission of the American Society for Biochemistry and Molecular Biology.] *B:* ATPase activity of a brain homogenate is plotted against the concentration of K$^+$ added to the medium; Na$^+$ was the counterion in neutralizing the 5 mM ATP present and perhaps with the 100 mM glycylglycine buffer. [Plotted from the data in Gore (1951), Table 2. Reprinted by permission of the Biochemical Society.]

interest lay in ATP hydrolysis as an energy source for brain "activity." Gore's results are the complement of Utter's: ATPase activity increased with the K$^+$ concentration, from 0 to 19 mM (Fig. 9.3*B*); Na$^+$ was stated to be present with ATP and presumably was with the 100 mM glycylglycine buffer also.[26] She thus described a K$^+$-stimulated ATPase in the presence of Na$^+$.

Gore's K$^+$-stimulated ATPase did not appear on Glynn's list of cation-sensitive enzymes, and no one else put together Utter's and Gore's findings either. Of the 22 papers citing Gore's report prior to 1962, only four—all preceding Skou's paper— commented on the K$^+$ stimulation. Henry McIlwain, in whose group Gore worked, twice mentioned it without suggesting any physiological role or noting that Na$^+$ was present in Gore's reaction medium (he also cited Utter's paper but did not mention the Na$^+$ stimulation).[27] Muriel Findlay and associates in London, Ontario also merely cited Gore's finding K$^+$-stimulated ATP hydrolysis.[28] On the other hand, Oliver Lowry in St. Louis reported in 1954 that "small amounts of K . . . stimulat[ed ATPase activity], as Gore had found";[29] he used 5 mM ATP as the Na$^+$ salt, did not cite Utter, made no suggestion about the functional significance of such an ATPase, and did not return to the problem.

SUCCESSORS

Skou first announced his results at an international neurochemistry meeting in Aarhus in July 1956, where he described ATPase activity that he linked to Na$^+$ extrusion from nerve.[30] Also present[31] at this meeting was Alfred Pope, a neuropathologist from Harvard University who had been measuring the activities in cerebral cortex of several enzymes, including ATPases, which he studied with Helen Hess. They too had noticed cation effects.[32] Hess and Pope then corroborated Skou's work with their studies on rat brain, described in an abstract for the 1957 spring FASEB meeting:[33] adding 145 mM Na$^+$ plus 10 mM K$^+$ doubled ATPase activity. But they did not refer to cation transport, and their title was merely "Effect of Metal Cations on Adenosinetriphosphatase Activity of Rat Brain." (The full paper was not published until 1962;[34] their interests lay elsewhere.)

Post read Skou's 1957 paper when it was published, thinking at first that Skou had found at least part of the pump, and then thinking probably not (because of the strong inhibition by excess K$^+$ that Skou's reported).[35] Post had met Skou at Woods Hole in 1953 and had driven him to the International Physiological Congress in Montreal that summer. Post next came across Skou at the International Congress of Biochemistry in Vienna in 1958. Skou was presenting his work with local anesthetics, which was still his major interest, but mentioned to Post that he "thought he had found the pump."[36] Post asked, "Have you tried ouabain?" And Skou replied, "What is ouabain?"[37] Skou telephoned to Aarhus to arrange such experiments, and when Post followed Skou to Aarhus after the meeting the results were ready. Ouabain inhibited ATPase activity and Post was convinced of the coupling to transport. By 1958 Skou had extended his work on ATPase activity to experiments with brain and other tissues. Post, who was experienced in red blood cell transport studies, asked Skou if he could pursue that aspect. Skou graciously agreed and never published on red blood cells.

From Aarhus Post went to Cambridge. Glynn had completed his military service in 1957 and returned to the Physiological Laboratory. He was joined by Dunham, who had gone to Cambridge at Tosteson's recommendation, and together Dunham and Glynn were beginning studies on red blood cell transport and ATPase activity; they had by then seen Skou's paper. When Post visited, they outlined to him their plans. Then at the Red Cell Club meeting in October 1958, Post relayed Skou's findings to Tosteson and Hoffman.

Back in Nashville, Post set about characterizing the ATPase activity of red blood cell ghosts, comparing it to the pump activity of intact cells. His abstract for the 1959 spring FASEB meeting reported that ATPase activity was stimulated twofold when both Na$^+$ and K$^+$ were present but not with either alone, and that the concentrations of Na$^+$ and K$^+$ needed to activate both ATPase and pump agreed closely, as did the concentrations of ouabain needed to inhibit both.[38] His abstract concluded: "The correspondence between the actions of these agents on active transport in the intact cell and on ATPase activity in isolated membranes is evidence that the ATPase is part of the active transport system."

In 1960 Post's full paper elaborated on these findings.[39] Using an extensively washed ghost preparation, he distinguished between "(Na$^+$ + K$^+$)-dependent" and "(Na$^+$ + K$^+$)-independent" ATPase activities; the former required *both* Na$^+$ and K$^+$ (Table 9.1) and only it was sensitive to ouabain. The coupling between Na$^+$ and K$^+$ harked back to his demonstration of a fixed transport ratio (see Chapter 8). Now he interpreted that coupling in terms of Shaw's and Glynn's model of a sequential system: "on each side of the membrane the linked transport system must free itself of one [cation] as transported product before it can take on the other as transportable substrate."[40]

(Finding ATPases stimulated by (Na$^+$ + K$^+$) far more than by Na$^+$ alone became the norm. Stimulation by Na$^+$ in the nominal absence of K$^+$ could in some cases be accounted for by contaminating K$^+$,[41] and enzymes from different species had different requirements. The cation requirements of the crab nerve enzyme have not been reinvestigated; it is not a convenient preparation: Skou estimated that his studies consumed 25,000 crabs.)

Most important for convincing the world of a link between ATPase and pump was Post's compilation of nine similarities between the two. Both (1) were in the membrane; (2) used ATP but not an alternative nucleotide; (3) required Na$^+$ and K$^+$ together; (4, 5) depended on the concentrations of Na$^+$ and K$^+$ similarly; (6) responded to Na$^+$ as a competitor to K$^+$; (7) were inhibited by ouabain at similar concentrations; and (8, 9) would accept ammonium ion as a substitute for K$^+$ but not Na$^+$, at similar concentrations.

Post also reported the presence of ouabain-sensitive (Na$^+$ + K$^+$)–stimulated ATPase activity in kidney and considered that "a similar transport system may be present in the membrane of many kinds of cells [for it is] more widely distributed in animal tissues than has been thought."[42] Post did not mention Skou's work on other tissues or Skou's finding that ouabain inhibited the crab nerve ATPase. He did,

Table 9.1 Effect of Added Na$^+$ and K$^+$ on ATPase Activity of a Membrane Fraction from Human Red Blood Cells*

	ATPase Activity	
Na$^+$ or K$^+$ Added	*without Ouabain*	*with Ouabain*
None	1.2	1.1
72 mM Na$^+$	1.2	1.1
72 mM K$^+$	1.1	1.4
72 mM Na$^+$ and 72 mM K$^+$	2.6	1.3

*The red blood cell membrane fraction was incubated for 1 hour at 40° in media containing the added Na$^+$ and K$^+$ concentrations shown, 2 mM ATP (as the tris-hydroxymethylaminomethane salt), 2 mM Mg^{2+}, 40 mM imidazole plus 40 mM histidine (adjusted to pH 7.1), and in the absence or presence of 0.1 mM ouabain. The liberation of inorganic phosphate from ATP was measured, expressed as micromoles of phosphate produced per hour per milliliter of membranes. [From Post et al., 1960, Table II. Reprinted by permission of the American Society for Biochemistry and Molecular Biology.]

however, refer to Skou's 1957 paper at five points, including an acknowledgment in his second paragraph: "When Skou . . . reported such an ATPase in crab nerve, we were stimulated to investigate human erythrocyte ATPase for features already known to be characteristic of active sodium and potassium transport in intact cells."[43] The first paragraph, on the other hand, began: "The identification of any enzyme in broken cells as a participant in membrane transport has not been reported."[44]

Post's 1959 abstract surprised Dunham and Glynn, who had been proceeding similarly. They published an abstract in the spring of 1960 reporting that ATPase "activity was much increased if both Na and K were added, though either ion alone had no effect," and that strophanthidin "completely prevented the activating effect of Na + K."[45] Their full paper, which was submitted in November 1960 and published in 1961, considered two "components" of the ATPase activity: one stimulated by Na^+ plus K^+ and sensitive to cardiotonic steroids, and the other unaffected by Na^+, K^+ or the cardiotonic steroids.[46] They pointed out that inhibition by strophanthidin was antagonized by K^+, just as inhibition of the pump was, and that ATPase and pump responded similarly to chemical variants of the cardiotonic steroids. They cited Skou's 1957 paper but not his 1960 study; they cited Post's 1959 abstract but not his 1960 paper.

Tosteson, now in St. Louis, published an abstract for the 1960 spring FASEB meeting describing ATPase and pump activities of HK and LK sheep red blood cells (see Chapter 8): both $(Na^+ + K^+)$–stimulated ATPase and pump activities were three to four times higher in HK than LK cells, and both ATPase and pump activities were inhibited by strophanthidin.[47] Hoffman, still at NIH, described correlations between human red blood cell ATPase activity and transport at the 1960 Society for General Physiology meeting, and subsequently in papers published in 1961 and 1962.[48]

In 1960 Skou and Post described their results at a meeting in Prague that was attended by leaders in the transport field.[49] And by the end of 1962, Na^+- or $(Na^+ + K^+)$–stimulated ATPase activities had been described in brain, kidney, skeletal muscle, heart, liver, intestine, thyroid, frog skin, squid axon, and eel electric organ. Sjoerd Bonting and associates at NIH found $(Na^+ + K^+)$–stimulated ATPase activity in over a dozen organs of the cat.[50] They also described a tight correlation between their measured activities and reported values for active cation fluxes in six cell types or tissues spanning a 25,000-fold range.[51] This was convincing support for the link between ATPase and pump. Correspondingly, the number of papers on these ATPase activities multiplied dramatically (Fig. 9.4A), coming from scientists working in Denmark, Sweden, Finland, Hungary, the Netherlands, England, the United States, and Japan. Even more rapidly rose the citations to Skou's and Post's initial papers (Fig. 9.4B), indicating the spread of interest to those not themselves examining ATPase activity.

DEVELOPMENTS THROUGH 1965

Three further aspects of the Na^+/K^+-ATPase were quickly described.

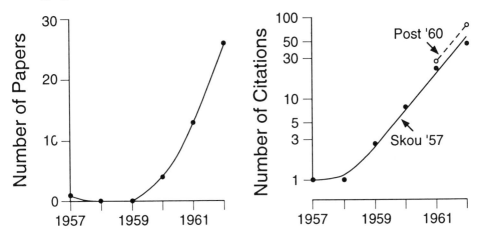

Fig. 9.4 Number of papers (*A*) on the Na$^+$/K$^+$-ATPase and the number of citations (*B*) to Skou's 1957 and Post et al.'s 1960 papers, for the years 1957–1962. The scale for citations is logarithmic. Citations were counted from the *Science Citation Index*, Institute for Scientific Information, Philadelphia.

Sidedness of Cation Activation

Particularly compelling evidence linking ATPase to pump came from studies published independently in 1962 by Glynn and by Whittam.[52] With resealed red blood cell ghosts containing various cation mixtures and incubated in media containing various cation mixtures, they showed that ATPase activity sensitive to cardiotonic steroids were activated by *extracellular* K$^+$ and *intracellular* Na$^+$, just as were cation fluxes sensitive to cardiotonic steroids. This parallel between asymmetric activation and asymmetric transport left little doubt about a common identity. (ATP was added intracellularly prior to resealing, so the site at which ATP activated the enzyme was also asymmetric, as shown previously for the pump.)

Stoichiometry of Pump/ATPase

Glynn found that "close to three Na ions leave by the [cardiotonic steroid] sensitive pathway for each inorganic phosphate formed by the [cardiotonic steroid] sensitive ATP-ase,"[53] which was in accord with Post's finding a 3:2 coupling ratio for Na$^+$/K$^+$ transport in red blood cells and with Keynes's finding a third-power dependence of Na$^+$ extrusion from frog muscle on intracellular Na$^+$ concentration.[54] Using resealed ghosts, Amar Sen and Post then described a ratio of three Na$^+$ extruded to two K$^+$ taken up to one ATP hydrolyzed.[55] They also emphasized the correspondence with Zerahn's and Leaf and Renshaw's studies on frog skin (see Chapter 8).

Phosphorylation of the Enzyme by ATP

Skou's second paper on crab nerve ATPase activity, in 1960, described not only inhibition by ouabain but also an ADP/ATP exchange activity:[56] when labeled ADP and unlabeled ATP were incubated with the membrane preparation, labeled ATP was formed. Since no adenylate kinase activity was detected, the likely means by which ATP was labeled was through a reversible phosphorylation of the enzyme (E):

$$E + ATP \Leftrightarrow E\text{-}P + ADP$$

where $-P$ represents a phosphate group linked covalently. If unlabeled ATP donated its terminal phosphate to the enzyme (reaction to the right), the phosphorylated enzyme thus formed could transfer that phosphate to a labeled ADP, forming a labeled ATP (reaction to the left). But Skou's ADP/ATP exchange reaction was not stimulated by Na^+ or K^+, nor was it affected by cardiotonic steroids. It could nevertheless be part of a ($Na^+ + K^+$)–stimulated ATPase, and Skou drew a two step scheme in which Na^+ and K^+ activated hydrolysis of $E\text{-}P$:

$$E + ATP \Leftrightarrow E\text{-}P + ADP$$
$$E\text{-}P + Na^+_{in} + K^+_{out} \Rightarrow E + P_i + Na^+_{out} + K^+_{in}$$

where P_i represents inorganic phosphate. But no critical evidence tied this exchange reaction to the ATPase.

By contrast, in 1963 J. S. Charnock and Post described phosphorylation of a kidney membrane preparation (using ATP labeled with ^{32}P in the terminal phosphate) that was stimulated by Na^+.[57] K^+ did not stimulate. Instead, K^+, when added after $E\text{-}^{32}P$ was formed, decreasing enzyme labeling and increasing P_i release (i.e., increased ATPase activity). These striking observations led them to a quite different two-step scheme of Na^+-stimulated phosphorylation followed by K^+-stimulated dephosphorylation:

$$E + ATP + Na^+_{in} \Rightarrow E\text{-}P \cdot Na^+ + ADP$$
$$E\text{-}P \cdot Na^+ + K^+_{out} \Rightarrow E + P_i + Na^+_{out} + K^+_{in}$$

Using better methods to reduce the background level of radioactivity in their phosphorylation experiments,[58] Post, Sen, and A. S. Rosenthal provided in 1965 clearcut evidence for Na^+-stimulated phosphorylation and K^+-stimulated dephosphorylation: the rates were consistent with the measured ($Na^+ + K^+$)–stimulated ATPase activity, and the amount of labeling was linearly related to the rate of that ATPase activity.[59] Moreover, ouabain strongly inhibited K^+-stimulated dephosphorylation (it reduced Na^+-stimulated phosphorylation less).

Wayne Albers (Fig. 9.5) at NIH independently reported compatible observations in 1963 from experiments using an eel electric organ ATPase preparation: Na^+ stimulated phosphorylation of the preparation, whereas added K^+ decreased that phosphorylation.[60] And after digestion of the phosphorylated preparation with a proteolytic enzyme, some of the resultant peptides, separated by electrophoresis, were labeled. ATP phosphorylated the protein itself, in contrast to some expectations that membrane phospholipids were involved.

Fig. 9.5 R. Wayne Albers. (Photograph courtesy of Wayne Albers.)

The next year, Albers and associates described Na$^+$-stimulated ADP/ATP exchange that they attributed to the ATPase.[61] Integrating that reaction with the phosphorylation and dephosphorylation steps would require a more complex scheme, one suggesting particular mechanisms for transport as well as for ATP hydrolysis (see Chapter 11).

Also in 1964, Glynn and J. B. Chappell devised a simple method for preparing ^{32}P-labeled ATP that became widely used by individuals as well as by commercial purveyors.[62] Glynn and associates then demonstrated, by careful calculations of the sufficiency of time and amount in their experiments, that no phospholipid served as a phosphorylated intermediate.[63]

And in 1965 Lowell Hokin at the University of Wisconsin examined the labeled peptide fragments (formed by proteolytic digestion of brain membranes phosphorylated with ^{32}P-ATP) and concluded, from the sensitivity to chemical and enzymatic modification, that Na$^+$-stimulated phosphorylation produced an acyl phosphate.[64] The significance of an acyl phosphate lay in its energy status: acyl phosphates are acid anhydrides of phosphoric acid and the acyl group of a carboxylic acid, and such anhydrides are—in Lipmann's terms—"energy rich" phosphates. That assignment thus was in accord with the Na$^+$-stimulated ADP/ATP exchange reaction, which required an "energy-rich" *E*-P. Independently, Makoto Nakao and associates in Tokyo published similar evidence and arguments.[65]

COMMENTARY

Places and People

During this period new investigators from across the globe joined familiar figures from the membrane transport field. Prominent among these were Glynn in Cambridge, Post in Nashville, and Skou in Aarhus. Although subsequent chapters will continue to celebrate their achievements, a brief comparison of these protagonists is appropriate now.

Ian Michael Glynn (b. 1928)

Glynn is a tall, elegant presence, impeccably courteous, honorably correct, comfortably literate, broadly informed, analytically precise. He radiates the full virtue of the English gentleman and Cambridge scholar, his archetypical accent modulated with assurance. He is also the descendant of Polish Jews and a tradition of dedication to learning and of reverence for the professions. He intones the Latin grace at High Table with the aplomb of his collegial agnostics. And his scientific papers are equally superlative: clear, thorough, meticulously reasoned. His data are fully warranted. The range of his experiments—from categorizing cation fluxes to measuring enzymatic responses—and the authority of his opinion have been beacons to those in the field.

   Glynn entered Trinity college with a goal of academic medicine. An uncle was a distinguished pathologist, an aunt progressed from biochemistry to medicine, and an older brother is a physician involved with research. Hodgkin was Glynn's tutor, however, and Glynn became enthralled with the glories of axonal conduction. After medical training at University College Medical School in London, he therefore returned to Cambridge as a research student with Hodgkin. Biochemical studies on squid were impractical since they were not always available, so Glynn studied transport in red blood cells instead, receiving his Ph.D. in 1956—with his course interrupted at that critical time by compulsory military service.

   Upon his return, Glynn faced heavy teaching responsibilities, to which were added—as he ascended from fellow to instructor to reader to professor (1975)—administrative duties, including Vice-Master of Trinity College (1980–1986), chairman of the editorial board of the *Journal of Physiology* (1968–1970), and chairman of the Physiological Laboratory (1986–1995). Both space and facilities were sharply restricted in Cambridge at that time. A further hindrance was the paucity of graduate students in physiology. Instead, Glynn played gracious host to a succession of accomplished visitors, including Dunham, Carolyn Slayman, Marcia Steinberg, and J. J. Grantham from the United States; Patricio Garrahan, V. L. Lew, José Cavieres, Luis Beaugé, and Donald Richards from Argentina; Ursula Lüthi from Switzerland; Y. Hara from Japan; and Steven Karlish from Israel. But a consequence of his relying on foreign visitors was the failure to establish a self-sustaining group in Cambridge to ensure a succession. He retired in 1995.

The academic remove from the contemporary world is well exemplified by Glynn's failure to patent his method for preparing ATP labeled with ^{32}P: "it seemed selfish to patent a method developed in a university laboratory and likely to be of value to colleagues involved in medical research."[66] In any event, academic honors followed. He was elected to the Royal Society, as an honorary member to the American Academy of Arts and Sciences and to the American Physiological Society, and he was awarded an honorary degree from the University of Aarhus.

But Glynn was edged out by Skou in finding the ATPase and by Post in characterizing the red blood cell enzyme. In his turn, Glynn categorized the diverse fluxes through the pump, contributed greatly to studies on the reaction mechanism, and demonstrated the cation occlusion proposed by Post (see Chapter 11).

He is married and the father of three.

Robert Lickely Post (b. 1920)

Post is slight, erect, formally correct in dress and manner, impeccably courteous and proper, generous with detailed advice and suggestions. He is a Philadelphia Quaker, the son of a professor of Greek at Haverford College. He is also striving, ambitious, and unquestionably successful, with a firm grasp of the difference between right research and wrong. (Perhaps he is excessively dichromatic; there are those who expressly request that he not referee their submitted manuscripts.) In Tennessee, he was an enthusiastic sailor, competing in races because sailing was then "much more fun." In conversation he is soft-spoken and engaging, in formal talks concise and orderly, capable of felicitous phrases with an amiable, dry wit. His memoirs are lively and entertaining. His scientific papers, on the other hand, are sometimes dense, complex, and difficult; a careful reading is rewarded with accounts of unexpected observations and imaginative insights.

Post attended Quaker schools in the United States and in England (the Downs School, which Hodgkin and Sanger also attended). He studied Latin for six years and Greek for three, but chose science in the end. After a year at Exeter, Post entered Harvard College in 1938, majoring in biochemistry with a research project under John Edsall. He preferred medical school to military service in 1942, completing the wartime accelerated program in 1945. After an internship, he migrated to academic science: as instructor in physiology at the University of Pennsylvania for 2 years and at Vanderbilt University for 5 before being promoted to assistant professor in 1953. He published no papers between 1948 and 1955. But as Chapter 11 will illustrate, the impressive beginnings he made flowered into a succession of papers defining the reaction mechanism of the Na^+/K^+-ATPase, dominating the field with new types of experiments and new interpretations of the data. Following the direction of his chairman, C. R. Park[67]—"Do it right the first time"—he made few errors, and those he did make he admitted candidly.

Post had no graduate students that followed him in the field, but he collaborated with a steady stream of visitors to his laboratory, including Charnock from Australia, Sen from India, Hermann Bader from Germany, Sven Mårdh from Sweden,

Csaba Hegyvary from Hungary, Peter Jørgensen from Denmark, and Shoji Kume, Gotara Toda, Kazuya Taniguchi, Yoshihiro Fukushima, Motonori Yamaguchi, and Kuniaka Suzuki from Japan. He had no colleagues in his department faculty working in the same field and was sensitive to his isolation in Nashville. On the other hand, he felt it beneficial not to have been in Cambridge, for, as he observes, "I would have been thinking the way they thought and I would have been less likely to do something different." Like Glynn, he left behind no descendants at his university when he retired in 1991, moving back to Philadelphia.

Despite his stellar accomplishments, Post received only one major recognition, the Cole prize of the Biophysical Society. He is married but childless.

Jens Christian Skou (b. 1918)

Skou may seem the least impressive personality of this trio at first meeting; he is tall but slim, now bearded, reserved, without pretensions, unassuming to a fault. He describes himself as shy, with a difficulty in meeting people. On closer acquaintance, Skou emerges as a contemplative person, with interests from architecture to sailing to music. He is an amusing teller of tales, entertaining, a warm and generous host. He is now fluent in English but hesitant in speech, occasionally meandering. The scientific papers, however, are straightforward and direct. His memoirs are quite enjoyable.

Skou comes from a business rather than an academic or professional family. Of the three, he was most attracted to medical practice and the least prepared for a scientific career; with characteristic modesty he admits to having "never been a scientist." Certainly, he was less prompt in publishing than his competitors, especially in the early days. When he joined the physiology department in Aarhus he was one of four faculty members, with a corresponding share of teaching obligations. Facilities were quite meager. And as the next senior member he soon had thrust upon him the departmental administrative duties when the chairman, S. L. Ørskov, became chronically ill. After Ørskov's death in 1963, Skou became chairman officially. Over the years the department expanded, with a new building and an enlarged faculty. By his considered decision never more than a third of the faculty worked on the Na^+/K^+-ATPase. Those who contributed significantly to our understanding of the ATPase include (with dates of appointment), Jørgen Jensen (1962), Jens Nørby (1965), Peter Jørgensen (1966), Otto Hansen (1969), Liselotte Plesner (1971), Paul Ottolenghi (1971), Irena Klodos (1972), and Mikael Esmann (1977).

In 1978, when Skou was appointed professor of biophysics, he was finally freed of his other tasks to do research. He retired in 1988.

Beyond discovering the Na^+/K^+-ATPase, Skou and his collaborators were prominent in purifying the enzyme, reconstituting it in artificial membranes, and studying its reaction mechanism by a range of chemical and physical techniques (Chapter 11). Skou has also been active in writing reviews and organizing symposia. He has made occasional mistakes, as in misidentifying Na^+-independent ADP/ATP exchange with the Na^+/K^+-ATPase.[68] But his candid modesty has contributed far more to a serious underappreciation of his extensive contributions. He looks back with regret on having abandoned his experiments with local anesthetics, which he recalls enjoying more than those on the ATPase.

He has, of course, received numerous awards, including membership in the Royal Danish Academy of Sciences, the Deutsche Akademie der Naturforscher Leopoldiana, the Leo Medical Prize, the Novo Prize, the Carlsen Prize, and the Anders Retzius Prize.

He is married with two daughters and is a firm believer in allocating time for family and holidays.

Skou's Achievement

In a memoir, Skou acknowledged with typical candor: "Had I not identified the pump, I am sure logical reasoning would soon have led someone in the transport field to the discovery; the time was ripe."[69] Most scientists assume that there is a world of "real" objects available for discovery. On this basis they could confidently agree that the Na^+/K^+-ATPase would have been discovered without Skou's efforts— eventually. But such assurance ought not to diminish the applause for Skou's achievement. A careful scrutiny of others' efforts through 1957 reveals no one else progressing along the necessary experimental route: looking for an ATPase stimulated by Na^+ *plus* K^+ and inhibited by cardiotonic steroids. Whether holding that key of "logical reasoning" would have sufficed for finding and turning the lock expeditiously is unclear as well. In any event, the first enunciation of such logical reasoning was not voiced until 1960, after the gate had been swung open. Only then did Post rationalize: "if the hydrolysis of ATP is stoichiometrically coupled to the linked transport of sodium and potassium across the membrane, then . . . there will be an ATPase [in membrane fractions] which requires both sodium and potassium ions together, not separately, as cofactors."[70]

Through 1957 the actual course followed by those prominent in the field had been worse than halting, at best a step or two in assorted directions. The one announced search for a linked ATPase, carried out by Dunham, recruited none of the criteria Post specified. And the observations published by Utter and Gore were not pursued by them or others. Even after Skou's work had appeared, the logical reasoning was not immediately revealed to all. Three more examples should suffice.

McIlwain, a leading neurochemist, also attended the 1956 Aarhus meeting. Nevertheless, his first paper on the Na^+/K^+-ATPase, published 5 years later, began by neglecting K^+:

> Expectation that the adenosine triphosphatases of neural tissue . . . might participate in the tissue's utilization of metabolically derived energy . . . was supported by the findings of Skou . . . and of Hess and Pope . . . that the adenosine triphosphatase . . . required the presence of sodium ions for maximal activity. The enzymes also showed the requirements for magnesium and potassium salts *which are more usual* in reactions involving adenosine triphosphate. . . ; the *sodium requirement* might connect the enzymes with the performance of osmotic work in neural tissue [emphases added].[71]

McIlwain failed to cite the K^+ stimulation found by Gore. Not until 1962 did he remark: "The metal activation . . . of cerebral enzymes forming adenosine diphosphate from adenosine triphosphate [was] examined by Gore . . . and by Lowry."[72]

In 1959 Julian Tobias dismissed Skou's results as "evidence . . . of the 'compatible with' sort, but were the substrate [ATP] somehow to bind some sodium, then such reactions could make sense."[73] That urge to impose responsibility for binding on the nucleotide apparently arose from a reluctance to endow proteins with such discrimination.

In another review, also published 2 years after Skou's 1957 paper, K. I. Altman surveyed the ATPase activities of red blood cell ghosts, concluding that "it is unknown whether there is a functional significance in the location of these enzymes at or near the cell surface."[74] He did not cite Skou.

Why Was it Skou who Discovered the ATPase?

Certainly Skou had some good luck. He was fortunate in studying the crab nerve ATPase, for its activity with $Na^+ + K^+$ was 10-fold that without, a magnitude that commanded attention. By contrast, only a twofold increase was found by Post and by Hess and Pope. Others reported even less,[75] but by then they knew something was to be found and what features to look for. Skou was fortunate in following an early paper implicating ATP in Na^+ extrusion from the squid axon, unaware that Hodgkin and Keynes had recently reported their failure to drive outflux with injected ATP.[76] He was fortunate in not knowing about cardiotonic steroids and thus did not rely on them as criteria for the pump, for the crab nerve ATPase turned out to be quite insensitive to ouabain when compared with other pump activities than being studied.[77] He was fortunate in not observing a common practice among enzymologists of that period, which was to use reagents solely as K^+ salts. And he was fortunate in receiving a fellowship to visit Woods Hole,[78] without which he probably would not have read Libet's abstract nor talked with Post in Vienna.

Budding scientists are constantly admonished with Pasteur's aphorism about chance favoring the prepared mind. Skou was prepared with an interest in nerve conduction and with a reverence for physiological relevance as a guiding light. And he was fortunate in possessing the wit and industry to exploit all of this.

Negotiations

Negotiation among scientists is currently a popular topic for commentators, and the range and nature of such interactions are indeed interesting. Here we saw Post negotiate successfully with Skou for a protected share of research on the Na^+/K^+-ATPase, and unsuccessfully with Dunham and Glynn. Not explicitly depicted are negotiations for faculty positions, laboratory space, research support, publication, symposium participation, and for esteem and rewards. But in the view of most scientists, negotiating does not extend to "matters of fact." Whether a phosphorylated enzyme participates in the reaction process of the Na$^+$/K$^+$-ATPase is not "negotiated" in that sense. Instead, the conclusion is based on experimental criteria: whether *E-P* is present in sufficient amounts, is formed and destroyed at sufficient rates, etc. Judgments about what is sufficient then depend on scientific *principles* of what a reaction intermediate must be, within the general framework of shared biochemical theory.

Incomplete Descriptions

Through the 1950s the monovalent cation content of incubation media was rarely specified in the papers cited here. That content was not deemed a significant consideration in the sense that incubation temperatures or substrate concentrations were. Anticipating the variables that will be pertinent in the future is obviously risky. Clearly reprehensible, however, are accounts where it is impossible to tell what was done, how, and to what.[79] The developing *ideal* of scientific practice is approximated by fallible experimenters, whose descriptions and interpretations are screened by fallible referees, accepted or rejected for publication by fallible editors, and finally contemplated (or overlooked) by fallible readers.

Chairmen

Finally, a sociological aspect of science departments in U.S. medical schools should be pointed out, for it deserves scholarly attention, namely, the power of the chairman. In medical schools the chairmen are frequently appointed for life, answerable to few, and endowed with despotic power to appoint and promote staff and to allocate funds and facilities. Even the scientific direction of their serfs may be dictated and the fruits of that toil taxed: the chairman affixing his name to papers he walks past, exacting his toll from the research grants of his minions, squandering space and students. Choosing chairmen on the basis of their scientific accomplishments and entrepreneurial skills is presumed to be an adequate basis for selecting wise, nurturing, and benevolent administrators as well. This is like judging the optimal development of the French nation under Louis XIV from the grandeur of Versailles.

One of the more beneficent examples of such influence, however, is the case of C. R. Park at Vanderbilt. Park was an accomplished scientist who made notable progress in the field of diabetes and sugar transport after his appointment as chairman in 1952. He told Post, who had been working (not very productively) on cardiac electrophysiology, that he, Park, was going to have a department that concentrated on hormones and transport. If Post wanted to stay he could work on transport. Post decided he wanted to stay. Initially, he collaborated with Park in studies on sugar transport. Park suggested that Post find the mechanism of the sodium pump. Although Post did not find it first—Skou beat him to discovering the Na$^+$/K$^+$-ATPase—Post led the way in determining how the enzyme worked. When asked why he suddenly became a productive scientist after a decade of meager achievement, Post replied: "I finally had a chairman who liked me."[80]

Scientific Summary

In 1957 Skou reported that including both Na^+ and K^+ in the assay media stimulated ATPase activity of crab nerve membranes 10-fold. He proposed that this ATPase activity was involved with active cation transport, and he subsequently showed that the cation-dependent increment was inhibitable by ouabain. Further studies by increasing numbers of investigators established compelling parallels between ATPase and pump, including similar activation by extracellular K^+ and intracellular Na^+. The ATPase was soon described in multitudes of cell types, with enzyme activity corresponding to transport flux across a wide biological range. By 1965 studies on the reaction process pointed to a Na^+-activated phosphorylation of the enzyme protein (forming an "energy-rich" acyl phosphate intermediate) and a K^+-activated dephosphorylation.

Notes to Chapter 9

1. Glynn (1957b), p. 297.
2. Ibid., p. 295.
3. Ibid., pp. 296, 297.
4. Clarkson and Maizels (1952). They discussed the activity as an "apyrase," meaning that the final product of ATP hydrolysis was AMP; by contrast, ADP is the final product with an ATPase. But an ATPase in the presence of adenylate kinase will also produce AMP from ATP. With hindsight I refer to ATPase activity.
5. Herbert (1956), p. 29. The Na^+ concentration equivalent to "0.4%" is 150 mM. He does not say whether or not K^+ was present.
6. Dunham (1957b).
7. The attempt is feeble because many enzymes have similar pH optima: it is not a discriminating characteristic.
8. Skou (1957a).
9. Ibid., p. 394. Skou's introduction implies that he undertook the experiments expressly to study cation transport. Skou later acknowledged that this presentation was a "subsequent rationalization" (1989, p. 436).
10. ATP is an acid, and neutralization with a base necessarily introduces some cation. Skou converted ATP (obtainable commercially as some salt) to the free acid—by passing it through an ion exchange column in the H^+ form—and then neutralized it with the organic base 2-amino-2-methyl-1,3-propanediol.
11. I refer to the ATPase activity of impure enzyme preparations due to Na^+ and K^+ as "$(Na^+ + K^+)$–stimulated ATPase activity"; I reserve "Na^+/K^+-ATPase" for the actual enzyme.
12. Skou (1957a), p. 401.
13. This account is based on Skou (1989), Post (1974, 1989), interviews and correspondence with Skou, Post, Hoffman, Tosteson, and Nørby (1993–1995), and Skou's examination of his laboratory notebooks (1993).
14. Libet (1954).
15. Abood and Gerard (1954).
16. Through 1964, Skou's 13 papers on this subject were cited nearly 150 times.
17. Skou's laboratory notebook for 28 June 1955 (the last workday before he left for vacation) asks: "ATP—is it Na^+ or K^+?". The reminiscences (Skou, 1989, p. 436, and Post,

1974, p. 6) imply, misleadingly, that the time spent untangling the sources of the erratic results was far longer.

18. Skou attributed the delay in publishing to his primary interest continuing to be local anesthetics, to his unfamiliarity with English, and to the absence—in his environment—of a sense of urgency, either from a fear of competitors or from a need to build a bibliography.

19. Utter (1950). Utter presented his results as apyrase activity, although he explicitly noted that the activity could result from an ATPase plus adenylate kinase.

20. Steinbach (1954). He also had been studying ATPases, but with Ca^{2+} as the divalent cation (Steinbach, 1949).

21. Glynn (1956, 1957b, 1959).

22. I have omited a few papers on bacteria, which are irrelevant to this story.

23. Muntz and Hurwitz (1951).

24. Novikoff et al. (1952), p. 160. Stimulation by even optimal concentrations of Na^+ and K^+ would not be large with liver homogenates; Emmelot and Bos (1962) found only 30% stimulation using a plasma membrane fraction.

25. Gore (1951).

26. With 5 mM ATP, the Na^+ concentration at pH 7.4 would be 15–20 mM. Since K^+ was being varied, it presumably was not also present with the glycylglycine (Na^+ and K^+ were then the standard cations for buffers). At pH 7.4, roughly 15 mM Na^+ would be present with 100 mM glycylglycine.

27. McIlwain (1952a,b).

28. Findlay et al. (1954).

29. Lowry et al. (1954), p. 28.

30. Skou (1957b). His description was during the discussion period following a paper by Keynes. By the time of the meeting, Skou had already submitted his paper.

31. Others present included Lowry, McIlwain, Steinbach, Tosteson, and Ussing.

32. According to Pope (telephone conversation, 1993), they had previously found effects of cations on ATPase activity but had been interpreting their results differently.

33. Hess and Pope (1957). "FASEB" is the abbreviation for the Federation of American Societies for Experimental Biology, which held large meetings each spring.

34. Hess (1962).

35. Post (1974).

36. Skou (1989), p. 437. Skou states his supposition that Post had not yet read his 1957 paper.

37. Ibid. Inhibition of transport in the squid axon by ouabain had not been reported by 1958, and studies on squid were Skou's link to the transport field.

38. Post (1959).

39. Post et al. (1960).

40. Ibid., p. 1801.

41. Aldridge (1962).

42. Post et al. (1960), pp. 1801, 1802.

43. Ibid. p. 1796.

44. Ibid.

45. Dunham and Glynn (1960), p. 62P.

46. Dunham and Glynn (1961).

47. Tosteson et al. (1960).

48. Hoffman (1961, 1962b). The meeting abstract of 1960, by Hoffman and H. E. Ryan, was not published.

49. Skou (1961); Post and Albright (1961).

50. Bonting et al. (1961).

51. Bonting and Caravaggio (1963).

52. Glynn (1962); Whittam (1962a,b). Whittam had moved from Cambridge to Oxford.

53. Glynn (1962), p. 18P.

54. Post and Jolly (1957); Keynes and Swan (1959).

55. Sen and Post (1964). Others proposed that far fewer Na$^+$ were transported for each ATP consumed, e.g., Caldwell et al. (1960a).

56. Skou (1960). It was submitted for publication in November 1959.

57. Charnock and Post (1963); a more complete account was published in Charnock et al. (1963).

58. The critical advance was including unlabeled ATP in the solution used to wash the precipitated enzyme after labeling with ^{32}P-ATP.

59. Post et al. (1965).

60. Albers et al. (1963).

61. Fahn et al. (1964).

62. Glynn and Chappell (1964).

63. Glynn et al. (1965).

64. Hokin et al. (1965).

65. Nagano et al. (1965).

66. Glynn (1989b).

67. Quoted in Post (1989).

68. Perhaps that early misidentification contributed to Skou's long disbelief in a phosphorylated enzyme participating in the ATPase reaction sequence (other evidence against E-P participating included an absence of stimulation by K$^+$ at 0°).

69. Skou (1989), p. 436.

70. Post et al. (1960), p. 1796. On the other hand, not all those active in the field were then convinced of a stoichiometric link between Na$^+$ outflux and K$^+$ influx.

71. Deul and McIlwain (1961), p. 246.

72. Schwartz et al. (1962), p. 626.

73. Tobias (1959), p. 306.

74. Altman (1959), p. 937.

75. For example, the increment in ATPase activity on adding Na$^+$ and K$^+$ was reported to be a mere 20% to 30% by Yoshida and Fujisawa (1962) and Emmelot and Bos (1962).

76. Hodgkin and Keynes (1955a, 1956). Skou (1989) acknowledged his ignorance of the latter paper at that time.

77. Skou (1960). Crab nerve ATPase required about 0.3 mM ouabain for half-maximal inhibition, several orders of magnitude more than needed for red blood cell ATPase. When an enzyme is inhibitable only by unusually high concentrations, there is a real likelihood that the effect is "non-specific," i.e., that the reagent is then acting in a different fashion.

78. At a party after a friend's thesis defense, Skou met Ejnar Lundsgaard, who was in Aarhus as an examiner; Skou mentioned to Lundsgaard his wish to prepare acetylcholinesterase and Lundsgaard arranged the visit with Nachmansohn, who was an old acquaintance.

79. For example, Kimura and DuBois (1947), in a paper on cardiotonic steroids and heart ATPase activity, failed to specify not only the monovalent cation content, but also the ATP concentration, the presence or absence of Mg^{2+}, the reaction pH, and temperature.

80. Post, interview (1993).

chapter 10

CONTEMPORARY EVENTS: 1953 – 1965

W HILE the experiments, analyses, formulations, and arguments described in the preceding chapters were underway, the fields of anatomy, physiology, and biochemistry were advancing at an accelerating pace. A few pertinent issues will be discussed here; parallel studies on oxidative phosphorylation will be covered in Chapter 16.

MEMBRANE STRUCTURE AND COMPOSITION

Electron microscopy progressed rapidly in the 1950s, employing better methods for fixing, staining, embedding, and sectioning specimens as well as improved optics. At the end of that decade J. D. Robertson, then in London, promulgated the "unit membrane" model based on micrographs showing two dark bands about 20 Å thick bordering a clearer zone about 35 Å thick (Fig. 10.1A).[1] Robertson interpreted these images as a lipid bilayer sandwiched between two protein sheets, stressing (1) universal occurrence, for it appeared in multitudes of cells examined and in their organelle as well as plasma membranes; (2) specific, consistent dimensions, defining a distinct structure; (3) a single bilayer of lipid; and (4) membrane asymmetry due to differences in the flanking sheets. He also correlated his observations with J. B. Finean's X-ray diffraction studies of myelin,[2] the only membrane system then amenable to such examination. Robertson's micrographs were of permanganate-fixed tissue. Discrepancies with osmium tetroxide–fixed tissue were soon rationalized by carefully matching experimental conditions,[3] and his proposal was the reigning model throughout the 1960s. Nevertheless, the unit membrane model was not universally accepted. Fritiof Sjöstrand detected globular elements in electron micrographs that he interpreted as lipid globules separated by protein strands, and David Green proposed lipid micelles embedded in protein shells, with membranes assembled from

A

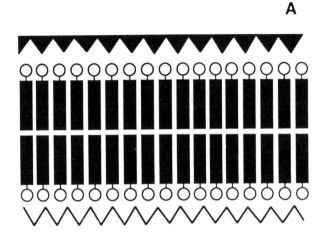

B

Polar Pore

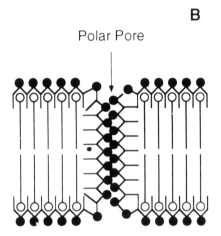

Fig. 10.1 Membrane models. *A:* The unit membrane consists of a bilayer flanked by unfolded protein sheets that are different on each surface. [From Robertson, 1964, Fig. 13, © 1964. Reprinted by permission of John Wiley & Sons.] *B:* An unfolded protein sheet extends across the bilayer to line a polar pore. [From Stein and Danielli (1956), Fig. 8. Reprinted by permission of the Royal Society of Chemistry.]

a single layer of such repeating units.[4] These formulations were supported by model lipid systems having interconvertable micellar forms as well as bilayers.[5] And direct measurements of protein structure did not indicate extended conformations.[6]

In 1956 Wilfred Stein and Danielli drew pores across the bilayer lined by single polypeptide chains (Fig. 10.1*B*), but without morphological or chemical evidence.[7] Their conjecture, however, emphasizes a disappointment with other models of this time: no routes for the passage of polar solutes were apparent.

On the other hand, some progress was made in isolating plasma membranes (from other than red blood cells), a prerequisite to determining their chemical composition. In 1960 David Neville in Rochester described procedures for liver cells based on controlled homogenization conditions and density-gradient centrifugation; he identified plasma membranes by electron microscopy.[8] P. Emmelot and associates in Amsterdam modified this procedure somewhat and showed that antibodies to their preparation bound preferentially to plasma membranes of whole liver cells.[9] Identification of membrane proteins could then proceed, although protein fractionation techniques were still rudimentary. Lipid analytical techniques, however, had advanced sufficiently to allow cataloging.[10]

Finally, two model systems that were developed subsequently should be introduced here. In 1962 Paul Mueller and Donald Rudin in Philadelphia described the preparation of planar bilayers.[11] They spread a lipid extract from brain across a 1 mm hole in a septum separating two aqueous solutions. The film then thinned spontaneously to bilayer dimensions and characteristics: 60–90 Å thick, with resistance of 10^7–10^8 ohm cm^2 and capacitance of 1 μfarad/cm^2. And in 1965 A. D. Bangham in Babraham formed multilaminated lipid vesicles: he added aqueous salt solutions to dried lipid films, creating concentric bimolecular lipid layers separating aqueous compartments.[12]

PROTEIN STRUCTURE AND DYNAMICS

At this time, X-ray crystallography provided the first pictures of protein structure on the atomic scale and the first images of how such structures underlay catalysis. In 1958 John Kendrew and associates in Cambridge and London described a three-dimensional model of myoglobin at 6 Å resolution. By 1960 they improved the resolution to 2 Å, defining the peptide chain's path (Fig. 10.2A) and identifying for the first time in a protein the α-helices proposed by Pauling and Corey.[13] They commented that "the most remarkable features of the model are its complexity and its lack of symmetry; [absent are] the kind of regularities which one instinctively anticipates, and it is more complicated than has been predicted by any theory of protein structure."[14]

Myoglobin, an 18 kDa protein with a single iron-containing heme group, functions as an oxygen carrier in muscle cells. Hemoglobin, the oxygen carrier in red blood cells, is a 67 kDa protein composed of two pairs of similar subunits, each of the four having a heme group. Max Perutz in Cambridge began crystallographic studies on hemoglobin in the 1930s (Kendrew was his student). By 1960, after devising necessary techniques and enlisting electronic computers, Perutz and associates reported a model of oxygenated hemoglobin at 5.5 Å resolution, showing two α- and two β-subunits, each closely resembling myoglobin.[15] It had been known since the nineteenth century that oxygenated and deoxygenated hemoglobin formed quite different crystals, and in 1963 Perutz described a model for deoxygenated hemoglobin as well; comparisons now revealed that with oxygenation the two β-subunits moved 7 Å closer together.[16] By the end of the decade a

A

B

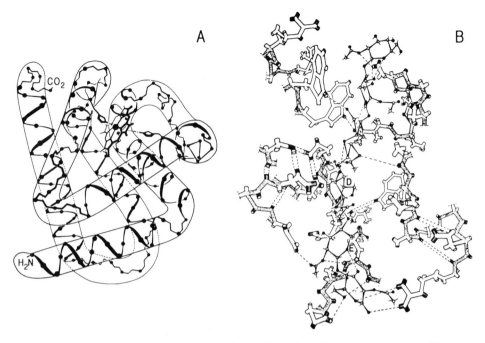

Fig. 10.2 Protein structure. *A:* Myoglobin. Straight cylinders contain α-helices that are disrupted at the turns; the circle at the upper center represents iron in its porphyrin ring. [Redrawn from Dickerson, 1964, Fig. 15. Reprinted by permission of Academic Press.] *B:* Lysosome structure around its active site cleft. The substrate, a sugar polymer, lies in the cleft so that cleavage between sugars D and E is achieved by glutamate-35 "donat[ing] a proton to the glycosidic oxygen, while the negatively charged [aspartate-52] stabilizes an intermediate carbonium ion at C(1) of sugar . . . D." [Redrawn from Phillips (1967), Fig. 7; p. 494. Reprinted by permission of David C. Phillips.]

gratifying model of oxygen uptake and release—and its modification by physiological regulators—appeared.[17]

Lysozyme, a 15 kDa protein that cleaves polysaccharides of bacterial cell walls, was the first enzyme whose detailed crystal structure was determined. In 1962 David Phillips and associates in London presented a three-dimensional model at 2 Å resolution, describing also the structure with a competitive inhibitor occupying a transverse crevice in the protein surface.[18] These models pointed to a palpable mechanism for catalysis (Fig. 10.2B). This was a conceptual triumph as well as a structural revelation. (Lysozyme contained not only α-helices but a β-sheet, demonstrating its occurrence as well.)

Although X-ray crystallography portrayed static structures, other approaches emphasized flexibility. In 1958 Daniel Koshland at Brookhaven National Laboratory proposed an "induced-fit" model for enzyme–substrate interaction.[19] Instead of the rigid image of lock and key specified at the close of the nineteenth century, Koshland considered that substrate binding caused "appreciable" structural changes in enzymes, promoting the requisite orientation for effective catalysis. Then in 1963 Jacques Monod, Jean-Pierre Changeux, and François Jacob in Paris described regulatory control of enzymes by substances binding at "allosteric sites," structurally separate from the active site where catalysis occurs.[20] An explicit, quantitative model in 1965 specified concerted, all-or-none transitions between alternative conformations, pictured as changes in subunit interactions of oligomeric proteins.[21] Since the early 1900s oxygen binding to hemoglobin had been known to be "cooperative": binding increased sharply with concentration in a sigmoidal fashion, as if the first oxygen bound increased the affinity of hemoglobin for the second oxygen, the second for the third, and the third for the fourth. Monod and associates accounted for such binding in terms of two states of the hemoglobin tetramer—T having low affinity and R having high—with oxygen binding favoring the concerted transition from T to R. In 1966 Koshland and associates formulated an alternative model with sequential changes in affinity.[22]

An even more dramatic instance of molecular motion is muscle contraction. In 1954 Andrew Huxley and Rolf Niedergerke in Cambridge, England, and Hugh Huxley and Jean Hanson in Cambridge, Massachusetts, proposed "sliding filament" models (from studies using interference microscopy, electron microscopy, and X-ray diffraction).[23] This scheme was developed vigorously throughout this period in terms of "cross-bridges" on myosin filaments repetitively attaching to, pulling, and detaching from actin filaments, so that the longitudinal fibers would slide past each other.

Two further developments during this productive time should also be cited. Walter Kauzmann in Princeton and Charles Tanford in Durham, North Carolina stressed the significance of hydrophobic association among nonpolar side chains of amino acids and the role of water in determining protein structure.[24] And Christian Anfinsen at NIH proposed a "thermodynamic hypothesis" relating protein structure to minimal states of free energy, an interpretation derived from his studies of proteins folding spontaneously to their native state.[25]

OTHER SCIENTIFIC ADVANCES

At the beginning of this period, James Watson and Francis Crick in Cambridge published their double-helix model of DNA and enunciated functions attributable to that structure.[26] By the end of this period, the "central dogma" of genetic information flow—from DNA to RNA to protein—was established. Enzymes replicating DNA and producing RNA from DNA templates were identified, as was the process of protein synthesis by ribosomes. The ribosomal synthesis was shown to involve enzymatic linking of amino acids tagged by specific transfer RNAs, in sequences prescribed by messenger RNAs, following the code of nucleotide triplets designating each amino acid.[27]

Also early in this period, while examining how certain hormones increase glucose production by the liver, Earl Sutherland and associates in Cleveland showed that these hormones raised the activity of the enzyme phosphorylase (which liberates glucose from its storage polymer). They then showed that the increased phosphorylase activity was due to its phosphorylation and that this phosphorylation was stimulated by cyclic AMP, whose production from ATP was increased by the hormone.[28] Sutherland termed cyclic AMP a "second messenger" since it carried information from the hormone (the "first messenger", which interacted with receptors at the cell surface) to affect multitudes of processes throughout the cell.[29] By 1965 a host of first messengers were shown to alter cyclic AMP levels, and in later years other second messengers were identified as well. In many instances second messengers activated protein kinases, enzymes catalyzing protein phosphorylation by ATP. The first of these to be distinguished, in 1956 by E. G. Krebs in Seattle, was the protein kinase phosphorylating phosphorylase.[30] These studies, as well as those on allosteric enzymes, reflected a growing interest in cellular control mechanisms that are essential during development and maturity.

SPONSORS

Arguably the most significant event fostering research in this and subsequent decades was the development of the National Institutes of Health as the premier patron.[31] In 1930 the Hygenic Laboratory of the Public Health Service in Washington, which focused its research on infectious diseases, became the National Institute of Health, and this was joined by the Cancer Institute in 1937. After World War II the resources of NIH increased rapidly. This growth was fed by perceived shortages of physicians, by financial plights of medical schools, and by a euphoric faith in the beneficent promise of science (Vannevar Bush's *Science—The Endless Frontier* appeared in 1945). In 1948 the National Institutes of Health was created from the prewar institutes plus the Mental Health Institute (1946) and the Heart Institute (1948). Importantly, it was charged not only with intramural research on its campus (now in Bethesda) but also with financing extramural research through direct grants to investigators. A crucial innovation at this time was the creation of "study sections," panels of scientists knowledgeable in the particular field who evaluated the scientific merit and feasibility of each application. This peer review system provided quality assurance. Moreover, awarding funds directly to those devising the projects avoided filtering support through layers of academic politics.

During the period considered here, the number of extramural grants increased sevenfold and the total dollars of those grants 26-fold (Fig. 10.3). The number of NIH employees increased from 3.9 thousand to 12.2 thousand and the total NIH appropriation from $81 million (1955) to $1,059 million (1965). This growth was in large part due to congressional support led by Rep. John Fogarty and Sen. Lister Hill (which consistently exceeded presidential recommendations), to the wise leadership of James Shannon (director of NIH, 1955–1968), as well as to the effective lobbying of Mary Lasker.

The National Science Foundation was established in 1950, and its Division of

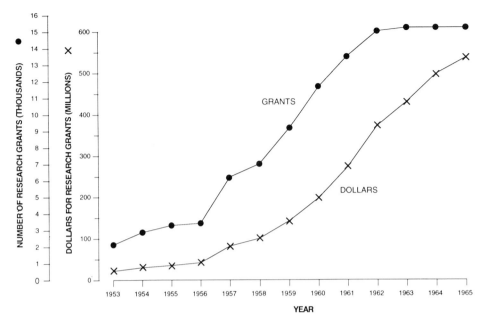

Fig. 10.3 Increases in number of research grants and research grant dollars over the years 1953–1965. Data are from *NIH Factbook* (1976).

Biological Sciences, which addressed basic research issues, grew in parallel (although its level of support was less than 5% of NIH's). The Soviet Union's *Sputnik* triumph in 1957 spurred increased funding, encouraging perceptions that scientific achievement and education in the United States were dangerously lagging and required urgent remediation.

Private, nonprofit support for biomedical research in the United States also increased during this time, although by comparatively less—only threefold.

Consequences of these benefactions are the studies displayed in these chapters. Working at NIH were Albers, Anfinsen, Dunham, Hoffman, and Tosteson. Recipients of NIH research grants included Hokin, E. G. Krebs, Koshland, Post, Sutherland, and Tanford in the United States, as well as Phillips in London, Skou in Aarhus, and Ussing in Copenhagen. The Rockefeller Foundation financed X-ray crystallography in Cambridge as well as Hodgkin's group in the Physiological Laboratory. The Medical Research Council of the United Kingdom became a major supporter of molecular biology in Cambridge, and the Nuffield Foundation replaced the Rockefeller Foundation in funding physiology.

Another consequence of increased funding was more university faculty; the number of full-time medical faculty in the United States increased from 3,500 (1951) to more than 17,000 (1966). Research money also prompted the development of enabling technologies (e.g., automated radioactivity detectors, such as liquid scintillation counters with sample-changers) and reagents (e.g., commercially available ATP, which freed researchers from having to extract their own). On the other hand, costs then grew so rapidly that the number of grants lagged behind total spending (Fig. 10.3).

All of this translated into an avalanche of publications. From 1953 to 1965 the number of biomedical journals grew 75%, with the introduction of such prominent periodicals as *Biochemistry* (1962), *Biophysical Journal* (1960), *Journal of Biophysical and Biochemical Cytology* (1955; it subsequently became *Journal of Cell Biology*), and *Journal of Molecular Biology* (1959). Established journals also expanded; the *Journal of Biological Chemistry* increased its number of pages by 50% between 1959 (when it doubled its page size) and 1965.

NOTES TO CHAPTER 10

1. Robertson (1959, 1964).
2. Finean and Robertson (1958).
3. Maddy (1966). Questions remained about what permanganate and osmium tetroxide actually labeled, however.
4. Sjöstrand (1963); Green and Fleischer (1963).
5. Luzzati and Huson (1962).
6. Maddy and Malcolm (1965). Their study, nevertheless, did not deal with native proteins. They measured infrared spectra of dried membranes and optical rotary dispersion of an extracted fraction.
7. Stein and Danielli (1956).
8. Neville (1960).
9. Emmelot et al. (1964). Antibody binding was to histological sections of liver cells.
10. See, for example, Takeuchi and Terayama (1965).
11. Mueller et al. (1962). This preparation is often called a "black lipid membrane" from its optical interference when illuminated at an angle. To form films efficiently they added to their extracts substances like α-tocopherol. By modern calculations, 60–90 Å is too thick for a bilayer, and there probably were inclusions between the layers.
12. Bangham et al. (1965).
13. Kendrew et al. (1958, 1960).
14. Kendrew et al. (1958), p. 665.
15. Perutz et al. (1960).
16. Muirhead and Perutz (1963).
17. Perutz (1970).
18. Blake et al. (1965); Johnson and Phillips (1965). Phillips had collaborated with Kendrew on myoglobin structure.
19. Koshland (1958). In 1957 Edna Kearney interpreted increases in enzyme activity by certain substances as activator-induced changes in protein "configuration."
20. Monod et al. (1963).
21. Monod et al. (1965).
22. Koshland et al. (1966).
23. A. F. Huxley and Niedergerke (1964); H. Huxley and Hanson (1964). The Huxleys are not related, although both are British (as was Hanson).
24. Kauzmann (1959); Tanford (1962). Perutz et al. (1965) stressed such interaction in determining the structure of hemoglobin.
25. Epstein et al. (1963). An alternative view imagined that templates assisted folding to complementary conformations.

26. Watson and Crick (1953).

27. See *Cold Spring Harb. Symp. Quant. Biol.* 28, 1963.

28. Rall et al. (1957); Sutherland and Rall (1958).

29. Sutherland et al. (1965).

30. Krebs and Fischer (1956).

31. Chronologies and statistics are from Bordley and Harvey (1976); Mushkin (1979); Strickland (1972); and *NIH Factbook* (1976).

chapter 11

CHARACTERIZING THE
Na⁺/K⁺-ATPase

B Y 1965 numerous approaches had converged, supporting proposals that (Na⁺ +
K⁺)–stimulated ATPase activity underlay the Na⁺/K⁺ pump of higher animals.[1] Later papers in vast profusion reported diverse studies of ATPase and pump,
describing details of details, but also demonstrating unequivocally the identity of
pump and ATPase. Here only a drastically condensed sketch can be achieved. My
aim is to illustrate the extensive branching of experimental endeavors that revealed
complexities in this entity—a multi-sited enzyme—and in the process—cation
transport by multiple modes linked to particular sequences of enzymatic reaction
steps.

MULTIPLE TRANSPORT MODES

Before we turn to reactions catalyzed by this ATPase, a quick survey of the various
ways Na⁺ and K⁺ pass through the pump is necessary, for tightly coupled Na⁺/K⁺
exchange turned out to be not the only way cations are moved. Characterizing these
alternatives was begun by Glynn, who extended his thesis research of a decade
earlier with the announced goal: "to get . . . further in understanding the processes
involved [one] approach is to look more closely at the overall behaviour of the
transport system, particularly under abnormal conditions, in the hope that knowledge of the kinds of behaviour of which the system is capable will indicate . . . the
kind of mechanism it must have."[2] Although Glynn dominated these efforts
through the next decade, others also made notable contributions, culminating in the
recognition of six transport modes (Fig. 11.1).

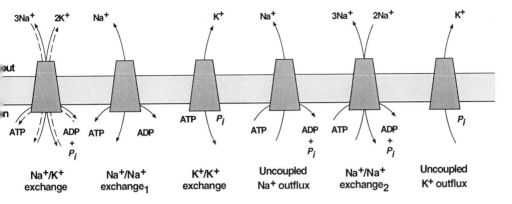

Fig. 11.1 Six transport modes of the Na^+/K^+-ATPase.

Na⁺/K⁺ Exchange Consuming ATP

This is the "normal" transport mode, moving Na^+ out and K^+ in, driven by ATP hydrolysis. In 1967, however, Glynn and Patricio Garrahan showed that this mode was reversible, that it could synthesize ATP.[3] They loaded red blood cells with K^+ and a high ratio of ADP plus P_i to ATP and then incubated them in media containing high Na^+ concentrations. The steep cation gradients (K^+ outward and Na^+ inward) drove $ADP + P_i$ to ATP.

Na⁺/Na⁺ Exchange Requiring ATP and ADP but not Consuming ATP

Also in 1967 they described a 1:1 exchange of intracellular Na^+ for extracellular Na^+ that did not involve ATP hydrolysis but nevertheless required that ATP be present.[4] Glynn, while visiting Hoffman (now at Yale University), used a system regenerating ATP from ADP to show that ADP was necessary.[5] Then, using an inhibitor of adenylate kinase (an enzyme present in these cells that converts ADP reversibly to ATP plus AMP), he and José Cavieres were able to demonstrate that ATP also was required.[6]

K⁺/K⁺ Exchange Requiring Pi and ATP but not Consuming ATP

In 1971 Glynn, together with Hoffman and Virgilio Lew, reported that red blood cells depleted of ATP and ADP could no longer exchange intracellular K^+ for extracellular K^+.[7] This was surprising, for ATP was then thought to be involved solely with enzyme phosphorylation, a Na^+-activated step occurring early in the reaction sequence. Timothy Simons, a graduate student in Cambridge, found that the 1:1 exchange required $P_i + ATP$ without consuming ATP; indeed, structural analogs of ATP unable to phosphorylate the enzyme were effective, and he concluded that binding alone was sufficient.[8]

Uncoupled Na⁺ Outflux Consuming ATP

Glynn and Garrahan also described Na^+ outflux when both Na^+ and K^+ were absent extracellularly.[9] In 1973 Lew and colleagues reported that neither ADP nor P_i was needed, and in 1976 Glynn and Steven Karlish reported that two to three Na^+ were extruded per ATP hydrolyzed.[10]

Na⁺/Na⁺ Exchange Consuming ATP

In 1980 Rhoda Blostein at McGill University discovered another transport mode: exchange of intracellular Na^+ for extracellular Na^+ (in the absence of extracellular K^+) that did not require ADP but consumed ATP.[11] She concluded that this Na^+/Na^+ exchange represented the Na^+/K^+ exchange mode but with extracellular Na^+ transported inward in lieu of extracellular K^+. She calculated that three Na^+ were extruded per ATP hydrolyzed; the ratio of outflux to influx was near 1:1, but it could have been 1.5 to 1. Michael Forgac and Gilbert Chin at Harvard University showed that this transport mode was electrogenic, in accord with a 3:2 stoichiometry of Na^+ pumped out to Na^+ pumped in.[12]

Uncoupled K⁺ Outflux not Requiring ATP

In 1986 John Sachs at the State University of New York at Stony Brook recognized that current reaction sequences (see below) depicted a loop, previously unnoticed, that effected K^+ outflux. This prediction he then tested, finding K^+ outflux in the absence of ATP into media free of Na^+ and K^+.[13]

Electrogenic Pumps and the Stoichiometry of Na⁺/K⁺ Exchange

As just noted, the pump can move one or the other ion alone. But the issue here is the stoichiometry of Na^+/K^+ exchange, by far the dominant mode under physiological conditions. Chapters 8 and 9 described several approaches using red blood cells that pointed to ratios of three Na^+ pumped out for two K^+ pumped in. Few other cells were amenable to such measurements, however, and one that was, the giant axon of squid, provided a different answer. In 1969 Lorin Mullins and Frank Brinley, at the University of Maryland and Johns Hopkins University, respectively, reported a variable stoichiometry sensitive to intracellular Na^+: at low concentrations the coupling was near 1:1, but at high concentrations it rose to three to four Na^+ per K^+.[14] Such variable coupling seemed quite reasonable for efficient and responsive functioning, and it was consistent with views of the cations as allosteric effectors.[15] But plausible as that formulation was, subsequent studies by Paul De Weer, at Washington University and later the University of Pennsylvania, demonstrated convincingly that in squid giant axons the coupling ratio was also a closely maintained 3:2.[16]

A pump with that 3:2 stoichiometry would be electrogenic if no other ions were transported concomitantly. The most thorough and compelling early study demonstrating such electrogenicity was by Roger Thomas at the University of London in

1969.[17] He injected Na$^+$ into large neurons of snail ganglia electrophoretically through Na$^+$-containing microelectrodes, measured the membrane potential with intracellular microelectrodes, and followed cytoplasmic Na$^+$ concentrations with Na$^+$-sensitive microelectrodes. Injecting 25 picomoles of Na$^+$ raised the potential about 20 mV, a rise acutely inhibited by ouabain, and both the increase in pump rate and in Na$^+$ extrusion were proportional to the increase in Na$^+$ concentration. Thomas calculated that one-third of the Na$^+$ extruded was not associated with the transport of another ion, which was in accord with a 3:2 exchange of Na$^+$ for K$^+$.

PURIFICATION AND RECONSTITUTION

Although links between ATPase and pump were apparent by the 1960s, significant questions persisted. Was the ATPase the whole pump or were other components, such as lipid carriers, involved? Was the ATPase a single enzyme or a system of enzymes? Was ATPase activity measured in the absence of Na$^+$ and K$^+$ catalyzed by the same enzyme? The obvious route to resolving such uncertainties was to purify the enzyme(s), identify the chemical structure(s) and reaction(s) catalyzed, and determine the functional adequacy of the isolated enzyme(s). This is the standard reductionistic approach. But realizing that approach was hindered by serious barriers: in the early 1960s there was no recognized approach to purifying membrane-bound enzymes and no method for demonstrating transport by isolated enzymes.

Isolation and Purification

One apparent strategy was to dissolve the membranes with detergents and then apply techniques developed over the preceding decades for "soluble" enzymes (those not membrane-bound). Beginning in the mid-1960s, Hokin followed this route, and by 1971 he described preparations from brain with specific activities of 8–12 units (micromoles of ATP hydrolyzed per minute per milligram of protein present).[18] He also applied a crucial new technique to assess purity, electrophoretic separation on a polyacrylamide gel of proteins solubilized with a detergent, sodium dodecyl sulfate (this technique, called "SDS-PAGE," separates according to molecular weight[19]). He found a major band migrating with an apparent molecular weight of 94 kDa. Only this band was labeled after phosphorylation with ^{32}P-ATP in the presence of Na$^+$ + Mg^{2+}.

Also in 1971 Jack Kyte published his doctoral research from Harvard University that represented the alternative strategy.[20] He extracted extraneous proteins from kidney membranes and then isolated fragments enriched with the ATPase, obtaining a specific activity of 13 units. But Kyte found *two* bands by SDS-PAGE, with apparent molecular weights of 87 and 54 kDa, that had copurified (they were present at the same ratio in successive steps of the purification).[21] (These became known later as the α- and β-subunits, respectively, and I will so refer to them now.) He also confirmed that the α-subunit was phosphorylated by ATP, showed that the β-subunit contained sugars (was "glycosylated"), and "cross-linked" the α- and β-

subunits, joining them *in situ* with a double-ended reagent to demonstrate a close association *in vivo*.[22]

A third attempt at purification was begun by Skou and completed by Peter Jørgensen. They used the outer medulla of kidney, which was known from physiological studies to represent the site of greatest Na^+ transport, and their 1969 paper described treating a microsomal fraction of the medulla with detergent, followed by centrifugation to separate membrane fragments. The specific activity was 15 units.[23] In 1971 they obtained a specific activity of 25 units, but they did not report electrophoretic analysis.[24] And in 1974 Jørgensen described a procedure yielding specific activities of 32–37 units with an estimated purity of 99%.[25] A simplified version gave specific activities of 20–26 units with an estimated purity of 50–70%. Both provided two prominent bands by SDS-PAGE with apparent molecular weights of 96 kDa and 35–57 kDa. Because of their high specific activities and relative ease, Jørgensen's two procedures soon became favorites. Other prominent purification procedures of that period from brain, kidney, eel electric organ, and shark rectal gland all confirmed the presence of two subunits.[26]

Uncertainties about the molar ratio of α- to β-subunits were initially conspicuous, attributable at least in part to difficulties in determining protein contents and molecular weights and to artifacts from aggregation and detergents. Most commentators came to accept a 1:1 stoichiometry.[27] More troubling and persisting was the question of whether the functional enzyme was a single $\alpha\beta$ heterodimer or a multiple of that, $(\alpha\beta)_n$ (where n > 1). The complex issues cannot be displayed here succinctly but the history can be condensed as: $(\alpha\beta)_n$ formulations were popular in the 1970s, but $\alpha\beta$ was favored in the 1980s when *almost* all the observations were revised and reinterpreted and when new data favoring the $\alpha\beta$ interpretation were reported.[28] Nevertheless, a universal consensus was not achieved, and the residual doubts represent a salient characteristic of scientific research: *some* issues resist clarification despite honest efforts, *some* interpretations must be held provisionally for long periods, and uncertainty about *some* issues can coexist with consensus about others closely related.

Reconstituting the Enzyme into Vesicles

Purifying the Na^+/K^+-ATPase demonstrated that α- and β-subunits were responsible for the $(Na^+ + K^+)$–stimulated ATPase activity. Still unanswered, however, was whether these peptides alone were the total machinery of the pump.[29] For that assurance transport must be demonstrated, requiring that the purified enzyme be in a membrane separating two compartments so cation movements can be measured. In the early 1970s Efraim Racker at Cornell University developed procedures for "reconstituting" membrane proteins, solubilized with detergents, into microscopic vesicles formed from phospholipids.[30] Adapting these procedures to the Na^+/K^+-ATPase was thus an obvious goal.

In 1974 Hokin, Shirley Hilden, and associates, using purified Na^+/K^+-ATPase from shark rectal gland and their modification of Racker's procedure, found that

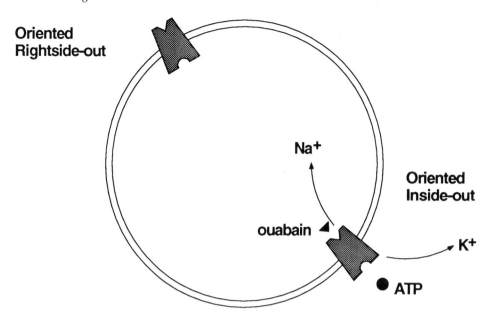

**Oriented
Rightside-out**

Na$^+$

**Oriented
Inside-out**

ouabain ◄

K$^+$

● ATP

Fig. 11.2 Orientation of ATPase molecules after reconstitution. ATPase mole-cules are reconstituted randomly in the phospholipid vesicles. Those oriented inside-out have their ATP-binding sites facing the medium and ouabain-binding sites facing the vesicle interior.

reconstituted vesicles took up ^{22}Na in the presence of ATP, attaining a 2-fold accumu-lation (both Na$^+$ and K$^+$ were within the vesicles and in the medium).[31] Ouabain trapped in the vesicles blocked ^{22}Na uptake but not ouabain added to the medium. This was consistent with Na$^+$ transport by ATPase molecules oriented in the vesicle membrane inside-out (Fig. 11.2). The ratio of Na$^+$ transported to ATP hydrolyzed was low, 0.3–0.4 Na$^+$/ATP, even with ouabain in the medium to block hydrolysis in leaky vesicles. They did not measure K$^+$ transport. On the other hand, they found ^{22}Na uptake when K$^+$ uptake was absent on both sides of the membrane, and they interpreted this as Na$^+$/Na$^+$ exchange. ATP was required and would provide ADP, although this was not examined.

   Earlier that year Stanley Goldin at Harvard University described similar experi-ments with Kyte's kidney enzyme, but—peculiarly—he found no ATP-dependent outflux of ^{42}K from the vesicles.[32] In 1975, however, Goldin and Kathleen Sweadner, using a partially purified brain enzyme, described outflux of ^{86}Rb (an analog of K$^+$), with the ^{22}Na:^{86}Rb stoichiometry a satisfying 3:2.[33] On the other hand, their en-zyme's specific activity was only 1 unit: other proteins were present in excess.

   Also in 1975 Hilden and Hokin reported ATP-dependent outflux of ^{42}K, with a Na$^+$:K$^+$ stoichiometry of 3:2.[34] The ratio of Na$^+$ influx to ATP hydrolysis was now 1.4:1. They also found ^{42}K outflux when K$^+$ was present in the medium but Na$^+$ was absent; this flux required ATP and P_i, and they equated it to K$^+$/K$^+$ exchange.

Since several percent of the "purified" enzyme preparations was neither the α- nor β-subunits, another protein could be present (during reconstitution selective enrichment might occur, up to molar equivalency with α and β). But in 1977 Goldin showed by SDS-PAGE that more than 95% of the reconstituted protein was the α- and β-subunits.[35] He again reported a 3:2 $Na^+:K^+$ stoichiometry, but now with 3.6 Na^+ taken up per ATP hydrolyzed.

REACTION MECHANISM

Efforts to determine the ATPase reaction mechanism and relate it to pump activity—to "explain" how the pump works—began in the 1950s and continued into the late 1980s when this account fades away. Here only a few lines of inquiry are sketched.

The Albers-Post Scheme

While comparing ADP/ATP exchange (see Chapter 9) with $(Na^+ + K^+)$–stimulated ATPase activity, Albers discovered reciprocal relationships: as he raised Mg^{2+} concentrations, the exchange fell but ATPase activity rose, whereas the inhibitors oligomycin or N-ethylmaleimide blocked ATPase activity but not exchange.[36] To explain these findings he presented a model with *two* phosphorylated enzyme forms in sequence, subsequently known as E_1-P and E_2-P. E_1-P was involved with exchange and its phosphate then had to be "energy-rich" since the phosphate could be donated to ADP to form ATP. Conversely, E_2-P was not "energy-rich" since its phosphate could not be donated to ADP; moreover, E_2-P was involved with K^+-activated dephosphorylation. Albers proposed that conversion of E_1-P to E_2-P was favored by higher concentrations of Mg^{2+} but blocked by oligomycin and N-ethylmaleimide, and he suggested that the conversion "may be [due to] a conformational change [in the enzyme] or an intramolecular migration of the phosphate."[37]

Meanwhile, Post continued studying enzyme phosphorylation, confirming an acyl phosphate linkage and reporting that with enzyme from 6 tissues and 11 species the same phosphorylated peptide fragments were formed by proteolytic digestion of the phosphorylated enzyme (enzymes were labeled with ^{32}P-ATP, digested with the proteolytic enzyme pepsin, and the fragments separated by electrophoresis; labeled fragments from various enzymes all migrated identically).[38] In 1969 he reported a further salient generality: phosphorylated fragments were the same from enzymes in the E_1-P and E_2-P states (enzyme phosphorylated under standard conditions was considered to be E_2-P, whereas enzyme treated with N-ethylmaleimide and then phosphorylated was considered to be E_1-P).[39] Since phosphate was bound covalently to the same amino acid in both, E_1-P and E_2-P had to differ in conformation, in the folding together of the peptide chains within the enzyme.

Granted that an acyl phosphate forms with the same amino acid in both E_1-P and E_2-P, which amino acid is it? A free carboxyl is required, but all are bound in peptide linkages except the "extra" carboxyls of dicarboxylic amino acids, aspartate

$$E_1 + MgATP \quad \overset{Na^+}{\rightleftharpoons} \quad E_1 \sim P + MgADP \qquad (1)$$

$$E_1 \sim P \quad \overset{Mg^{++}}{\rightleftharpoons} \quad E_2 - P \qquad (2)$$

$$E_2 - P \quad \overset{K^+}{\longrightarrow} \quad E_2 + P_i \qquad (3)$$

$$E_2 \rightleftharpoons E_1 \qquad (4)$$

Fig. 11.3 Albers's four-step scheme for the Na$^+$/K$^+$-ATPase. Enzyme phosphorylation and dephosphorylation occur at steps 1 and 3, and transport occurs at steps 2 and 4. [From Albers (1967).]

and glutamate (plus that at the end of the chain). In 1967 Hokin reported that ATP phosphorylated a glutamate; his analysis involved formation and separation of a radioactive derivative but the final yield was only 2%.[40] In 1973 Post reinvestigated enzyme phosphorylation, identifying the phosphorylated amino acid in a proteolytic fragment as aspartate.[41] This identification was confirmed by Boyer and by Hokin, using a third method.[42]

In an early version of his scheme, Albers included—for symmetry but without direct evidence—two forms of the unphosphorylated enzyme as well, E_1 and E_2, in a four-step sequence (Fig. 11.3).[43] He then linked Na$^+$ and K$^+$ transport to the two interconversions: "E_1 [forms] are enzyme molecules with inwardly oriented cation sites of high Na$^+$ affinity, [whereas] E_2 forms have outwardly oriented cation sites of high K$^+$ affinity."[44] Thus the E_1-P to E_2-P step drove Na$^+$ outflux and the E_2 to E_1 step drove K$^+$ influx. Post's subsequent demonstration that the E_1-P to E_2-P transition represented an enzyme conformational change supported this proposal.

But, as happens not infrequently, a scheme broadly correct in principle was subsequently shown to be incorrect in much of its detail.[45] (1) ADP, not MgADP, is the product of the first reaction.[46] (2) Mg^{2+} is not required for the conversion of E_1-P to E_2-P.[47] (3) Dephosphorylation, step 3, is easily reversible,[48] and (4) ready conversion of E_2 to E_1 requires ATP binding.[49] Over the following decades that four-step scheme was thus extensively modified as well as enlarged (see below).

Conformational Changes

Throughout the early 1970s various approaches suggested that Na$^+$ and K$^+$ bound to different enzyme conformations, but the most dramatic demonstrations were those by Jørgensen. In 1975 he described proteolytic digestion of the ATPase that produced strikingly different fragments depending on whether proteolysis was in Na^+ media or in K^+ media.[50] He used trypsin, which cleaves specifically at peptide bonds between lysine or arginine and another amino acid; the lysine-X and arginine-X bonds must also be accessible to the active site of trypsin, however, and this depends on how the protein is folded. The experiments clearly showed that when the enzyme bound K^+ one particular peptide bond of the α-subunit was exposed to attack by trypsin, which was different from those exposed when the enzyme bound Na^+. These alternative conformations were soon equated with the E_2 and E_1 conformations, respectively, of the Albers-Post scheme.

Jørgensen also showed that adding ATP to media containing K^+ converted the cleavage pattern to that of the Na^+ form (interpreted as ATP, like Na^+, favoring E_1), whereas adding ATP to media containing Na^+ plus Mg^{2+} converted the cleavage pattern to that of the K^+ form (interpreted as enzyme phosphorylation producing E_2-P).

Such experiments required static enzyme forms during which trypsin could act, and this approach obviously was not suitable for following conformational changes occurring while the ATPase reaction progressed. To achieve this time resolution, Karlish recorded changes in the "intrinsic" fluorescence of the enzyme, the fluorescence from tryptophans in the enzyme whose fluorescent yield was a function of their local environments and hence changed with changing enzyme conformations. In 1978 he described an increased fluorescence when K^+ was added in the absence of Na^+, and a decrease when Na^+ was added next; ATP greatly accelerated the fluorescence change when Na^+ was added. This Karlish interpreted as ATP accelerating the transition from E_2 to E_1.[51] Unfortunately, the magnitudes of these fluorescence changes (only a few percent) were too small for measuring reaction rates accurately. So Karlish, who was now at the Weizmann Institute, turned next to labeling the α-subunit covalently with a fluorescent indicator, fluorescein isothiocyanate (FITC). In 1980 he reported that adding K^+ to the FITC-labeled enzyme produced a 15–20% change in fluorescence and adding Na^+ reversed that change. These fluorescence changes indicated different local environments around the FITC label and thus conformational changes in the α-subunit.[52] The rates were measurable but not the effects of ATP on them, for FITC blocked ATP binding.

Another fluorescent label, eosin, bound tightly to the enzyme, and the eosin–enzyme complex also responded to added Na^+ and K^+ with large changes in fluorescence.[53] But, also like the FITC-labeled enzyme, the eosin-labeled enzyme did not bind ATP. Nevertheless, Skou and Mikael Esmann added an important detail: the "K^+ form" of the enzyme was favored not only by adding K^+ but also by lowering the ionic strength or the pH of the media, whereas the "Na^+ form" was favored by some other cations, such as choline$^+$, that could not activate catalysis. These and related studies thus demonstrated that the E_1 and E_2 conformations represented intrinsic alternative conformations *selected* by Na^+ and K^+ rather than *induced* by their binding.

To obtain a labeled enzyme that retained full activity, Marcia Steinberg at the State University of New York in Syracuse tried fluorescent derivatives of iodoacetate, since iodoacetate bound without inhibiting.[54] The successful reagent, iodo-acetamidofluorescein (IAF), did not affect ATPase activity when bound covalently to a specific cysteine of the α-subunit, and there its fluorescence yield changed by 10–15% between Na$^+$ and K$^+$ media. Using this reagent Steinberg and others were then able to record rates of conformational change occurring over the entire reaction sequence as well as selected segments of it.[55]

Cardiotonic Steroid Binding

As noted in Chapter 8, interest in cardiotonic steroids was spurred by their therapeutic significance and by the persisting ignorance of how they acted. That interest will not be portrayed here, but a few points deserve celebration. In 1961 Kurt Repke at the German Academy of Science in (East) Berlin proposed that the Na$^+$/K$^+$-ATPase was the pharmacological receptor for these drugs and assembled impressive correlations between therapeutic responses and ATPase inhibition.[56] These studies were soon extended by, among others, Arnold Schwartz, who was first at Baylor University and then at the University of Cincinnati, and by Tai Akera and Theodore Brody at Michigan State University.[57] How such inhibition might stimulate cardiac contraction was suggested independently by Repke and by Glynn in 1964, who proposed that elevated intracellular Na$^+$ concentrations (from inhibition of the pump) increased intracellular Ca^{2+} concentrations (Ca^{2+} was then recognized as a crucial determinant of contractile force).[58]

But how do cardiotonic steroids inhibit? Using radioactive digoxin, Schwartz and associates in 1968 identified two sets of ligands promoting its binding:[59] ATP + Na$^+$ and Mg^{2+}, and P_i + Mg^{2+}. The first set was explained in terms of E_2-P uniting with digoxin in a stable complex. The second set suggested formation of a stable E-P·digoxin complex from P_i, and they demonstrated this phosphorylation by using ^{32}P-P_i.[60] In 1968 Albers reported similar binding (using ^{3}H-ouabain) as well as ouabain-stimulated phosphorylation of the enzyme by P_i.[61] Moreover, in 1969 Albers, Post, Sen, and colleagues showed that the chemically same E_2-P was formed from P_i + Mg^{2+} in the presence of ouabain as by ATP + Na$^+$ and Mg^{2+} in its absence.[62] In 1974 Kyte, who was now at the University of California in San Diego, showed that a reactive derivative of cardiotonic steroids bound covalently to the α-subunit.[63] Inhibition was thus due to cardiotonic steroids stabilizing a phosphorylated form of the enzyme and to the enzyme–steroid complex being resistant to phosphorylation by ATP.

Phosphorylation by Phosphate

In 1973 Post and associates reported a surprising result: they labeled the enzyme with ^{32}P-P_i in the absence of Na$^+$ and K$^+$ and of cardiotonic steroids.[64] They had formed E_2-P (as they confirmed) not with ATP but with P_i. Since E_2-P bears an acyl phosphate and ordinary acyl phosphates cannot be formed in appreciable yield from P_i, the energy differential must be balanced by conformational changes in the protein.

Two years later Post and Kazuya Taniguchi described formation of ATP from P_i and ADP.[65] In contrast to Glynn's reversing the pump with transmembrane ion fluxes, these experiments relied not on energy from ion gradients but on binding energy provided by the sequential additions of cations. Their two-step reaction involved (1) forming E_2-P from P_i (as above), followed by (2) adding ADP + 1,200 mM Na^+. The high concentration of Na^+ then drove the reaction sequence backward, from E_2-P to E_1-P. E_1-P, with its "energy-rich" acyl phosphate, could then form ATP from ADP.

Analogous interactions were discovered from quite different courses. Lewis Cantley at Harvard University traced a peculiar kinetic response of the enzyme to an impurity in commercial ATP preparations and identified the contaminant as vanadate.[66] He attributed inhibition by vanadate, VO_4^{3-}, to its structural similarity with that assumed by phosphate, PO_4^{3-}, during phosphorylation–dephosphorylation reactions. Another inhibitor, fluoride, had long been known to inactivate the ATPase, and subsequently Joseph Robinson at the State University of New York in Syracuse reinterpreted its actions similarly: an aluminum fluoride complex, AlF_4^-, mimicked phosphate structurally (that complex was formed from aluminum ions leached from the glassware).[67] As with vanadate, inhibition required Mg^{2+} and was stimulated by K^+.

ATP Binding

Direct measurements of radioactive ATP binding to the enzyme were described in 1971 by Post and Csaby Hegyvary and by Jens Nørby and Jørgen Jensen in Aarhus.[68] Both groups found a dissociation constant in the range 0.1–0.2 μM, indicating extremely tight binding. K^+ decreased the enzyme's affinity for ATP; Post and Hegyvary found that Na^+ antagonized this effect and proposed that "Na^+ and K^+ induce alternative conformations" of the enzyme.[69] They also concluded that the low apparent affinity in the ATPase reaction "suggests that ATP ordinarily combines with a K^+ form of the enzyme and that the dissociation of K^+ is a rate limiting step" in the overall ATPase reaction.[70] This was a crucial interpretation, as we shall see in the next section.

Before these descriptions of ATP binding were published, Yuji Tonomura at Osaka University reported in 1970 that ATPase activity increased with ATP concentration in a biphasic fashion, which he interpreted in terms of *two* substrate sites with K_ms of 3 μM and 0.2 mM.[71] The nature of these two classes of sites was then debated vigorously.[72] In 1980, however, Cantley published a model having a single substrate site that changed its affinity during the reaction sequence.[73] That simpler formulation also agreed with Post's findings (below) and became widely accepted.

Cation Occlusion and Deocclusion

Customary models for cation transporters had depicted two states—with cation sites facing either the cell interior or the exterior—but Post and Sen pointed out in 1965 the necessity for a third intervening state: with the cation occluded and accessible to neither.[74] Moreover, as noted above, Post and Hegyvary suggested in 1971 that ATP, through binding to low-affinity sites, might accelerate K$^+$ release, which they proposed to be the rate-limiting step in the ATPase reaction. Post and associates brought these themes together in 1972.[75] They first added ^{32}P-ATP, Na$^+$, and Mg^{2+}; within seconds the enzyme was maximally labeled. Next they added either Rb$^+$ or Li$^+$ as analogs of K$^+$ to stimulate dephosphorylation.[76] These additions led to far lower steady-state levels of enzyme labeling within a few tens of seconds. Finally, they added oligomycin (to block the E_1-P to E_2-P transition and thus any further effect of Rb$^+$ and Li$^+$ on dephosphorylation) together with a high concentration of Na$^+$ (to overcome competition from Rb$^+$ or Li$^+$ at the Na$^+$ sites activating phosphorylation). They then measured the rate at which labeling increased, which represented the rate of enzyme *re*phosphorylation.

Post and associates identified three important results. (*1*) The rate of rephosphorylation was far slower after dephosphorylation was stimulated by Rb$^+$ than by Li$^+$. (*2*) Adding Rb$^+$ or Li$^+$ with oligomycin and Na$^+$ did not affect rephosphorylation; the crucial consideration, therefore, was which cation was present during dephosphorylation. And (*3*) adding a high concentration of ATP (0.4 mM—the ATP concentration for the initial phosphorylation was only 0.02 mM) with oligomycin increased the rate of rephosphorylation significantly. They concluded that Rb$^+$ (or K$^+$)—but not Li$^+$—became trapped or occluded after activating dephosphorylation of E_2-P, caught in a closed configuration of the enzyme, and that ATP promoted release or deocclusion of that Rb$^+$ (or K$^+$). In this formulation ATP at high-affinity sites phosphorylated the enzyme, whereas ATP at low-affinity sites modulated activity by accelerating the rate-limiting step, K$^+$ deocclusion.

Karlish, by measuring protein fluorescence changes, confirmed in 1978 the slow conversion from the E_2 conformation bearing K$^+$ to the E_1 conformation.[77] And in 1979 Glynn and Luis Beaugé directly demonstrated occlusion and deocclusion.[78] They added ^{86}Rb to the enzyme, running the deocclusion reaction in reverse,[79] and then forced the mixture—with a motor-driven plunger—rapidly through a cation-exchange column that removed any free, exchangeable ^{86}Rb. Radioactivity emerging from the column with the protein thus represented nonexchangeable, occluded Rb$^+$. And the amount of radioactivity associated with the enzyme was of the appropriate magnitude. Moreover, including ATP with the enzyme drastically reduced the amount of ^{86}Rb emerging with the enzyme, consistent with ATP accelerating Rb$^+$ release. Subsequently, Glynn and Donald Richards measured occlusion formed in the forward direction.[80] In this case, they mixed enzyme and ^{86}Rb with Na$^+$, Mg^{2+}, and a low concentration of ATP to generate E_2-P; again, a corresponding amount of ^{86}Rb accompanied the enzyme through the column. By collecting multiple, successive samples rapidly, Bliss Forbush at Yale University better defined the time course of ^{86}Rb release; with these data he then formulated a "flickering gate" model for K$^+$ deocclusion.[81]

Glynn and associates similarly demonstrated occlusion of Na$^+$ in the presence of ATP + Mg^{2+},[82] while Skou and Esmann showed that oligomycin promoted Na$^+$

occlusion even in the absence of enzyme phosphorylation,[83] accounting neatly for oligomycin's ability to block ATPase activity but spare ADP/ATP exchange.

And Karlish found that after prolonged proteolytic digestion of the Na^+/K^+-ATPase, during which the extramembrane domains were removed, the remaining 19 kDa fragment in the membrane could still occlude cations.[84] Peptide strands within the membrane were thus responsible for cation occlusion.

REACTION SEQUENCE

These studies (plus a host of others not cited) can be accommodated in reaction schemes like that in Fig. 11.4.

1. Na^+/K^+ exchange comprises steps 1 through 21 and back to 1. Reversal will synthesize ATP. With low (nonphysiological) ATP concentrations in the micromolar range, ATP will not bind to the low-affinity site, and that reaction sequence comprises steps 34 plus 1 through 18, and then steps 41, 42, and back to 34. This route is far slower, reflecting the absence of ATP at the low-affinity site that accelerates K^+ deocclusion, the rate-limiting step.
2. Na^+/Na^+ exchange requiring ATP and ADP comprises steps 34 plus 1 through 12, with the exchange accomplished by shuttling back and forth over this segment. ADP/ATP exchange reflects steps 34 plus 1 through 7 and back; oligomycin allows this reaction to proceed while inhibiting Na^+/Na^+ exchange.
3. K^+/K^+ exchange comprises steps 12 forward through 1, and back.
4. Uncoupled Na^+ outflux comprises steps 34 plus 1 through 12, and then steps 31 back to 34.
5. Na^+/Na^+ exchange consuming ATP comprises steps 34 plus 1 through 18, and then steps 41, 42, and back to 34, but with extracellular Na^+ binding at steps 13 and 14 in place of extracellular K^+.
6. Uncoupled K^+ outflux comprises steps 34 backward to 42 through step 18, continuing backward through step 12, and then to steps 31 through 34.

COMMENTARY

Places and People

The diversity represented in these experimental results is matched by that of the scientists characterizing the Na^+/K^+-ATPase (Figure 11.5).[85]

Robert Wayne Albers was born in 1928 in Hebron, Nebraska. After receiving an undergraduate degree from the University of Nebraska, he became a laboratory technician for Oliver Lowry at Washington University. When Albers asked about the possibility of earning a master's degree, Lowry told him it must be a Ph.D. or nothing. So he became Lowry's first graduate student, receiving a Ph.D. in pharma-

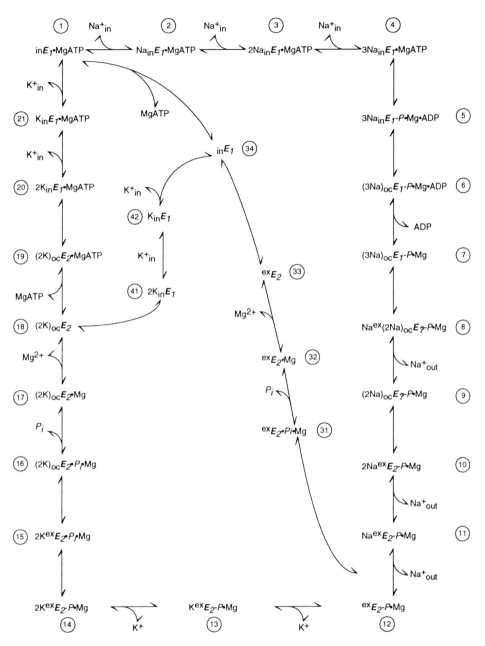

Fig. 11.4 Reaction sequence of the Na$^+$/K$^+$-ATPase. Superscripts and subscripts after the cation symbols indicate that cations occupy sites exposed to extracellular (ex) or intracellular (in) media or are occluded (oc). Subscripts after E indicate the conformation ("?" indicates uncertainty). Not all alternative pathways are shown. [Adapted from Robinson and Pratap (1993), Table 1.]

A

Fig. 11.5 Participants. *A:* Amir Askari; *B:* Rhoda Blostein; *C:* Peter Jørgensen; *D:* Steven Karlish; *E:* Makoto Nakao; *F:* Jens Nørby; *G:* Kurt Repke; *H:* Arnold Schwartz; and *I:* Amar Sen. (Photographs courtesy of these individuals.)

cology in 1954 with studies on brain enzyme activities. That year he moved to NIH, where he has remained, working first on the biosynthesis of gamma-aminobutyric acid in brain. When a colleague down the hall imported electric eels from Brazil, Albers took some of the electric organ to survey enzyme activities. His pursuit of the Na^+/K^+-ATPase was motivated by curiosity about cation-enzyme interactions. The naming of the Albers-Post scheme rightly celebrates his salient contributions to delineating the phosphorylation and ADP/ATP exchange reactions. Astute and informed but retiring and soft-spoken, Albers has broad scientific interests; he has continued to pursue neurochemical investigations (he is an editor of a prominent neurochemistry text) as well as the Na^+/K^+-ATPase.

Amir Askari was born in 1930 in Ahwaz, Iran, and came to the United States in 1949 as 1 of 10 recipients of a competitive governmental award established to educate physicians abroad. After receiving his undergraduate degree from the University of Dubuque, however, he switched to science, obtaining a master's degree in organic chemistry from New York University and a Ph.D. in biochemistry from Cornell University Medical College in 1960, with thesis research on the biosynthesis

B

of ergothionine. He moved directly to a faculty position in the Pharmacology Department at Cornell, beginning independent studies on choline transport by red blood cells. From these experiments he eased into research on cation-activated enzymes, K^+-activated AMP deaminase and the Na^+/K^+-ATPase. In 1975 he moved to the Medical College of Ohio as chairman of pharmacology, where he attracted a number of able investigators. Over the years, Askari has studied a wide range of problems using diverse techniques, from cardiotonic steroid effects on the heart to cross-linking the α-peptides to delineating divalent cation interactions, and he has advanced reasoned, thoughtful, imaginative, and occasionally dissenting viewpoints. He is a gregarious, cheerful, and engaging individualist.

Rhoda Blostein was born in 1936 in Montreal, Quebec, and received her undergraduate and doctoral degree (1960) from McGill University. Her thesis research in

C

biochemistry concerned enolase activity of red blood cells. After postdoctoral years at the University of Illinois studying another glycolytic enzyme, aldolase, she returned to McGill where she began independent research on red blood cells, initially approaching the Na^+/K^+-ATPase as a link to cellular energy consumption. For a number of years she pursued in careful detail the processes of enzyme phosphorylation and dephosphorylation using red blood cells and inside-out vesicles of their membranes, defining at which surface of the enzyme various reactants achieved their effects. More recently she has studied properties of the purified Na^+/K^+-ATPase, its modification chemically and its function when reconstituted into lipid vesicles. She is an enthusiastic and charming colleague as well as an original and dedicated scientist.

Peter Leth Jørgensen was born in 1938 in Copenhagen and received his medical

D

degree from the University of Aarhus in 1964, having worked with Skou on the Na^+/K^+-ATPase during his student years. After clinical training and a year in the army, he joined the physiology department in Aarhus in 1966. In 1989 he became Professor of Molecular Physiology at the August Krogh Institute of the University of Copenhagen. Jørgensen has been outstandingly successful in purifying the Na^+/K^+-ATPase, in studying its conformational states and transitions, and in exploring its chemical and physical structure. He is industrious, ambitious, focused, and productive, a major contributor to the field; he is formally polite and responsive, but he can be somewhat prickly on occasion and difficult to deal with.

Steven J. D. Karlish was born in 1944 in Lahore but grew up and was educated in England, receiving his undergraduate degree in biochemistry from Oxford University in 1966. He then emigrated to Israel and received his doctoral degree from the Weizmann Institute in 1970 for research on cation transport associated with photosynthesis. After 2 years with Wilfred Stein at the Hebrew University, which included studies of glucose transport by red blood cells, Karlish moved on to studies of the Na^+/K^+-ATPase with Glynn in Cambridge. In 1975 he returned to the Weizmann Institute as a faculty member, where he has remained aside from numerous professional visits to New York, Los Angeles, Cambridge, Bristol, and Copenhagen. Karlish has been an extremely important contributor to transport and enzymatic

E

studies, as this chapter illustrates. More recently he has pursued questions of how protein structure affects ATPase function. He is confident, assured, assertive, and articulate, and is an interested and willing collaborator.

Makoto Nakao was born in 1924 in Miyazaki, Japan. Following the interruptions of World War II, he received his medical degree from the University of Tokyo in 1947. After holding positions in biochemistry at the Universities of Gunma and Yokohama, he returned to the department of biochemistry at the University of Tokyo in 1963 where he remained until retirement in 1989. His early research was on red blood cell metabolism, but he was a pioneering investigator of the Na^+/K^+-ATPase. He has examined such issues as the acyl phosphate nature of the reaction intermediate, explored cation interactions with the enzyme, devised purification techniques, and estimated functional size by radiation-inactivation. Nakao is friendly although reserved, and he is an active participant at meetings despite the hindrance of language differences. In 1972 he organized a significant meeting in Japan that was attended by among others Blostein, Hoffman, Post, and Tosteson.

Jens Gregersen Nørby was born in 1932 in Copenhagen and received his degree in chemical engineering from the Danish Technical University in 1956. He then proceeded to the biochemistry department in Aarhus, where he studied polyunsaturated fatty acids in atherosclerosis and co-authored a textbook of biochemistry. In 1965 he moved to the physiology department, beginning studies on the Na^+/K^+-

F

ATPase that included measuring ATP binding, determining rates of phosphorylation and dephosphorylation (deriving detailed quantitative models from these data), and assessing the evidence for subunit interactions. Nørby is a thoughtful and serious scientist. He is also a prominent personality: generous, outgoing, entertaining, and humorous. Nørby has traveled widely and is a popular and esteemed visitor.

Kurt Robert Hermann Repke was born in 1919 in Friesach, Germany. His education was disrupted by military service (he was also taken prisoner by the United States Army), but he received his medical degree in 1945 from the University of Rostock. After clinical training in internal medicine he began research in pharmacology. In 1955 he moved to the Institute of Biochemistry of the German Academy of Sciences in (East) Berlin and worked under Karl Lohmann. From 1971 until his retirement in 1984 he headed the Biomembrane Section of the Central Institute of Membrane Biology. Repke's initial interest in steroids evolved into studies of the metabolism of cardiotonic steroids. But after his recognition that the therapeutic effect of these steroids was mediated through the Na$^+$/K$^+$-ATPase, he concentrated on these interactions and on the reaction mechanism of the ATPase. Although immersed in administrative duties and hindered by political restraints, Repke participated actively at international meetings. He is a formal, reserved, polite, and determined contributor.

Arnold Schwartz was born in 1929 in New York City and received an under-

G

graduate degree in pharmacy from Long Island University in 1951. After 4 years in the Air Force he received a master's degree from Ohio State University. This was followed by a Ph.D. in pharmacology in 1961 from the State University of New York in Brooklyn, where he studied cardiotonic steroids. Schwartz next spent a year with McIlwain in London working on the Na^+/K^+-ATPase and six months with Skou in Aarhus. He began his independent career at Baylor University and moved to Cincinnati in 1977 as chairman of pharmacology. His studies on the Na^+/K^+-ATPase included examinations of the reaction mechanism and identification of the amino acid sequence, but his major focus was on the effects of cardiotonic steroids. Indeed, he has been a major investigator in several areas of cardiac function where he has led new and fruitful endeavors. He is exuberant, energetic, accomplished, entertaining, organized—and organizing.

Amar Kumar Sen was born in 1927 in Calcutta to a medical family. His grandfather introduced the root of *Rauwolfia serpentina* from Indian medicine to Western medicine for the treatment of schizophrenia and hypertension and then supplied the root from which reserpine is extracted. After receiving his medical degree in 1955 and practicing medicine briefly, Sen went to the University of London to further his scientific education on behalf of the family firm. Sen received his Ph.D. in 1960,

H

working with W. F. Widdas on sugar transport by red blood cells, and Widdas recommended him to C. R. Park at Vanderbilt University. There Sen instead worked with Post, from 1960 to 1961 and again (after an absence due to visa restrictions) from 1963 to 1966. Sen then moved to the department of pharmacology at the University of Toronto where he continued work on cardiotonic steroid interactions and on the reaction mechanism of the Na$^+$/K$^+$-ATPase, as well as studying regulation of this enzyme and extending his interests to other transport and pharmacological issues. Sen is a patient and careful investigator; he is also a warm, gracious, and dignified gentleman.

These and many others formed a web of cordial relationships that underpinned the web of experimental evidence they amassed. These relationships are documented by joint authorships of collaborative studies, and they are evidenced as well in long-standing friendships. For example, Karlish published with Glynn, Jørgensen, and Steinberg, and acknowledged help from H.-J. Apell and Post, among others. Blostein thanked in various papers Glynn, Hoffman, Post, Karlish, and Philip Dunham, whereas Askari thanked Schwartz, Post, Nørby, Irena Klodos, and Earl Wallick. Hoffman is a personal friend of Glynn, Post, and Skou, and Post and Skou have maintained a warm relationship over the decades. Some investigators built large collaborative and international groups, as did Schwartz in Houston and Cincinnati.

I

Others' groups were more isolated, as was Nakao's in Tokyo by geography and Repke's in Berlin by Cold War partitionings. Some of these scientists are more gregarious, as are Askari, Nørby, and Schwartz, while others are more reserved, as are Albers and Sen.

Meetings

Scientific meetings helped to form such collaborative and personal associations. In this field the most praised meetings have been the international conferences on the Na^+/K^+-ATPase. Askari organized the first one in New York in 1973, which assembled scientists with varied but overlapping interests from four continents, many of whom had never met before. The second meeting was in Sønderborg in 1978, the third in New Haven in 1981, the fourth in Cambridge in 1984, the fifth in Fuglsø in 1987, and the sixth in Woods Hole in 1990.

There is less enthusiasm and more disparate opinion about broader, more general meetings. Skou felt that meetings were for him, a "poor listener", largely "a waste of time"; on the other hand, his trip to Montreal with Post in 1953 and his encounter with Post in Vienna in 1958 were clearly beneficial, and he has described how his lunch with Hodgkin and Keynes in Buenos Aires in 1959 encouraged him to consider the Na^+/K^+-ATPase a worthwhile topic.[86] Post feels that meetings are important for turning a name on paper into a personality; he advises young investigators "to take an interest in how [their] competitors' minds work."[87]

New Techniques

Many of the central results described here relied on new methods and approaches. Particularly important were new separation techniques, notably SDS-PAGE for proteins, high-voltage paper electrophoresis for peptides, and various chromatographic methods for a range of needs. In addition, density-gradient and zonal centrifugation techniques were necessary for enzyme preparation. Essential also were methods for reconstituting proteins in phospholipid vesicles, for making inside-out vesicles from red blood cell membranes, and for dialyzing squid giant axons to control intracellular constituents. Crucial analytical techniques included using ion-specific intracellular electrodes and labeling proteins with fluorescent reporter groups. Important reagents ranged from inhibitors such as oligomycin to nonhydrolyzable analogs of ATP. Of all these, none was developed specifically for studying the Na^+/K^+ pump or ATPase.

Scientific Summary

Purification of the Na^+/K^+-ATPase established its biochemical identity as a dimer of α- and β-subunits having approximate molecular weights of 100 and 40 kDa. Reconstitution of the purified enzyme into lipid vesicles established through transport experiments the physiological identity of this enzyme as the Na^+/K^+ pump. Concurrently, efforts to understand the reaction mechanism of this ATPase and pump included characterizing intermediate stages that bound ATP, became phosphorylated, underwent a conformational change, were dephosphorylated, and returned to the original conformation: while binding, occluding, and releasing first three Na^+ and then two K^+. These stages became discrete steps in a reaction sequence that was correlated with the various transport modes of the pump.

NOTES TO CHAPTER 11

1. Through discovery of alternative mechanisms and failure to find the Na^+/K^+ pump, it became clear that bacteria, fungi, plants, and lower animals do not possess the Na^+/K^+-ATPase described in molluscs, arthropods, and chordates.

2. Garrahan and Glynn (1967a), p. 160.
3. Garrahan and Glynn (1967e). These experiments showed incorporation of ^{32}P from P_i into ATP; later experiments showed net synthesis of ATP (Lew et al., 1970).
4. Garrahan and Glynn (1967a,c,d).
5. Glynn and Hoffman (1971).
6. Cavieres and Glynn (1979).
7. Glynn et al. (1971).
8. Simons (1974, 1975).
9. Garrahan and Glynn (1967a,b).
10. Lew et al. (1973); Glynn and Karlish (1976).
11. Lee and Blostein (1980); Blostein (1983).
12. Forgac and Chin (1982).
13. Sachs (1986).
14. Mullins and Brinley (1969).
15. For example, Robinson (1967, 1969) described cooperative binding of cations to the ATPase, following the allosteric models of Monod et al. (1963, 1965). Activation could then occur without all sites being filled, in contrast to models requiring that three Na$^+$ and two K$^+$ necessarily be bound.
16. Abercrombie and De Weer (1978); De Weer and Geduldig (1978); Rakowski et al. (1989).
17. Thomas (1969).
18. Uesugi et al. (1971).
19. However, further experience demonstrated that membrane-bound enzymes, which necessarily have extensive hydrophobic surfaces, may migrate differently from "soluble" proteins of equal molecular weight.
20. Kyte (1971a).
21. In photographs of Hokin's gels, bands are visible not only at 94 kDa, but also at 64, 53, and 35 kDa.
22. Kyte (1971b, 1972).
23. Jørgensen and Skou (1969).
24. Jørgensen and Skou (1971); Jørgensen et al. (1971). By that time (and before Kyte's paper appeared), Skou had detected two bands by SDS-PAGE, but after some disagreement, they chose not to describe the finding (Skou, interview, 1993).
25. Jørgensen (1974a,b).
26. Nakao et al. (1974); Lane (1973); Dixon and Hokin (1974); Hokin et al. (1973).
27. See, for example, Robinson and Flashner (1979); Jørgensen (1982); Glynn (1985).
28. See, for example, Repke and Schön (1973) and Askari et al. (1980) vs. Brotherus et al. (1981, 1983) and Jørgensen and Andersen (1986). A similar issue, whether two ATP-binding sites coexist on the enzyme, was strongly advocated by, among others, Askari (1987), Scheiner-Bobis et al. (1987), and Schuurmans-Stekhoven et al. (1981); however, Nørby (1987) could find only one ATP-binding site per α-subunit.
29. A reasonable concern was whether subunits necessary for transport might be selectively lost during the purification procedure.
30. Kagawa and Racker (1971); Racker (1972, 1973). The phospholipid vesicles are often termed "liposomes."
31. Hilden et al. (1974).
32. Goldin and Tong (1974). This paper was submitted (and appeared) two and a half months before the paper by Hilden et al.
33. Sweadner and Goldin (1975).
34. Hilden and Hokin (1975).
35. Goldin (1977).
36. Fahn et al. (1966a,b).

37. Fahn et al. (1966a), p. 1888.

38. Bader et al. (1966, 1967, 1968).

39. Post et al. (1969).

40. Kahlenberg et al. (1967, 1968).

41. Post and Kume (1973).

42. Degani et al. (1974); Hokin (1974). Degani et al. noted that their study was prompted by Post's finding, and Hokin noted that his reinvestigation was prompted by Degani and Boyer's development of their technique.

43. Albers (1967); Siegel and Albers (1967).

44. Albers (1967), p. 743.

45. This is not paradoxical since the schemes remembered are those having some general validity, and schemes are necessarily modified in detail as further studies reveal intricacies characteristic of biological processes.

46. Beaugé and Glynn (1979a).

47. Robinson (1974a,b); Klodos and Skou (1975).

48. Post et al. (1973, 1975).

49. Post et al. (1972).

50. Jørgensen (1975, 1977).

51. Karlish and Yates (1978).

52. Karlish (1980). He reported that only the α-peptide was labeled, with one to two FITC per peptide.

53. Skou and Esmann (1980, 1981).

54. Kapakos and Steinberg (1982, 1986a,b); Tyson et al. (1989).

55. Steinberg and Karlish (1989); Stürmer et al. (1989); Pratap et al. (1991).

56. Repke (1963); the meeting was held in 1961.

57. See, for example, Schwartz et al. (1969); Akera et al. (1969).

58. Repke (1964); Glynn (1964). This proposal was developed by Langer (1972), who included the Na$^+$/Ca^{2+} exchanger linking Na$^+$ and Ca^{2+} fluxes.

59. Schwartz et al. (1968). Charnock et al. (1963) had previously proposed that the cardiotonic steroids reacted with the phosphorylated enzyme.

60. Lindenmayer et al. (1968).

61. Albers et al. (1968).

62. Siegel et al. (1969); Sen et al. (1969); Post et al. (1969).

63. Ruoho and Kyte (1974).

64. Post et al. (1973, 1975).

65. Taniguchi and Post (1975).

66. Cantley and Josephson (1976); Josephson and Cantley (1977); Cantley et al. (1977, 1978). Vanadate was a previously unrecognized contaminant of ATP prepared from horse muscle.

67. Robinson et al. (1986). Arguments for the participation of Al^{3+} from glassware and for the resemblance of AlF$_4^-$ to phosphate were presented by Sternweis and Gilman (1982) and by Bigay et al. (1985).

68. Hegyvary and Post (1971); Nørby and Jensen (1971); Jensen and Nørby (1971).

69. Hegyvary and Post (1971), p. 5239.

70. Ibid.

71. Kanazawa et al. (1970).

72. See, for example, Robinson (1976); Askari (1987).

73. Smith et al. (1980). Moczydlowski and Fortes (1981) made the same argument.

74. Post and Sen (1965); they attributed the notion of an intermediate stage to the analysis by Rosenberg and Wilbrandt (1957). Post and Sen repeated their proposal in 1969 (Post et al., 1969) with the argument that otherwise a leak could occur.

75. Post et al. (1972).

76. Post et al. selected Rb^+ and Li^+ empirically for their extreme effects. By that time Rb^+ was known to be a good substitute for K^+ in the Na^+/K^+-ATPase reaction and Li^+ a poor substitute.

77. Karlish and Yates (1978).

78. Beaugé and Glynn (1979b).

79. K^+ was thought to be binding from the cytoplasmic surface and from there moving to the occluded state. In the forward direction, K^+ binds from the extracellular surface (when the enzyme is in the E_2-P state) and from there moves to the occluded state. Release from the occluded state would then be from the cytoplasmic surface.

80. Glynn and Richards (1982).

81. Forbush (1987).

82. Glynn et al. (1984).

83. Esmann and Skou (1985).

84. Karlish et al. (1990).

85. Biographical information is from interviews and correspondence (1993–1995).

86. Interview (1993); Skou (1989).

87. Interview (1993).

chapter   12

STRUCTURE AND RELATIVES
OF THE Na⁺/K⁺-ATPase

P REVIOUS chapters described how a physiological process, the Na^+/K^+ pump, became identified with a biochemical entity, the Na^+/K^+-ATPase, an enzyme consisting of α- and β-subunits and coupling ATP hydrolysis to Na^+ and K^+ transport. Unanswered, however, were essential questions: What is the chemical identity of the subunits and what is the chemical mechanism of hydrolysis and transport? Partial answers to the first question will be described next; satisfactory answers to the second were merely being approached as the 1980s closed. But a further realization emerged that was unanticipated initially. The Na^+/K^+-ATPase belongs to a family of transport ATPases that are related structurally and functionally and are present across the biological realm, from bacteria and fungi to plants and animals.

CHEMICAL IDENTITY

Sanger first succeeded in determining the "primary structure" of a lengthy peptide when he sequenced the B chain of insulin in 1951 (see Chapter 7). Since identifying the next amino acid becomes increasingly uncertain as one proceeds further along a chain, he cleaved the peptide into various short fragments, separated and sequenced these, and then pieced together the overall amino acid sequence from those of overlapping fragments.

For the Na^+/K^+-ATPase, Post and associates identified in 1973 several amino acids flanking the phosphorylated aspartate (see Chapter 11), but the first full-scale attempt to apply Sanger's approach to the α-subunit was begun in the later 1970s by John Collins, working with Schwartz in Cincinnati. That effort, however, was frustrated by a problem common to membrane-bound enzymes: hydrophobic fragments were insoluble and thus could not be separated for analysis. Still, by 1982 Collins and associates had sequenced a number of hydrophilic stretches,[1] and this information was essential for an alternative method that did prove successful. For in 1977 Sanger—and independently Allan Maxam and Walter Gilbert at Harvard University—described a quite different approach based on amino acids being encoded in DNA by successive nucleotide triplets: they determined the sequence of DNA nucleotides (made easy by clever techniques) and from that deduced the amino acid sequence.[2]

In 1981 Jerry Lingrel moved from the biochemistry chair in Cincinnati to the microbiology chair.[3] One of the microbiology faculty, Dennis Lang, was studying bacterial ATPases and mentioned to Lingrel the physiological importance of the Na^+/K^+-ATPase and that its amino acid sequence was unknown. Lingrel, who had a background of pioneering studies on hemoglobin biosynthesis from gene regulation to messenger RNA (mRNA) processing, recognized the advantages of deducing amino acid sequences from DNA sequences. Together, Lingrel and Lang submitted in 1981 a grant application to sequence the Na^+/K^+-ATPase with this method. By the time the grant was awarded in 1982, however, Lang had left Cincinnati, and Lingrel proceeded with an able postdoctoral fellow, Gary Shull. Lingrel knew Schwartz as a fellow chairman at Cincinnati, and Lingrel and Shull adopted a strategy dependent on the sequences that Collins, Schwartz, and associates had determined.

To find the relevant DNA they first synthesized "oligonucleotide probes" corresponding to two of the peptides that Collins and associates had sequenced: deoxyribonucleotide polymers encoding a five amino acid sequence of one peptide and a seven amino acid sequence of the other. Because more than one nucleotide triplet may encode a single amino acid, they had to synthesize 32 separate probes for the first peptide and 256 for the second to cover all possible ways each could be represented by DNA.

Their other starting material was complementary DNA (cDNA) formed from sheep kidney mRNA, the RNA the kidney cells use to direct their protein syntheses.[4] This cDNA thus encoded the amino acid sequences of all peptides that the kidney cells were making. They cut the cDNA enzymatically into appropriate lengths, inserted these enzymatically into plasmids,[5] and incorporated the plasmids randomly into bacteria. A given bacterium then reproduced a single plasmid containing a single stretch of added cDNA. Lingrel and Shull thus produced a "cDNA library" of 50,000 colonies of the bacterium *Escherichia coli*, each colony representing a "clone"—cells genetically identical to a single ancestor—and each clonal colony containing a particular stretch of rat kidney cDNA.

To find which of the 50,000 colonies contained cDNA for the Na^+/K^+-ATPase, they "hybridized" the DNAs with each probe: radioactively labeled probes were incubated with DNA of each colony under strictly controlled conditions, and any probe not complementary to the DNA of a colony was washed away. Lingrel and Shull found cDNA from 12 of the 50,000 colonies hybridized with both probes.

Finally, they determined the nucleotide sequence of the hybridizing cDNA and from that sequence they deduced the amino acid sequence. All 16 of Collins's peptides were represented in the 1,016 amino acid sequence, which gave a calculated molecular weight of 112,177. That was a straightforward application of established techniques, but it was nonetheless a significant accomplishment on which many further achievements were based.

Shull, Lingrel, and Schwartz's paper appeared in *Nature* in August 1985.[6] That same issue included a report from Shosaku Numa's group in Kyoto.[7] Unknown to those in Cincinnati, Numa and associates had been proceeding similarly with the ATPase of an electric fish, the ray *Torpedo californica:* they sequenced tryptic peptides and then prepared nucleotide probes based on these sequences to screen a cDNA library. Their 1,022 amino acid sequence from fish was nearly identical to that from sheep.[8] Similar reports appeared rapidly thereafter. In 1986 sequences were published for the α-subunit of pig kidney, rat brain, and human tumor cells, and by the end of the decade for the α-subunit from such diverse species as brine shrimp and fruit flies.

On the basis of hydrophobic stretches of amino acids,[9] Shull and colleagues proposed that the amino acid chain passed across the membrane eight times. This value was disputed over the next decade, with eight or ten favored by the 1990s (Fig. 12.1). Soon localized along this sequence were the sites for phosphorylation by ATP and for covalent binding of the fluorescent labels FITC and IAF. The bonds cleaved by trypsin in the presence of Na^+ or K^+ (see Chapter 11) were localized also.[10]

(With the amino acid sequence defined, the goal of attributing functional roles to individual amino acids of the chain was greatly facilitated as well. A traditional approach, through chemical modification of particular amino acids, suffered from limitations in attaining the requisite specificity.[11] But a new technique of enormous potential, "site-directed mutagenesis," was just beginning to be applied to the Na^+/K^+-ATPase at the end of the 1980s. With this method any amino acid of interest can be deleted or transformed into another amino acid by changing the nucleotide triplet of the DNA encoding that amino acid (either removing the triplet or substituting another triplet encoding a different amino acid), and then "expressing" the modified DNA to produce the modified enzyme. This latter step involves synthesizing mRNA enzymatically from the modified DNA and using that mRNA to direct protein synthesis, e.g., by injecting the mRNA into oocytes of the frog *Xenopus laevis*, which readily synthesize protein from exogenous mRNA. Thus, Masaru Kawamura and associates in Kitakyushu, Japan changed the aspartate that is phosphorylated by ATP to the other dicarboxylic amino acid, glutamate.[12] ATPase activity was lost, consistent with a change in location of the free carboxyl—glutamate has an additional $-CH_2-$ group in the side chain—and in its orientation, since the peptide backbone zigzags and thus successive $-CH_2-$ groups point in a different directions.)

In 1986 Shull and associates reported the sequence for the sheep kidney β-subunit, specifying 302 amino acids with a calculated molecular weight of 34,937.[13] They proposed one transmembrane segment (Fig. 12.1) and pointed out three "consensus sites"[14] for covalent binding of sugars (glycosylation). Independently, sequences were reported that same year for the β-subunits of electric ray, human tumor cells, rat brain, and pig kidney by groups in Japan, the United States, and the Soviet Union.[15]

By injecting mRNA for α- and/or β-subunits into *Xenopus* oocytes, Numa and

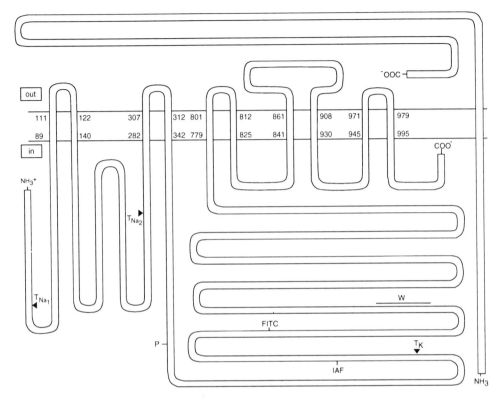

Fig. 12.1 Structural diagram of the α- and β-subunits. Shown for the α-subunit are the locations of the Walker motif (W), the phosphorylation site (P), the sites binding FITC and IAF, and those at which trypsin cleaves in Na^+ media (T_{Na1} and T_{Na2}) and in K^+ media (T_K). Ten proposed transmembrane segments are labeled with the amino acid sequence numbers. The β-subunit with a single transmembrane segment is shown on the right. [Redrawn from Lingrel and Kuntzweiler (1994), Fig. 2. Reprinted by permission of the American Society for Biochemistry and Molecular Biology.]

associates showed that both α- and β-subunits of the Na^+/K^+-ATPase were required for expressing ouabain-sensitive ATPase activity and ^{86}Rb uptake.[16] This experiment also demonstrated that these two peptides were sufficient.

ISOFORMS[17]

In 1962 Skou described how the $(Na^+ + K^+)$–ATPase activities of rabbit kidney and brain differed in their responses to Na^+ and K^+ and in their sensitivities to ouabain, implying that the enzymes in the two organs were not identical.[18] But the dramatic demonstration came in 1979 when Sweadner reported that purified Na^+/K^+-ATPase preparations from brain gave two bands by SDS-PAGE in the region of the

α-peptide.[19] One band corresponded to that found in other organs; the second, with an apparent molecular weight 2 kDa greater, was found in brain and neural tissue from a wide range of animals. She named this $\alpha+$. (That $\alpha+$ appeared larger than α was important for convincing other investigators, since proteases, which are abundant and active when cells are broken apart, can shorten proteins during purification procedures. Previous investigators may have overlooked $\alpha+$ since detecting that small difference by SDS-PAGE requires careful technique lest the two bands smudge together.)

In 1986 Shull and Lingrel screened cDNA libraries from rat brain and kidney and found cDNA for three different isoforms, subsequently named $\alpha1$, $\alpha2$, and $\alpha3$ (with α corresponding to $\alpha1$, and $\alpha+$ to $\alpha2 + \alpha3$).[20] Calculated molecular weights from the deduced amino acid sequences were, respectively, 112,573, 111,736, and 111,727 (the greater apparent molecular weight of $\alpha+$ was thus an artifact of SDS-PAGE). A second form of the β-subunit ($\beta2$) was identified in 1989.[21]

THREE-DIMENSIONAL STRUCTURE

The amino acid sequence represents only one aspect of a protein's chemical structure, its primary structure. How that chain folds locally into α-helices and β-sheets and how these in turn produce the full three-dimensional form, its "secondary" and "tertiary" structures, is the crucial concern: this folding brings together the reactive sites in functional order. The standard approach to learning such structure is from X-ray diffraction of three-dimensional crystals, but while membrane-bound enzymes may form two-dimensional crystals in their membrane, general procedures for stacking these in register to create three-dimensional arrays were not forthcoming. (The initial success at crystallizing membrane-embedded proteins was with a bacterial complex, the photosynthetic reaction center. This structure was then solved by X-ray crystallography at a resolution of 3 Å, a feat meriting a Nobel prize.[22] But that achievement did not immediately establish techniques successful for membrane-bound proteins generally.)

An alternative approach to defining a protein's three-dimensional conformation employs nuclear magnetic resonance (NMR) spectroscopy, although solving the structure of 100 kDa proteins had not been achieved during the 1980s. Still, Charles Grisham at the University of Virginia used proton NMR relaxation measurements in 1988 to determine the conformation of ATP bound to the enzyme, as well as to calculate the distances to a bound manganese ion (Mn^{2+}, used as an analog of Mg^{2+}).[23]

Gross structural features of the ATPase were discernable by various electron microscopic techniques. In 1973 Hokin and associates published electron micrographs of negatively-stained shark rectal gland Na⁺/K⁺-ATPase preparations, revealing knobs projecting from the membrane surface[24] that they likened to similar images of the mitochondrial ATP synthase (see Chapter 16). Four years later Repke and associates published similar micrographs, localizing the knobs to the cytoplasmic surface and proposing that fuzzy blurs on the opposite (extracellular) surface represented sugar polymers of the β-subunit.[25] They also published freeze-fracture electron micrographs showing particles *within* the membrane, in agreement with reports the previous year by Schwartz and associates and that year by Arvid Maunsbach and associates in Aarhus.[26] These images thus depicted a cytoplasmic mass, an intramembrane domain, and an extracellular region.[27]

In 1982 Maunsbach described two-dimensional crystals having unit cells compatible with $\alpha\beta$ complexes (induced by incubating with Mg^{2+} plus vanadate) or with $(\alpha\beta)_2$ complexes (induced by $Mg^{2+} + P_i$).[28] In 1985 Maunsbach published a three-dimensional model, calculated from tilted projections of two-dimensional crystals, that featured a large cytoplasmic mass extending 40 Å from the membrane, an intramembrane domain, and an extracellular portion extending 20 Å extracellularly.[29] The resolution of these images, however, was only 20–25 Å.

REACTION MECHANISM

A *qualitative* understanding of the reaction mechanism requires a model depicting: (1) how ATP binds among the amino acids so their side chains facilitate phosphorylation of one particular aspartate and then its dephosphorylation; (2) how the side chains form sites accepting three Na^+ from the intracellular medium, block egress (occlude), allow discharge extracellularly, then form sites accepting two K^+ from the extracellular medium, block egress, and allow discharge intracellularly; and (3) how the structural changes for ATP catalysis correlate with those for cation transport, all within the successive steps of the reaction sequence. Understanding the reaction mechanism thus requires knowing the structure in atomic detail at each step, with that structure then interpreted by the rules of chemical reactions. For transport, plotting the route of cations through the enzyme has no established paradigm, although Mitchell's "mobile barrier" depiction (Figure 8.10F) has been the favored scheme for some decades. No firm answers had emerged by the later 1980s.

THE FAMILY OF P-TYPE ATPases

By 1980 a number of transport ATPases were known to share common characteristics, notably reaction mechanisms involving enzyme phosphorylation, and these ATPase were designated P-type on that basis.[30] With the development of cDNA methods, similarities in amino acid sequences became apparent also, and from these data evolutionary as well as structural relationships were deduced (Fig. 12.2). By the early 1990s over 50 P-Type ATPases had been sequenced. Here I will point to a few prominent examples.

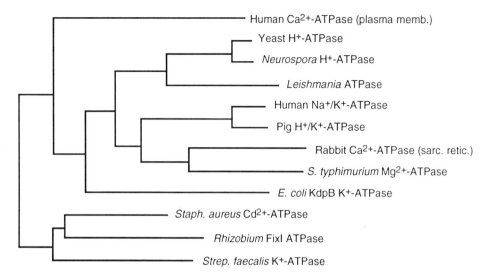

Fig. 12.2 Evolutionary relationships among P-type ATPases based on amino acid sequences. [From Maguire (1992), Fig. 3. Reprinted by permission of Plenum Publishing Corp.]

Sarcoplasmic Reticulum Ca^{2+}-ATPase[31]

In 1950 muscle action potentials were known to trigger contraction through a causal chain involving Ca^{2+}. Sydney Ringer had reported in 1883 that omitting Ca^{2+} from the medium blocked muscle contraction and George Mines reported in 1913 that omitting Ca^{2+} blocked contraction without affecting the action potential. And in 1947 Victor Heilbrunn and Floyd Wiercinski showed that injecting Ca^{2+} into muscle caused contraction.[32]

On the other hand, relaxation did not result simply from electrical activity ceasing, for contraction could persist long after the muscle action potential passed. What then caused relaxation? An answer arose, however, from different interests. B. B. Marsh in Cambridge noted that he "commenced [this project] to study the effect of ATP on the water retention of a muscle homogenate . . . suggested by the familiar observation that when fresh muscle is minced, the sarcoplasmic fraction becomes more extractable."[33] What he found was that "there appears to be a substance . . . in muscle, easily removable by dilute potassium chloride solution, . . . which determines the effect [that ATP] will produce" on the contractile state.[34] He centrifuged muscle minced in potassium chloride, monitoring the volume of the sedimented pellet. Over a period of several minutes the pellet shrank (likened to contraction), but adding ATP could then cause the pellet to expand (likened to relaxation). Yet if the potassium chloride supernatant fraction were removed and the pellet resuspended in pure potassium chloride, adding ATP made pellets shrink further instead of expanding. Thus the potassium chloride extract he removed contained a "factor" causing muscle relaxation with ATP rather than contraction. Moreover, the extract slowed muscle ATP hydrolysis 10-fold.

The following year Marsh interpreted these observations as the factor suppressing ATPase activity, with the consequent increase in ATP producing relaxation. Contraction, he concluded, was due both to an increased energy release from ATP hydrolysis and to the absence of ATP from a region of myosin where it favored relaxation.[35] Marsh also noted that added Ca^{2+} antagonized the factor's effect, which he explained as Ca^{2+} favoring ATP hydrolysis.

After Marsh returned to New Zealand, J. R. Bendall continued these studies in Cambridge. Initially, he confirmed the ability of Ca^{2+} to inhibit the lengthening of muscle fibers. He pointed out that the relaxing effect required time to develop, time "for the formation of the active substance or complex," and proposed that the active substance prevented "access of [ATP] to the contractile sites during relaxation and rest."[36] Consequently, Bendall set about to isolate and identify the active substance, concluding from standard approaches that it was a heat-labile, nondialyzable macromolecule and that Mg^{2+} was also necessary; the factor was "a functional protein, activated by magnesium."[37]

Bendall also described ATPase activity of his preparation, which was increased by 0.2 mM Ca^{2+} in the presence of millimolar Mg^{2+}. He concluded, however, that this ATPase "play[ed] no part in the factor activity."[38] That judgment arose from Bendall's focus on relaxation achieved through preserving muscle ATP content and his realization that this (Ca^{2+} + Mg^{2+})–stimulated ATPase activity represented only a small component of the total ATPase activity present. (Although Marsh[39] cited Kielley and Meyerhof's description of a muscle ATPase in addition to myosin,[40] Bendall did not. In any case, Kielley and Meyerhof measured ATPase activity either with Ca^{2+} alone or with Mg^{2+} alone or with Ca^{2+} + Mg^{2+} at equimolar ratios, concluding that their ATPase was inhibited by Ca^{2+}.)

In the mid-1950s Setsuro Ebashi in Tokyo began with a similar protein-purification approach: he separated two active fractions, one of which contained a high concentration of lipid and a Mg^{2+}-activated/Ca^{2+}-inhibited ATPase that he likened to the Kielley-Meyerhof ATPase.[41] But in 1958 Ebashi reported that relaxing activity was due to "granules," to particles sedimented from muscle extracts by centrifugation.[42] He also noted that ATPase activity of the fraction was sedimented similarly and that when ATPase activity was lost so was relaxing ability.

Ebashi turned next to the antagonism between Ca^{2+} and relaxation.[43] In 1960 he reported correlations between the abilities of a series of substances to bind Ca^{2+} and their abilities to relax muscle, concluding that "since Ca is required for contraction . . . removal of Ca has a direct connection with the the onset of relaxation and . . . may be the mechanism . . . of the natural particulate relaxing factor."[44]

Although Ebashi was a fellow at the Rockefeller Institute in 1959–60, his important experiments from that visit, with Fritz Lipmann as co-author, were not published until 1962.[45] They then described ATP- and Mg^{2+}-dependent "retention" of ^{45}Ca by a muscle membrane fraction—attaining a 1,400-fold concentration over that in the medium—with that ^{45}Ca released when ATP levels fell: there was "a dynamic transfer in or across a membrane [representing] an ATP-dependent osmotic concentration of calcium ion."[46] (At that time the sarcoplasmic reticulum, a network of membrane-bounded cisterns and tubules in the muscle cell cytoplasm, was being described by electron microscopists at the Rockefeller Institute.[47] Ebashi and Lipmann's electron micrographs showed vesicles presumably formed from that reticulum during homogenization.)

Their membrane fraction had both ATPase and ADP/ATP exchange activity. The latter was interpreted as being part of the ATPase reaction, thus indicating "a reversible phosphorylation of the membrane."[48] Peculiarly, Ebashi and Lipmann found that Ca^{2+} *inhibited* both ATPase and ADP/ATP exchange activities; however, they did not specify the concentration ranges examined.

The third fruitful approach was by Wilhelm Hasselbach in Heidelberg, who modified experimental conditions for studying the Marsh-Bendall relaxing factor in a fortunate way. Attributing varying efficacy to varying amounts of contaminating Ca^{2+}, Hasselbach added to his media millimolar concentrations of oxalate, a calcium-binding compound, noting that "oxalate does not affect . . . contraction [but] increases the activity of relaxing factors."[49]

Then, in 1960, along with two visitors from Japan, Torao Nagai and Madoka Makinose, Hasselbach reported that a "granule fraction" from muscle inhibited contraction in the presence of ATP, Mg^{2+}, and oxalate.[50] Since the effect was rapidly lost when granules were removed, they concluded that "granules release a labile inhibitor."[51]

The next year, however, Hasselbach and Makinose described a quite different process[52] independently of Ebashi and Lipmann but examining issues raised in that still unpublished paper. Hasselbach and Makinose now reported that relaxing factor granules took up large quantities of ^{45}Ca in the presence of ATP, Mg^{2+}, and oxalate, with Ca^{2+} precipitated as the insoluble calcium oxalate salt within the granules. In addition to the massive uptake—6,000-fold due to the presence of oxalate—they described a striking relationship with ATPase activity, one opposite to that of Ebashi and Lipmann: adding Ca^{2+} *increased* ATPase activity eightfold over the "basal-ATPase" activity, and this "extra-ATPase" activity ceased once the granules reached their maximal Ca^{2+} content. They calculated that 1.2 Ca^{2+} were transported per "extra" ATP hydrolyzed. In 1962 they reported that the "extra-ATPase" ceased as soon as the Ca^{2+} concentration in the medium fell below 1 μM, suggested that "the calcium-activated step . . . takes place on the outer side of the membrane," and noted that the free Ca^{2+} concentration in the vesicles reached 0.2 mM, in equilibrium with insoluble calcium oxalate.[53] The Ca^{2+} pump thus established an enormous concentration gradient across the membrane. (The ability of the pump/ATPase to operate at such low Ca^{2+} concentrations accounts for failures of previous workers to notice that requirement: micromolar Ca^{2+} is present as a contaminant unless stringent precautions are observed.)

Hasselbach and Makinose also described ADP/ATP exchange correlated with the "extra-ATPase" and Ca^{2+} transport, and they suggested that "calcium-induced 'extra'-splitting is initiated by the transfer of the phosphate of . . . ATP into an 'energy-rich' bond of an unknown substance [which] becomes a carrier by phosphorylation," a carrier that then "diffuses to the inner surface [where] the phosphate group is split off [diminishing] the calcium affinity."[54] Their 1963 paper corrected the stoichiometry to two Ca^{2+} per ATP,[55] a value that is still proclaimed.

By the mid-1960s the concept of a Ca^{2+} pump in the sarcoplasmic reticulum membrane was thus established, a pump that transported Ca^{2+} from the cytoplasm to halt muscle contraction and hence promote muscle relaxation, a pump manifested as an ATPase requiring millimolar Mg^{2+} plus micromolar Ca^{2+} and known as the sarcoplasmic reticulum Ca^{2+}-ATPase. This ATPase is therefore identifiable with Kielley and Meyerhof's ATPase, but with their assessment of divalent cation requirements corrected. Indeed, high concentrations of Ca^{2+}, as used by Kielley and Meyerhof, inhibit the sarcoplasmic reticulum Ca^{2+}-ATPase.

(The Kielley and Meyerhof ATPase is of historical interest in another context. Because it was labeled a lipoprotein,[56] Skou considered incorporating it into his lipid monolayer system for studying anesthetics instead of Libet's neural ATPase.[57] Since Skou has a penchant for clarifying details, he probably would then have characterized the $(Ca^{2+} + Mg^{2+})$–ATPase of muscle instead of the $(Na^+ + K^+)$–ATPase of nerve.)

Further steps, parallels with studies of the Na^+/K^+-ATPase, will be listed selectively and briefly. (1) In 1967 Anthony Martonosi in St. Louis and Tonomura described a Ca^{2+}-activated labeling of their sarcoplasmic reticulum fractions by ^{32}P-ATP.[58] (2) In 1971 Hasselbach and Makinose showed incorporation of ^{32}P from P_i into ATP driven by transmembrane Ca^{2+} gradients, and in 1972 Leopoldo de Meis in Rio de Janeiro showed enzyme phosphorylation by P_i in the absence of a gradient.[59] (3) In 1973 Boyer demonstrated that an aspartyl group was phosphorylated. Independently, Post and associates showed that the phosphorylated aspartate was contained within the same sequence of three amino acids as in the Na^+/K^+-ATPase.[60] This was the first evidence for a *structural* as well as a *functional* similarity between the two transport ATPases. (4) In 1967 Giuseppi Inesi in San Francisco and Tonomura reported biphasic substrate-velocity plots, indicating high- and low-affinity sites for ATP.[61] (5) In 1976 de Meis formulated a reaction sequence closely resembling the Albers-Post scheme.[62] (6) Makinose demonstrated Ca^{2+} occlusion in 1981.[63] (7) In 1980 Inesi argued for an electrogenic Ca^{2+}/H^+ exchange.[64] (8) David MacLennan in Toronto published a purification procedure in 1970 that provided a single peptide of 102 kDa, and Racker reported in 1972 that when reconstituted into phospholipid vesicles it transported Ca^{2+}.[65] (9) Michael Green in London sequenced extensive hydrophilic regions of the ATPase by Sanger's protein-cleavage approach, but MacLennan, in collaboration with Green, applied cDNA techniques to describe the complete sequence in 1985.[66] Their paper appeared in the same issue of *Nature* as Shull et al.'s report on the Na^+/K^+-ATPase, with each paper pointing to striking similarities between these sequences. The Ca^{2+}-ATPase model featured 10 transmembrane segments, a number that endured (and has influenced later proposals for the Na^+/K^+-ATPase). (10) Martonosi, now in Syracuse, and colleagues obtained two-dimensional crystals of the Ca^{2+}-ATPase for structural determinations in the 1980s;

this work was extended by Green to a resolution by electron microscopy of 6 Å in 1990. The resolution was still too low, however, for disclosing the reaction–transport process.[67]

Finally, I should point out that muscle endoplasmic reticulum (i.e., sarcoplasmic reticulum) is not alone in accumulating Ca^{2+}. Analogous experiments reflecting other interests demonstrated Ca^{2+}-ATPase activity of and Ca^{2+} transport into endoplasmic reticulum of numerous cell types. This mechanism became recognized as an essential component of intracellular signalling systems mediated through transient changes in cytoplasmic Ca^{2+} concentrations (see Chapter 13).

Plasma Membrane Ca^{2+}-ATPase

For cells with a 90 mV membrane potential, the equilibrium distribution of Ca^{2+}, according to the Nernst equation, should be a thousand-fold *greater* inside the cell than out. However, Keynes reported in 1956 that Ca^{2+} concentrations in squid giant axons were 20-fold *less*.[68] Indeed, Keynes and Hodgkin found the concentration of free, diffusable Ca^{2+} to be below 10 μM when the extracellular concentration was 11 mM, a thousand-fold less.[69] They also demonstrated that axon membranes were permeable to Ca^{2+}; consequently, a Ca^{2+}-pump had to be functioning.

Characterizing this pump began, however, with studies on red blood cells. This was a practical choice because these cells contain no endoplasmic reticulum that with its own Ca^{2+} pump could confound attributions (good techniques for separating plasma and endoplasmic reticulum membranes had not yet been developed). But the first step came during studies of the Na$^+$/K$^+$-ATPase. In their 1961 paper on ATPase activity of red blood cells, Dunham and Glynn described in passing a several-fold increase in ATPase activity with 2 mM Mg^{2+} when 0.1 to 1 mM Ca^{2+} was added; they called this stimulation "remarkable," adding that it "may be significant that red cells normally contain no appreciable calcium."[70] Caught up in experiments on the Na$^+$/K$^+$-ATPase, they did not pursue this further.

In 1966 Schatzmann described ATP-dependent extrusion of Ca^{2+} from red blood cells, and inferred from Dunham and Glynn's finding that a Ca^{2+}-stimulated ATPase "might be connected to active outward transport of Ca^{2+} in the same way as the sarcotubular ATPase in muscle is related to the Ca-accumulation into these structures."[71] Using resealed red blood cells, Schatzmann showed ATP-dependent losses of Ca^{2+} to levels well below those in the medium, indicating an active transport against a chemical gradient.[72] The following year Schatzmann confirmed Dunham and Glynn's Ca^{2+}-stimulated ATPase activity and linked it to Ca^{2+} transport.[73]

Further developments in this area can be condensed to a sequence paralleling those for the preceding ATPases. (*1*) In 1974 Philip Knauf and Hoffman reported the Ca^{2+}-activated phosphorylation of a red blood cell membrane protein identified with this Ca^{2+}-ATPase. Its molecular weight according to SDS-PAGE was about 150 kDa (considerably larger than that of the sarcoplasmic reticulum Ca^{2+}-ATPase or the α-subunit of the Na$^+$/K$^+$-ATPase), and its chemical properties were consistent with those of an acyl phosphate.[74] Alcides Rega and Garrahan in Buenos Aires confirmed these observations and calculated from the level of labeling that about 700 Ca^{2+}-ATPase molecules were present per red blood cell, which was several times the number of Na^+/K^+-ATPase units.[75] (2) In 1978 Rega and Garrahan described a Ca^{2+}- and ADP-activated dephosphorylation that was equivalent to Ca^{2+}-activated ADP/ATP exchange.[76] That year they also described a reversal of the whole ATPase reaction that was driven by Ca^{2+} gradients.[77] (3) Also in 1978, Rega and Garrahan published biphasic substrate-velocity plots, indicating high- and low-affinity sites for ATP.[78] Previously they had formulated a reaction sequence similar to the Albers-Post scheme.[79] (4) In 1973 Schatzmann calculated a stoichiometry of one Ca^{2+} transported per ATP hydrolyzed.[80] This is half that for the sarcoplasmic reticulum ATPase, which does not have to pump against an electrical gradient. In 1982 Ernesto Carafoli in Zurich reported that the ATPase transported H^+ in the direction opposite to Ca^{2+}.[81] (5) In 1979 Carafoli described a purification procedure for the red blood cell enzyme; one peptide was demonstrable by SDS-PAGE, with an apparent molecular weight of 125 kDa.[82] Soon afterward, this Ca^{2+}-ATPase was purified by similar methods from plasma membranes of various other cell types.[83] (6) In 1988 Shull, Carafoli, and their associates independently reported amino acid sequences of Ca^{2+}-ATPases from brain and tumor cells using cDNA approaches.[84] The greater molecular weight (compared with other P-type ATPases) was due to an extension near the carboxyl end for binding a regulatory peptide, calmodulin, that when bound increased activity.

Gastric H^+/K^+-ATPase

The stomach secretes acid, achieving an astonishing million-fold gradient of H^+. As noted in Chapter 8, Conway and Davies independently proposed redox mechanisms for gastric acid formation in 1948. They envisioned an extracellular release of H^+ by an oriented system converting hydrogen, which is extracted during cellular oxidation of carbohydrates, to a transported H^+ and an electron. The electron then reacted with oxygen plus water to form hydroxide, which was retained intracellularly:

$$4\,H \Rightarrow 4\,e^- + 4\,H^+_{out}$$

$$4\,e^- + 2\,H_2O + O_2 \Rightarrow 4\,OH^-_{in}$$

The maximal stoichiometry of such a system is thus $4\,H^+$ per O_2.

Davies, however, soon reported higher values, forcing him to rely on supplemental systems involving "metabolic energy (perhaps as high energy phosphate bonds)," and by 1955 he concluded that "any mechanism limited to a ratio of 4 is quantitatively inadequate and either supplementary processes must be involved . . . or, perhaps most likely, [ATP is] able to drive a single process."[85] In the early 1960s Davies, who was then in Philadelphia, described experiments analogous to Zerahn's on frog skin (see Chapter 8). Along with John Forte, a postdoctoral fellow, he compared shortcircuit current across gastric mucosal sheets with H$^+$ and Cl$^-$ movements and with oxygen consumption.[86] Again the stoichiometry exceeded four and they concluded that "a simple redox pump . . . cannot be the only operating mechanism."[87] Forte and Davies also correlated changes in ATP content with transport, calculating 1.5 H$^+$ secreted per ATP consumed.[88]

In 1967 Forte, who was now an independent investigator in Berkeley, noted that K$^+$ was an established cofactor for gastric acid secretion and that (Na$^+$ + K$^+$)-stimulated ATPase preparations also had K$^+$-stimulated phosphatase activity (operationally defined as the ability to hydrolyze *mono*phosphate substrates such as nitrophenyl phosphate). With that published rationale, he described membrane-bound phosphatase activity that was stimulated 10-fold by K$^+$, although he admitted that a "precise role . . . in [providing] gastric H$^+$ is, at present, purely speculative."[89] Although compelling links to H$^+$ transport were not immediately forthcoming, Forte tied K$^+$-stimulated phosphatase activity to K$^+$-stimulated ATPase activity, reporting in 1971 that ^{32}P-ATP labeled this membrane preparation (with characteristics of an acyl phosphate); that Na$^+$ did not stimulate such labeling (distinguishing it from the Na$^+$/K$^+$-ATPase); and that K$^+$ diminished labeling (which was compatible with K$^+$-stimulated dephosphorylation and K$^+$-stimulated phosphatase activity).[90] In 1973, after further separating K$^+$-stimulated activity from a background of K$^+$-independent activity, Forte described K$^+$-stimulated ATPase activity.[91]

The convincing link to H$^+$ transport was provided by Peter Scholes and associates at ICI Pharmaceuticals in Macclesfield, England. In 1974 they described the loss of H$^+$ from the medium when gastric microsomes were incubated with ATP.[92] This H$^+$ uptake was stimulated by K$^+$ *within* the vesicles and they argued for an electroneutral H$^+$/K$^+$ exchange. But a satisfying link between H$^+$ and ATPase activity came considerably later: in 1981 George Sachs and associates at the University of Alabama described H$^+$-stimulated phosphorylation of the enzyme.[93]

Further characterization will again be condensed to a mere listing. (*1*) In 1985 Post and associates reported phosphorylation of an aspartyl residue, in a paper describing a common amino acid sequence around the phosphorylation sites of the Na$^+$/K$^+$-, H$^+$/K$^+$-, and plant H$^+$-ATPases.[94] (*2*) Forte described high- and low-affinity sites for ATP in 1976 and formulated a reaction scheme with $\boldsymbol{E_1}$-P and $\boldsymbol{E_2}$-P intermediates.[95] (*3*) Sachs and associates achieved a 40-fold purification in 1977, and in 1985 they reconstituted a solubilized enzyme preparation into lipid vesicles to demonstrate transport.[96] The molecular weight by SDS-PAGE was reported in 1975 as 102 kDa.[97] (*4*) In 1986 Shull and Lingrel reported the amino acid sequence using cDNA methods;[98] similarity to the Na$^+$/K$^+$-ATPase sequence was particularly striking (see Fig. 13.2). Meanwhile, arguments for a β-subunit being part of this ATPase were accumulating. These were based not only on analogy with the Na$^+$/K$^+$-ATPase but also on a sugar-containing peptide being tightly associated with the catalytic peptide. In 1990 both Forte's and Sachs's groups published characterizations of an isolated glycopeptide.[99] That same year, Shull and three other groups announced the amino acid sequence;[100] the β-subunit too was similar to that of the Na$^+$/K$^+$-ATPase.

Fungal H^+-ATPase

Conway became interested in cation transport by yeast about the same time he began studying gastric acid production, and he soon proposed a redox H^+ pump for yeast also.[101] Identifying the functional mechanism for H^+ transport by fungi began, however, with investigations of a bread mold, *Neurospora crassa*, and from a quite different perspective.

Since "genes have been shown to influence ion permeability," Carolyn Slayman, a graduate student working with Tatum at Rockefeller University, began around 1960 searching for K^+ transport mutants of *Neurospora* with the goal of "correlat[ing] growth and cation-uptake studies on such a mutant."[102] These studies were in the mainstream of genetic exploration: matching a loss of function with a genetic mutation and then identifying the enzyme, specified by the mutated gene, responsible for the function. She identified a K^+ transport mutant but did not pursue that quest further.[103] Instead, the project took a different turn, due in part to the interests of her husband Clifford Slayman a graduate student in electrophysiology. The tubular filaments of *Neurospora* are as much as 15 μm in diameter—large enough to impale with microelectrodes. The membrane potential they then recorded had three striking characteristics.[104] (1) It was huge by nerve or muscle standards, averaging 127 mV when the K^+ concentration of the medium was 37 mM. (2) It was far too large for a Nernstian potential based on the ratio of intracellular to extracellular K^+, although of all the extracellular cations examined the potential was most sensitive to K^+. (3) It dropped abruptly—in less than 2 minutes—when cellular respiration was blocked by inhibitors such as cyanide. Subsequently, Clifford Slayman described potentials exceeding 200 mV and explicitly attributed them to an electrogenic pump.[105]

Then in 1970 at Yale University Clifford Slayman specified an electrogenic pump extruding H^+, in part from studies with Carolyn Slayman on K^+/H^+ exchange.[106] Also that year, he argued against a redox mechanism for the pump: cellular redox processes were halved by inhibitors well before the membrane potential declined, whereas the fall in potential paralleled closely the fall in ATP content.[107]

Meanwhile, Gene Scarborough in Denver had been studying sugar transport by *Neurospora* and described procedures for separating a plasma membrane fraction.[108] In 1976 he joined the Slaymans' quest with the explicit rationale: "*In vitro* biochemical proof that the electrical potential across the *Neurospora* plasma membrane is generated . . . by a plasma membrane ATPase requires isolation of plasma membrane vesicles in a pure form, demonstration of the existence of an ATPase in the isolated . . . vesicles, and finally, demonstration that ATP hydrolysis catalyzed by the . . . ATPase leads to the generation of an electrical potential."[109] Using plasma membrane vesicles, he showed that adding ATP generated electrical potentials (measured by the distribution of an ion that easily crosses membranes) that could be dissipated by chemicals facilitating H^+ diffusion through membranes ("H^+ ionophores"). And the next year he described some kinetic properties of the ATPase activity.[110]

By modifying Scarborough's membrane preparation procedure, Carolyn Slayman also described ATPase activity in 1977, and the next year she concluded from responses to inhibitors as well as from similarities in molecular weights that the *Neurospora* enzyme was related to Na⁺/K⁺- and Ca²⁺-ATPases.[111] In 1980 Clifford Slayman proposed a transport stoichiometry of one H⁺ per ATP hydrolyzed.[112]

Now in Chapel Hill, Scarborough used membrane vesicles to demonstrate ATP-dependent formation of pH gradients as well as electrical potentials that were inhibited in parallel:[113] these results strongly supported notions of a H⁺-transporting electrogenic pump. Also in 1980 he reported labeling with ³²P-ATP a protein having a molecular weight by SDS-PAGE of 105 kDa.[114] The next year he identified the product as an acyl phosphate.[115] In 1984 both his and Carolyn Slayman's groups described reconstitution of the partially purified enzyme into phospholipid vesicles capable of forming ATP-dependent pH and electrical gradients.[116]

Then in 1986 both Randolph Addison (formerly a postdoctoral fellow with Scarborough and now at Rockefeller University) and Carolyn Slayman and colleagues reported the amino acid sequence, noting similarities to Na⁺/K⁺- and sarcoplasmic reticulum Ca²⁺-ATPases.[117] Earlier that year Ramón Serrano, while visiting at the Massachusetts Institute of Technology from Madrid, reported a similar amino acid sequence for yeast H⁺-ATPase, comparing it to the sequences of animal ATPases[118] (that work arose from parallel studies on yeast electrogenic H⁺ transport by Serrano and others, notably the group of André Goffeau in Louvain[119]).

Bacterial Kdp K⁺-ATPase

The first amino acid sequence of a P-type ATPase published was, however, that of a K⁺-transporting enzyme from the bacterium *E. coli*.[120] This 1984 report by Wolfgang Epstein, Karlheinz Altendorf, and colleagues at the Universities of Chicago and Osnabrück, stressed similarities with sarcoplasmic reticulum Ca²⁺-ATPase sequences (as then known from Green's peptide-cleavage approach), particularly around the phosphorylation site of the muscle enzyme. Thereafter, as sequences of other P-type ATPases were announced, similarities with this bacterial enzyme were routinely drawn.

Why the bacterial enzyme was the first to be sequenced reflected both the technical advantages of working with bacterial systems and the interest and expertise of the investigators. Epstein started working on bacterial transport in Solomon's laboratory in the 1960s, collaborating with Stanley Schultz (who also became a leader in the transport field, although in another sector: see Chapter 14). They demonstrated that *E. coli* accumulated K⁺ by a saturable mechanism.[121] By 1976 Epstein, who was now in Chicago, showed that three separate systems for K⁺ accumulation were present. One of these, known as Kdp, was a high-affinity system expressed when bacteria were grown in K⁺-deficient media.[122] The following year Epstein reported that ATP drove the Kdp transport system, and in 1978 he described an associated K⁺-stimulated ATPase activity, noting that "bacteria have transport systems that are at least superficially similar to the ATP-driven systems of animal cells."[123]

Also in 1978 Epstein, Altendorf, and colleagues identified three peptides comprising the Kdp system.[124] These, known as KdpA, KdpB, and KdpC, had molecular weights by SDS-PAGE of 47, 90, and 22 kDa, respectively (when the sequences were determined, however, the calculated weights were actually 59, 72, and 20 kDa). KdpB had a sequence similar to that of other P-type ATPases, but neither KdpA nor KdpC was structurally related to known animal, plant, or fungal peptides associated with transport. Well after the sequence was determined, Altendorf reported in 1989 that KdpB was phosphorylated by ATP with the characteristics of an acyl phosphate, and that K^+ activated dephosphorylation of E-P.[125] This was similar to the reaction process of the Na^+/K^+- and H^+/K^+-ATPases.

COMMENTARY

Places, People, and Techniques

By the mid-1980s, powerful new techniques of molecular biology dominated the field, providing novel and sometimes unexpected insights. One consequence, of course, was that similar studies were similarly performed. Thus, Lingrel in Cincinnati, Numa in Osaka, MacLennan in Toronoto, Carafoli in Zurich, and Yuri Ovchinnikov in Moscow all addressed the same issues with the same newly available methods. New technologies established a homogeneous scientific culture of common approaches that yielded complementary if not identical findings and supplanted local or national research traditions and styles. Another consequence, one reflecting the labor-intensive demands of these techniques, was the proliferation of large groups collaborating on a single experimental project. The resulting publications were mega-authored papers (one cited here listing 17 names[126]). Still another consequence was the contemporaneous application of the same approach to different enzymes (Table 12.1). Obviously, the transfer of successful methods, like the spread of enlightening concepts, has been a major facilitator of scientific advances. Social factors also hastened or slowed the spread of new techniques as well as promoted or impeded the formation of efficient groups.

The similarity of approaches charted here should not, however, be interpreted as an effortless and thoughtless copying from case to case. Often considerable ingenuity was required to meet individual challenges. For example, to purify the Ca^{2+}-ATPase from red blood cell membranes, where few copies of this enzyme are present, Carafoli resorted to "affinity chromatography": he added the membranes (dissolved with detergent) to a chromatography column containing the regulatory peptide calmodulin covalently linked to a supporting gel.[127] Since this ATPase bears a binding site for calmodulin, the enzyme interacted with the calmodulin in the column, retarding its flow and thereby separating it from other proteins that did not bear such calmodulin-binding sites.

Table 12.1 Sequence of Reports*

	ATPase Activity					
Property	Na^+/K^+	Ca^{2+} (sarcoplasmic reticulum)	Ca^{2+} (plasma membrane)	H^+/K^+	H^+ (Neurospora)	K^+ (Kdp)
ATPase activity	1957	1961	1967	1973	1976	1978
Labeling with ^{32}P-ATP	1963	1967	1974	1975	1980	1989
Molecular weight by SDS-PAGE	1970	1971	1974	1975	1980	1978
Sequencing	1985	1985	1988	1986	1986	1984

*The dates of publication for reports of specific ATPase activity, of labeling the enzyme preparation with ^{32}P-ATP, of molecular weight determined by SDS-PAGE, and of amino acid sequencing through DNA sequencing are compared for the six P-type ATPases described here. (In several cases, reports in meeting abstracts appeared earlier than the papers whose dates are listed here.)

Comparisons

The proliferation of sequence data encouraged comparisons that in turn suggested common structural and functional domains. For example, John Walker in Cambridge described in 1982 an amino acid sequence associated with ATP binding sites in such diverse enzymes as adenylate kinase, phosphofructokinase, and ATP synthase. The "Walker motif" consisted of a basic amino acid (arginine or lysine), which, after an intervening two or three amino acids was followed by a glycine, followed by three other amino acids, and then a leucine plus two or three further hydrophobic amino acids.[128] In the sheep kidney α-subunit, residues 543–553 are: *arginine*, valine, leucine, *glycine*, phenylalanine, cysteine, histidine, *leucine, methionine, leucine*, and *proline*. In a similar manner, common structural domains and folding patterns were recognized. In 1991 Gilbert proposed a new experimental paradigm rooted in recognizing such structural-functional motifs.[129]

Scientific Summary

In the mid-1980s the chemical identity of the Na^+/K^+-ATPase was defined in terms of its amino acid sequence. Related studies provided some clues to the three-dimensional structure and to the chemical mechanisms of ATP hydrolysis and cation transport, although most details were still lacking in 1990. A number of transport systems for several different cations, which had been approached initially from diverse interests, were shown at this time to be ATPases with analogous mechanisms (including enzyme phosphorylation by ATP as part of the reaction sequence). These enzymes also had similar amino acid sequences, reflecting a common ancestry as well.

NOTES TO CHAPTER 12

1. Collins et al. (1983); it was submitted in 1982.
2. Sanger et al. (1977); Maxam and Gilbert (1977).
3. Historical information is from interviews with Lingrel and Schwartz (1994).
4. The DNA is complementary to mRNA, i.e., the sequences are related by the rules of "base pairing," and cDNA is formed from mRNA by the enzyme reverse transcriptase (it copies DNA from RNA, the reverse of the direction for protein synthesis).
5. A plasmid is an "accessory chromosome" that is physically independent from the single bacterial chromosome. The plasmid is replicated and transmitted to the daughter cells when the bacterium divides, but the plasmid may also be multiplied several-fold without cell division. The enzymes used for these particular cuttings and splicings are essential tools of these techniques.
6. Shull et al. (1985).
7. Kawakami et al. (1985).
8. There was 87% similarity between the amino acid sequences despite the evolutionary distance between fish and mammals ("identity" refers to identical amino acids, whereas "similarity" refers to the same amino acid or a conservative substitution with an amino acid of similar size, shape, and chemical properties.)

9. Their identification of intramembrane hydrophobic regions was based on the pioneering studies of Kyte and Doolittle (1982).

10. The aspartate and the FITC-site were identified from short sequences previously identified: Bastide et al. (1973); Farley et al. (1984); Kirley et al. (1984). The IAF-site was determined later (Tyson et al., 1989), as were the sites of tryptic cleavage (Jørgensen and Collins, 1986).

11. See discussion by Pedemonte and Kaplan (1990).

12. Ohtsubo et al. (1990).

13. Shull (1986b).

14. A consensus sequence for protein glycosylation, consisting of a glycosylated asparagine followed after an intervening amino acid by a serine or a threonine, had been established previously for a series of proteins.

15. Noguchi et al. (1986); Kawakami et al. (1986); Mercer et al. (1986); Ovchinnikov et al. (1986).

16. Noguchi et al. (1987).

17. "Isoform" refers to the same enzyme (performing the same catalytic function) but with a different although similar amino acid sequence.

18. Skou (1962).

19. Sweadner (1979).

20. Shull et al. (1986a). They used the sheep kidney enzyme probes, measuring hybridization under less stringent conditions, which was the standard approach to finding similar but not identical sequences.

21. Martin-Vasallo et al. (1989).

22. Deisenhofer et al. (1985). Johann Deisenhofer, Robert Huber, and Hartmut Michel received the Nobel Prize in chemistry in 1988.

23. Stewart and Grisham (1988).

24. Hokin et al. (1973). For negative staining, an electron-opaque substance (such as phosphotungstic acid) is applied to fill all the interstices of a specimen, leaving the structures electron-lucent.

25. Vogel et al. (1977).

26. Van Winkle et al. (1976); Deguchi et al. (1977). The freeze-fracture technique results in cleavages between the two leaflets of the lipid bilayer, and can reveal an intramembrane particle as an inclusion on one leaflet and a hole on the other (see Chapter 15).

27. Kyte (1975) had argued convincingly for an enzyme extending through the membrane, from experiments showing the simultaneous binding of strophanthidin from the exterior and an antibody from the cytoplasmic surface.

28. Hebert et al. (1982).

29. Hebert et al. (1985).

30. Pedersen and Carafoli (1987).

31. For a masterful history of efforts to understand muscle function see Needham (1971). I am also indebted to Anthony Martonosi for access to his unpublished history of the field.

32. Heilbrunn and Wiercinski (1947).

33. Marsh (1952), p. 247.

34. Marsh (1951), p. 1065.

35. Marsh (1952). This rationalization recognized that ATP was needed for contraction and was consumed by the myosin ATPase, but that ATP prevented rigor, a contracted state resisting extension.

36. Bendall (1952), p. 1060.

37. Bendall (1953), p. 252.

38. Ibid., p. 253.

39. Marsh (1952).

40. Kielley and Meyerhof (1948). See also Chapter 7.

41. Kumagai et al. (1955). They reported merely the "inactivat[ion] by a small amount of calcium ions" (p. 166).

42. Ebashi (1958). Independently, Hildegard Portzehl (1957) in Heidelberg showed the involvement of a "granule" fraction.

43. By this time, Alexander Sandow (1952) had proposed a role for Ca^{2+} in "excitation-contraction coupling" in muscle, Emil Bozler (1954) had showed that removing Ca^{2+} produced relaxation, and Annemarie Weber (1959) had demonstrated effects of Ca^{2+} on actomyosin ATPase activity (she suggested that the Marsh-Bendall factor might act by binding Ca^{2+}).

44. Ebashi (1960), p. 151.

45. Ebashi and Lipmann (1962). That paper was submitted in April 1962, although the work was completed in 1959; the delay was due to Lipmann's misgivings about such a role for Ca^{2+} (Ebashi, correspondence, 1995). Earlier, Ebashi, as sole author, reported other studies from that visit, noting that the vesicular fraction was able to "strongly bind calcium," a process that "depends on the presence of ATP," so that "calcium binding represents the physiological action . . . of the relaxing factor" (Ebashi, 1961, p. 236).

46. Ebashi and Lipmann (1962), p. 393.

47. Bennett and Porter (1953); Porter and Palade (1957).

48. Ebashi and Lipmann (1962), p. 395.

49. Hasselbach and Weber (1957), p. 108, citing unpublished observations.

50. Nagai et al. (1960).

51. Ibid., p. 223.

52. Hasselbach and Makinose (1961). This paper was submitted in October 1960.

53. Hasselbach and Makinose (1962), p. 133.

54. Hasselbach and Makinose (1962), p. 135.

55. Hasselbach and Makinose (1963).

56. Kielley and Meyerhof (1950).

57. Skou, interview (1993).

58. Martonosi (1967); Yamamoto and Tonomura (1967). Arguments that the phosphorylated intermediate was an acyl phosphate were advanced by Yamamoto and Tonomura (1968) and by Makinose (1969).

59. Barlogie et al. (1971); Makinose (1971); Makinose and Hasselbach (1971); Masuda and de Meis (1973). These last experiments thus paralleled those on the Na^+/K^+-ATPase at the same time by Post et al. (1973).

60. Degani and Boyer (1973); Bastide et al. (1973).

61. Inesi et al. (1967); Yamamoto and Tonomura (1967).

62. Carvalho et al. (1976). The scheme was clarified in de Meis and Vianna (1979).

63. Takisawa and Makinose (1981).

64. Chiesi and Inesi (1980).

65. MacLennan (1970); MacLennan (1971); Racker (1972).

66. Allen et al. (1980); MacLennan et al. (1985).

67. See, for example, Dux et al. (1987); Pikula et al. (1988); Stokes and Green (1990).

68. Keynes and Lewis (1956).

69. Hodgkin and Keynes (1957).

70. Dunham and Glynn (1961), p. 291.

71. Schatzmann (1966), p. 364.

72. Transport was also against an electrical gradient, which Schatzmann evaluated by measuring the ratio of Cl^- concentration within cells to that in the medium. Schatzmann also eliminated any potential role of a Na^+ gradient by measuring Ca^{2+} outflux with equal concentrations of Na^+ inside the cells and outside, and he blocked any effect of the Na^+/K^+-ATPase by adding ouabain.

73. Vincenzi and Schatzmann (1967).

74. Knauf et al. (1974).

75. Rega and Garrahan (1975).

76. Rega and Garrahan (1978).

77. Rossi et al. (1978).

78. Richards et al. (1978).

79. Rega and Garrahan (1975).

80. Schatzmann (1973).

81. Niggli et al. (1982).

82. Niggli et al. (1979).

83. E.g., heart (Caroni and Carafoli, 1981) and brain (Hakim et al., 1982).

84. Shull and Greeb (1988); Verma et al. (1988).

85. Crane and Davies (1948), p. xlii; Davies (1957), p. 291. The publication data of the latter is two years after the meeting it reports.

86. Forte and Davies (1963, 1964).

87. Forte and Davies (1963), p. 816.

88. Forte et al. (1965).

89. Forte et al. (1967), p. 303.

90. Tanisawa and Forte (1971).

91. Ganser and Forte (1973).

92. Lee et al. (1974). That transport rather than simple binding had occurred was supported by experiments showing a loss of H$^+$ from the microsomes when H$^+$-carrying compounds ("H$^+$ ionophores") or membrane-disrupting detergents were added.

93. Stewart et al. (1981).

94. Walderhaug et al. (1985).

95. Ray and Forte (1976).

96. Saccomani et al. (1977); Rabon et al. (1985).

97. Saccomani et al. (1975).

98. Shull and Lingrel (1986).

99. Okamoto et al. (1990); Hall et al. (1990).

100. Shull (1990); Reuben et al. (1990); Toh et al. (1990); Canfield et al. (1990).

101. Conway and O'Malley (1945); Conway (1953).

102. C. W. Slayman and Tatum (1965), p. 184. From studies on *Neurospora* Tatum and Beadle had enunciated the dictum "one gene, one enzyme" (see Chapter 7).

103. Ibid.

104. C. L. Slayman and Slayman (1962). These studies were on the native "wild type" *Neurospora crassa*; they concentrated on this rather than on the mutant after all.

105. C. L. Slayman (1965a,b).

106. C. L. Slayman (1970); C. L. Slayman and Slayman (1968). The H$^+$ gradient established by the pump could drive K$^+$ transport through K$^+$/H$^+$ exchange.

107. C. L. Slayman et al. (1970).

108. Scarborough (1975).

109. Scarborough (1976), p. 1485.

110. Scarborough (1977).

111. Bowman et al. (1978).

112. Warncke and Slayman (1980), p. 231.

113. Scarborough (1980).

114. Dame and Scarborough (1980).

115. Dame and Scarborough (1981).

116. Scarborough and Addison (1984); Perlin et al. (1984).

117. Addison (1986); Hager et al. (1986).

118. Serrano et al. (1986).

119. See reviews by Goffeau and Slayman (1981) and Serrano (1984).
120. Hesse et al. (1984).
121. Epstein and Schultz (1965).
122. Rhoads et al. (1976).
123. Rhoads and Epstein (1977); Epstein et al. (1978), p. 6668.
124. Laimins et al. (1978).
125. Siebers and Altendorf (1989).
126. Toh et al. (1990).
127. Niggli et al. (1979).
128. Walker et al. (1982).
129. Gilbert (1991).

chapter 13

ALTERNATIVES

I N earlier chapters we have seen how the Na^+/K^+-ATPase became established as the biochemical basis for the Na^+/K^+ pump. That triumphal course was traced against formulations of an early rival, the membrane redox pump. Here we go back to the 1950s to consider two further alternatives that attracted serious interest and consideration into the 1960s. These two, differing in the qualities of their proposals and the evidence adduced—and in their ultimate fates—were developed by three active and imaginative scientists: Lowell Edward Hokin, Mabel Ruth Hokin, née Neaverson, (Fig. 13.1), and Gilbert Ning Ling. The Hokins explored a model for a Na^+ pump that they later abandoned because of contrary evidence, in large part their own. Nevertheless, the basic phenomenon they described was amply confirmed; indeed, it was shown subsequently to have major functional significance in its own right. Ling, on the other hand, championed a tradition denying the existence of membrane pumps. Over the years, as evidence in favor of his views was not forthcoming while support for pumps proliferated, interest in Ling's hypotheses waned. And his intransigence undermined the initial respect for his insights and imagination and dissipated the sympathy felt for his stalwart iconoclasm. So this chapter also illustrates different fates that can befall "mistaken" formulations.

THE HOKINS' PHOSPHATIDIC ACID CYCLE

Birth

Mabel Hokin, a native of Sheffield, was drafted into agricultural work in the Women's Land Army after finishing school during World War II, but then, in answer to Hans Krebs's call for laboratory technicians, worked at the University of Sheffield for Henry McIlwain.[1] Krebs encouraged her to continue her education after the war, and she received a bachelor's degree from Sheffield in 1949 and a Ph.D. in 1952, studying acetate metabolism in pigeon muscle. Lowell Hokin, the

Fig. 13.1 Lowell E. Hokin and Mabel R. Hokin. (Photographs courtesy of Lowell Hokin and Mabel Hokin-Neaverson.)

same age as Mabel, received his M.D. degree from the University of Louisville in 1948 after a peripatetic wartime course beginning at the University of Chicago in 1942, passing through Dartmouth College and the University of Illinois, and including formative research experience on gastric acid secretion with Warren Rehm in Louisville. Following a year of clinical training, he entered the University of Sheffield in 1949, at age 25, for graduate study with Krebs. His thesis research dealt with the synthesis and secretion of digestive enzymes by slices of pigeon pancreas (he met Mabel through using pancreases from her pigeons). He received his Ph.D. in 1952, the year they were married.

As part of his thesis research, Lowell Hokin demonstrated that adding acetylcholine stimulated *secretion* of amylase by pigeon pancreas but not its *synthesis*.[2] (*In vivo* the pancreas synthesizes several enzymes that it stores and then secretes into the intestine, where they break down proteins, lipids, and carbohydrates for absorption into the bloodstream; amylase breaks down certain carbohydrates. *In vivo* secretion is stimulated by nerves releasing the neurotransmitter acetylcholine onto pancreas cells.) Then, shortly before he and his wife left for postdoctoral fellowships with J. H. Quastel at McGill University, Lowell began studying effects of acetylcholine on the labeling of "nucleoproteins" by ^{32}P-phosphate added to pigeon pancreas slices.[3] An experiment begun in Sheffield and completed together in Montreal, however, started them on a brand new course. By measuring the labeling of phospholipids as well as "nucleoproteins,"[4] the Hokins found that adding acetylcholine to pigeon pancreas slices increased the ^{32}P content of their lipid extract by five- to ninefold.

An effect of such magnitude is an experimenter's delight, as well as a likely indicator of significant physiological processes. The Hokins, working independently of Quastel, pursued their phenomenon vigorously. In 1953 they reported four crucial characteristics of the stimulated labeling.[5]

1. Atropine, a known antagonist at acetylcholine receptors, blocked the increased labeling, so acetylcholine was acting through specific receptors.
2. Acetylcholine did not affect oxygen consumption by the slice appreciably, so it was not merely boosting cell metabolism generally.
3. Acetylcholine did not increase the "specific activity" (radioactivity per gram of phosphorus present) of soluble, nonlipid compounds in the slices, so increased labeling of lipids was not secondary to a common process, such as pancreas cells becoming more permeable to ^{32}P-phosphate.
4. The specific activities of the labeled lipids rose, so increased labeling was not simply due to the presence of more phospholipids.

The following year the Hokins reported that the amount of phospholipids was unchanged during their experiments, so labeling represented a "turnover" of phospholipids, degradation and resynthesis in equal amounts.[6] They found, moreover, that not all of the phospholipid molecule (see Fig. 13.2) was broken down in this turnover: radioactivity from ^{14}C-glycerol was also incorporated into phospholipids but labeling was not increased by acetylcholine. In 1955, further sharpening their focus, they showed that not all species of phospholipids were turning over equally rapidly in response to acetylcholine.[7] With slices of both pigeon pancreas and guinea pig brain, the specific activities after labeling with ^{32}P-phosphate were greatest in phosphatidylinositol and phosphatidic acid, and increased most in these phospholipids when acetylcholine was added (Table 13.1). By contrast, the specific activities after labeling with ^{14}C-glycerol were about the same in all species of phospholipids.

But what physiological process did such turnover reflect? In 1953 the Hokins suggested that the labeled lipids might be part of an enzyme–lipid complex involved in secretion; in 1955 that turnover reflected acetylcholine interacting with a receptor lipoprotein; in 1956 that it was associated with active transport of proteins from cells; in 1958 that it accompanied secretion of smaller molecules as well; and in 1959 that "phosphatidic acid . . . participate[s] in a carrier mechanism . . . formed at the inner surface of the membrane [to make] lipophilic complexes with the hydrophilic secretory materials [that then] diffuse across the membrane to its outer surface, where . . . phosphate [is cleaved] from phosphatidic acid [leaving the] diglyceride."[8] This cycle was based also on their recent demonstration of an enzyme removing phosphate from phosphatidic acid (phosphatidic acid phosphatatase) to form diglyceride, and an enzyme phosphorylating diglyceride (diglyceride kinase) to form phosphatidic acid,[9] as well as on the notion of lipid carriers then pervasive. Indeed, the Hokins likened phosphatidic acid to coenzymes carrying electrons or chemical groups, with the associated enzymes determining the specificities.[10]

Diglyceride

Phosphatidic Acid

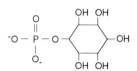

Inositol-*mono*phosphate

Inositol-*bis*phosphate

Cleavage by
Phospholipase-C **Phosphatidylinositol**

Inositol-*tris*phosphate

Fig. 13.2 Phospholipid structure. Glycerophospholipids consist of a three-hydroxyl alcohol, glycerol (represented by the E-shaped line on the left), with two of those hydroxyls esterified by long-chain fatty acids (the zigzag lines) and the third by phosphoric acid. This is the simplest phospholipid, phosphatidic acid. The phosphoric acid may, however, be esterified with a second alcohol, as with choline to form phosphatidylcholine, with ethanolamine to form phosphatidylethanolamine, with inositol to form phosphatidylinositol, etc. Phospholipase C cleaves to form diglyceride and phosphate (or phosphorylated alcohol). On the right are structures of three inositol phosphates formed by such cleavage (having no, one, or two additional phosphates esterified with inositol hydroxyls).

Table 13.1 Effect of Acetylcholine on Specific Activity of Phospholipids*

| | Specific Activity | | | |
| | Pancreas | | Brain | |
Lipid	Control	+Acetylcholine	Control	+Acetylcholine
Phosphatidylcholine	41	55	76	126
Phosphatidylethanolamine	236	433	9	12
Phosphatidylserine	9	25	2	3
Phosphatylinositol	440	7480	710	1650
Phosphatidic acid	754	1250	1020	2180

*Slices of pigeon pancreas or guinea pig brain were incubated with ^{32}P-phosphate in the absence or presence of acetylcholine. The phospholipids were extracted, the hydrolytic products separated, and the specific activity calculated, i.e., radioactivity (counts/min) expressed relative to the phosphorus content (μg P). [From L. E. Hokin and Hokin (1955b), Tables I and III. Reprinted with kind permission of Elsevier Science-NL, Amsterdam, The Netherlands.]

In 1958, however, Knut Schmidt-Nielsen published a series of papers that turned the Hokins' course decisively: he reported that nasal glands of marine birds secreted concentrated sodium chloride, up to 0.8 M, and that this secretion was stimulated by acetylcholine.[11] These observations caught the Hokins' attention, resonating with Lowell Hokin's early studies of gastric cation transport. So in 1959 the Hokins obtained albatrosses from Guam, made slices of the nasal salt glands, and measured phospholipid labeling by ^{32}P-phosphate. Added acetylcholine increased labeling of phosphatidic acid by a spectacular 15-fold (acetylcholine increased labeling of phosphaditylinositol only threefold, and they considered that increment secondary to labeling of phosphatidic acid).[12] The Hokins concluded that "it appears likely . . . phosphatidic acid is the sodium carrier,"[13] and they provided a concise scheme (Fig. 13.3A).

In an extensive paper published in 1960, the Hokins argued that phosphatidic acid was bound to proteins. They localized to a membrane fraction both the rapidly labeled lipids and the pertinent enzymes, phosphatidic acid phosphatase and diglyceride kinase.[14] And in 1961 they reported that ^{32}P-ATP labeled phosphatidic acid of red blood cell ghosts, confirming expectations that "If the phosphatidic acid-cycle is a general mechanism for active transport [it should be present] in all membranes which actively transport sodium."[15] Furthermore, the necessary enzymes were not only present in the membranes, but K^+ stimulated the kinase and Na^+ the phosphatase, each about twofold. Together, the kinase plus phosphatase are equivalent to an ATPase, and the Hokins suggested an identity with the $(Na^+ + K^+)$–stimulated ATPase, citing Skou and Post.

Death

By 1961 evidence associating the phosphatidic acid cycle with sodium transport was intriguing—and attracting considerable attention.[16] It was also tenuously indirect.

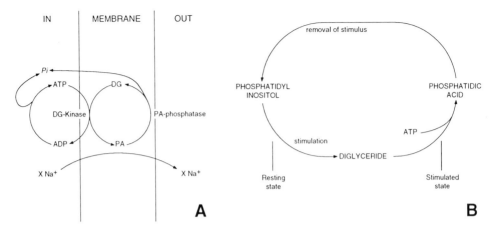

Fig. 13.3 Schemes for stimulated phosphatidic acid metabolism. *A:* Phosphatidic acid cycle. Transport of X Na$^+$ is coupled to the cyclical interconversion of phosphatidic acid (PA) and diglyceride (DG) by the corresponding phosphatase and kinase, consuming one ATP per cycle. [Redrawn from L. E. Hokin and Hokin (1963a), Fig. 1. Reprinted by permission of L. E. Hokin] *B:* Phosphatidic acid–phosphatidylinositol interconversions. Stimulation activates interconversion, but the consequences of that cycling are not specified. [From M. R. Hokin and Hokin (1964b), Fig. 3. Reprinted by permission of John Wiley & Sons, Ltd., copyright 1964.]

Acetylcholine stimulated Na$^+$ secretion *in vivo* and activated phospholipid turnover remarkably. But no link between transport and either labeling or its stimulation was demonstrated *in vitro*. And activation of the pertinent membrane-associated enzymes by Na$^+$ and K$^+$ was modest. Moreover, accumulating evidence tying the (Na$^+$ + K$^+$)–stimulated ATPase to the Na$^+$/K$^+$ pump shifted the Hokins' focus to incorporating the phosphatidic acid cycle into that formulation. Four concerns about such an integration were soon expressed, however.

The first concern was that ouabain blocked Na$^+$ secretion by the salt gland,[17] just as it inhibited Na$^+$/K$^+$ pumps elsewhere, but it increased labeling of phosphatidic acid rather than decreasing it.[18] In 1961 the Hokins also reported that ouabain stimulated labeling in the presence as well as in the absence of acetylcholine; however, they pointed out that inhibiting the phosphatase portion of their cycle could increase the steady-state level of labeling while decreasing activity around the cycle.[19] Nevertheless, ouabain and K$^+$ have antagonistic effects on the pump, whereas the Hokins' scheme specified ouabain inhibiting the Na$^+$-activated phosphatase.[20]

The second concern was that in the Hokins' cycle, K$^+$ activated the step for outward transport and Na$^+$ for inward transport (Fig. 13.3*A*), opposite to the direction of transport.[21]

Third, as the Hokins acknowledged, the kinase did not have a sufficiently high selectivity for Na$^+$ to discriminate under physiological conditions.[22]

Finally, as the Hokins also acknowledged, "one of the more important questions . . . is whether the rate [around the cycle] is compatible with the . . . rate of Na$^+$ transfer."[23] They calculated that rates for kinase and phosphatase were "sufficient to account for the Na$^+$ + K$^+$–dependent ATPase,"[24] and they showed that labeling of phosphatidic acid in salt gland slices reached its plateau within a minute or two after adding acetylcholine.[25] Labeling, however, must reach its plateau within a second if turnover of the entire phosphatidic acid pool represented a cycle of the pump, but the prolonged time course could represent slow diffusion of added acetylcholine to its receptors deep within the slice. On the other hand, Johan Järnefelt incubated brain microsomes with ^{32}P-ATP for 6 to 120 seconds and found that the rate of phosphatidic acid labeling was a thousand-fold slower than the rate of ATPase activity in the presence of Na$^+$ and K$^+$.[26] And in 1964 the Hokins agreed: after incubating red blood cell ghosts or homogenates of salt gland with ^{32}P-ATP, they found that rates of labeling were orders of magnitude too slow to be an intermediate in the ATPase reaction.[27] That decisive result terminated their advocacy of the phosphatidic acid cycle as a Na$^+$ pump.

Reincarnation

If the phosphatidic acid cycle was unrelated to the Na$^+$/K$^+$-ATPase and pump, what function did that prominent turnover represent? In 1964 the Hokins incorporated phosphatidylinositol and phosphatidic acid into a cycle with "stimulated" and "resting" phases (Fig. 13.3B), but just how were these phases manifested? The revelation will here be compressed into nine citations. (1) In 1968 Jack Durell reported that the initial effect of acetylcholine when added to brain homogenates was cleavage of phosphatidylinositol to diglyceride plus inositol-phosphate.[28] (2) In a massive review article published in 1975, Robert Michell marshalled arguments for phosphatidylinositol turnover being associated with increased levels of cytoplasmic Ca^{2+},[29] (3) which by then was recognized as an important "second messenger" regulating cellular functioning.[30] (4) Michell subsequently summarized evidence indicating that a rise in Ca^{2+} *followed* phosphatidylinositol breakdown.[31] (5) So far I have not distinguished between three forms of phosphatidylinositol (identified in 1961 by Hans Brockerhoff and Clinton Ballou[32]) having zero, one, and two additional phosphates esterified to inositol (Fig. 13.2). These represent phosphatidylinositol proper, phosphatidylinositol-phosphate, and phosphatidylinositol-*bis*phosphate. Cleavage by phospholipase-C (Fig. 13.2) will yield, respectively, inositol-phosphate, inositol-*bis*phosphate, and inositol-*tris*phosphate. (6) In 1983 Michael Berridge described inositol-*tris*phosphate production after stimulating blowfly salivary glands with the neurotransmitter serotonin,[33] and (7) that same year he showed that adding inositol-*tris*phosphate to pancreas cells released Ca^{2+} from intracellular stores.[34] (8) Meanwhile, Yasutomi Nishizuka described in 1979 a protein kinase ("protein kinase-C") activated by lipids in the presence of low concentrations of Ca^{2+},[35] and (9) later that year he demonstrated that diglycerides were most effective.[36]

These processes have been extensively studied in terms of their physiological, pharmacological, and pathological consequences, but here I will summarize only the basic mechanism depicted in current models. A host of neurotransmitters and hormones have receptors on cell surfaces that, when occupied, activate a coupled phospholipase-C specific for phosphatidylinositol-*bis*phosphate. Each of the two products of that enzymatic hydrolysis is a second messenger. Inositol-*tris*phosphate releases Ca^{2+} into the cytoplasm from stores within the endoplasmic reticulum (Ca^{2+} as a second messenger then activating a distinct class of protein kinases). And diglycerides also activate their own particular class of protein kinases.

LING'S ASSOCIATION–INDUCTION HYPOTHESIS

Selective binding of cations within cells is another means for creating asymmetric distributions, one I have reserved for discussion until this point. Before and through the 1950s several scientists, notably A. S. Troshin in Leningrad,[37] developed *equilibrium* models of cellular composition in which cytoplasmic binding or adsorption dictated cellular contents, with membranes playing negligible roles. The preeminent advocate, however, was Gilbert Ling: knowledgeable, imaginative, prolific, persistent.

When he was a graduate student at the University of Chicago in the later 1940s, Ling developed intracellular microelectrodes, a notable contribution cited in Chapter 5 and one leading also to his interpretation of ion distributions. Ling was struck by the persistence of membrane potentials in frog muscle when energy supplies were blocked (glycolysis by iodoacetate and oxidative metabolism by cyanide or anoxia).[38] He showed that potentials persisted as long as cationic asymmetries persisted, but how were those asymmetries maintained without metabolic energy?

In 1952 Ling, then at Johns Hopkins University, proposed that K^+ was adsorbed at fixed anionic sites within cells, sites selecting K^+ over Na^+ on the basis of charge density and size of hydrated ion.[39] By 1962 Ling, now at the Eastern Pennsylvania Psychiatric Institute in Philadelphia, formulated his association-induction hypothesis, adding cooperative–inductive interactions.[40] This formulation ranged across all cellular function, but here I will focus on ion distributions.

Ling asserted that free carboxyls of aspartates and glutamates in cytoplasmic proteins could selectively adsorb K^+ ("association") when ATP, through binding to "cardinal sites" on proteins, induced conformational changes that spread cooperatively throughout three-dimensional lattices of structured proteins ("induction"). These organized proteins also ordered adjacent water structure, excluding Na^+.[41] (Although K^+ was bound, it was not immobilized and could hop from site to site, in accord with Harris's measured diffusion of ^{42}K in frog muscle.[42]) Conversely, when ATP levels fell, protein conformations reverted to those no longer selecting K^+. Ling developed quantitative models with three classes of protein K^+ sites, although they were not identified with actual structures. In these models, water also was distributed passively, with solvation changes mimicking those predicted by membrane theories of osmosis, and electrical potentials arose at boundaries between phases (between ordered cellular structure and extracellular medium). The hypothesis thus specified equilibrium distributions of solutes and water dependent on ATP binding.

Metabolism was needed only to maintain ATP levels, and net ATP hydrolysis abolished selectivity.

Rationale

Foremost was Ling's conclusion that cellular energy supplies were insufficient for maintaining nonequilibrium ion distributions, that they were insufficient for powering membrane pumps. He cited Conway, Ussing, Harris, and Keynes, noting calculations that Na^+ transport consumed 20% or more of cellular energy production, a share he deemed unacceptable, a priori.[43] Maintaining other solutes at nonequilibrium levels would, he argued, require still more energy for their pumps, compounding the energy shortfall.

The primary critique arose from his experiments with frog muscles poisoned with iodoacetate and cyanide: at 0° they retained their ionic asymmetries for hours.[44] Ling measured ATP and creatine phosphate content; the latter fell somewhat, providing 22 calories per kilogram of muscle per hour. By contrast, the calculated outflux of ^{22}Na from muscle would require 343 calories per kilogram per hour. Therefore, insufficient energy was available for moving so much Na^+. Ling judged the pump hypothesis to be refuted.

Second, Ling noted that Conway's double-Donnan hypothesis was inadequate and that pumps transporting Na^+ alone did not account for K^+-distributions, without acknowledging that current proposals for Na^+/K^+ pumps addressed both issues.[45] But he did admit that active transport occurred across epithelia such as frog skin and kidney tubules[46] (correlations between epithelial Na^+ transport and Na^+/K^+-ATPase activity he ignored, however).

Third, Ling was motivated by preferences for simple, encompassing hypotheses. He referred to "living protoplasm" with a focus on the "*entire* cell," and disparaged pump hypotheses as "piecemeal" and "ad hoc."[47]

Criticisms

Surprisingly, no experiments comparable to Ling's on metabolically-poisoned frog muscle were published in the 1960s, and his data appeared only in a book, without full descriptions. Still, at first glance those results seemed impressive, although omissions and contradictions lurked among the arguments. A major concern was Ling's rate of ^{22}Na outflux from poisoned muscle, for it was 10-fold that in studies by Levi and Ussing and by Harris (see Chapter 3). Moreover, this discrepancy arose from Ling's approach to calculating outflux, which was internally inconsistent.[48] Another concern was the contribution to the observed ^{22}Na outflux of "exchange diffusion," the process described by Ussing that achieves net *isotope* outflux without consuming metabolic energy (see Chapter 3). Strangely, Ling did not refer to exchange diffusion in his reviews of this period even though Keynes had by then published strong evidence for it in frog muscle.[49] If outflux were comparable to that reported earlier and if a sizable fraction were due to exchange diffusion, then Ling's argument based on energy needs collapsed. Nevertheless, neither Ling nor his critics assessed these issues explicitly at that time.

Overshadowing these data, to many observers, were more serious flaws in Ling's proposal. Arguments against rival formulations were no substitute for evidence in favor of Ling's hypothesis, and selective binding could not be demonstrated *in vitro*. Ling attributed this failure to a loss of organization when cells were broken apart. Examining parts of a complex whole is a common challenge in biological research, but one that has generally been surmounted. In this case, uniquely, necessary structure might be irretrievably lost. But where was direct, independent evidence for that loss? What if there were no necessary structure and the hypothesis were wrong? Instead of pondering such possibilities, Ling blithely asserted that "the failure to demonstrate selective adsorption by protein solutions implies that the proteins within cells are not in the same state."[50] (Oddly, some philosophers have been enamored of Ling's arguments.[51])

The hypothetical adsorption of K^+ and ATP to cellular proteins was, moreover, wildly implausible. Ling calculated that these proteins contained 260 mmol of anionic sites (from free carboxyls) per kilogram of muscle,[52] more than enough to adsorb 100 mmol of K^+/kg. But cells contain thousands of different species of proteins in a wide range of structures forming distinctive associations. Contemporary views of protein structure lent no support to notions that anionic groups in diverse environments could rearrange into a few classes of K^+-specific sites. And there was no evidence for, and considerable reason to doubt, that sites capable of distinguishing ATP from ADP existed commonly among cellular proteins, or that by doing so they could induce the characteristic and comprehensive changes required.

These shortcomings contrasted with the blossoming of proposals for Na^+/K^+- and Ca^{2+}-ATPases. Formulations of these enzymes were not without problems either. But these ATPases were capable of experimental exploration and the consequent models open to further development.

Continuations

From the 1960s through the 1970s (and beyond), Ling and his fellow enthusiasts (1) reasserted their formulae; (2) measured adsorption to model systems such as wool and ion-exchange resins, which did not display the requisite selectivity; (3) used nuclear magnetic resonance spectroscopy (NMR) to try to show that water and Na^+ in cells were partially "bound" (their motions restricted),[53] and used Na^+-sensitive microelectrodes to show that Na^+ was partially "bound" (its chemical activity decreased);[54] and (4) criticized work on the Na^+/K^+-ATPase.

Interpretations of NMR results, however, were challenged;[55] one advocate was forced to retreat to the astonishing admission that "the data are compatible with some amount of complexing of sodium, between 1 and 99%, but we don't know how much it is."[56] And even if water were organized and Na^+ restricted, Ling's hypothesis was neither quantitatively established nor the pump hypotheses refuted. Furthermore, experiments with K^+-sensitive microelectrodes revealed relatively little binding of K$^+$.[57] Ling's rebuttal was that inserting microelectrodes disrupted the protein lattice locally, freeing K$^+$.[58] Any contrary evidence Ling rebuffed with claims about unobserved structure: "evidence" for that structure was its absence, for all searches had to perturb the native state.

A series of exchanges between critics and Ling in 1976/1977 further illustrates the controversies. Lawrence Palmer and Jagdish Gulati, who were former associates of Ling, published "a test of the binding hypothesis," showing that frog muscle in the presence of high extracellular K$^+$ *in vitro* accumulated 580 mM K$^+$ intracellularly, which was far in excess of anionic sites, while Na$^+$ changed little.[59] Ling responded by fitting the accumulation to his model of three classes of sites, with the residual K$^+$ free.[60] Gulati and Palmer then pointed out that Ling's fit was based on (1) choosing a distribution coefficient for K$^+$ that was 100-fold greater than that for Na$^+$, whereas Ling had previously used a coefficient only two-fold greater, and (2) using nine arbitrary constants.[61]

Criticisms of the Na$^+$/K$^+$-ATPase were marshalled in 1980 in a review that concentrated on purported transport by the purified enzyme reconstituted in phospholipid vesicles.[62] These early experiments were not without shortcomings (see Chapter 11). But Ling disingenuously noted flaws in one paper that were remedied in another. For example, unequal initial volumes of distribution for ^{22}Na and ^{42}K were interpreted by Ling as ATP preventing ^{22}Na loss preferentially during separation of vesicles from medium; but the volumes were equal in another paper.[63] He ignored evidence that was inconsistent with his explanation (e.g., a nonhydrolyzable analog of ATP that bound equally well and differed in only three atoms could not drive Na$^+$ gain/K$^+$ loss[64]). He also did not acknowledge evidence strongly favoring the pump hypothesis (e.g., the 3:2 exchange of ^{22}Na for ^{42}K and the time course and stoichiometry of ^{22}Na gain with ATP hydrolysis[65]). Ling likened K$^+$-binding by the Na$^+$/K$^+$-ATPase to K$^+$-binding of his association-induction hypothesis, but the Na$^+$/K$^+$-ATPase uses (probably) only *two* out of 130 aspartates and glutamates for that purpose—hardly the fraction of total anionic groups required by Ling's balance sheet.

Conclusions

In contrast with characterizations of the transport ATPase family, Ling's proposal developed negligibly. Not only was the necessary specificity for K$^+$ not demonstrated, but numerous opportunities remained unexplored. Two further examples should suffice. First, Ling argued that ouabain binding at the cell surface induced a cooperative change throughout the protein lattice such that K$^+$ was lost and Na$^+$ gained.[66] Ling's formulation thus challenged him to provide evidence for such ouabain-induced changes in protein structure and ordered water.[67] Second, dog red blood cells have high Na$^+$ and low K$^+$ contents, differing sharply from human red blood cells. Since hemoglobin is the predominant protein in red blood cells and so must provide most of the hypothesized cation binding sites, Ling's formulation challenged him to show how dog hemoglobin differs functionally from human hemoglobin—based on the known structures of these hemoglobins—to account for the contrast in cation contents.[68] (The orthodox explanation is that dog red blood cells lack the Na^+/K^+-ATPase and regulate Na^+ by a Na^+/Ca^+ exchanger, driven by the Ca^{2+} gradient, with K^+ distributed passively.[69])

Finally, I should emphasize that advocates of membranes and pumps do not deny that binding occurs (indeed, the standard view depicts most intracellular Ca^{2+} as tightly bound), but they point to overwhelming evidence for enzymes throughout the biological kingdoms consuming ATP to transport specific ions stoichiometrically.

COMMENTARY

Places and People

This seems an appropriate point for recognizing influences of class and gender. Mabel Hokin was from a working-class background, and she would not have attended a university had not the war and Hans Krebs intervened. She expresses gratitude for her acceptance by a less class-conscious United States. But in Montreal she was denied a faculty position when Lowell Hokin was appointed assistant professor, and finding a university in the United States whose nepotism rules would allow her to work with a husband was difficult. Even at the liberal University of Wisconsin she was relegated to the rank of research associate, although she was a partner in planning, experimenting, and writing (the absence of faculty status, among other deprivations, prohibited her from supervising graduate students). Not until 1965 was she appointed an assistant professor; by then her husband was a full professor. Avoiding such real vicissitudes would undoubtedly have made her life and science more pleasant and productive. But it would not have made the measured rate of phosphatidic acid labeling any faster.

Conflict Resolution

Excluding the phosphatidic acid cycle as part of the Na^+/K^+ pump was straightforward: experiments demonstrated that measured rates of a proposed part were far slower than measured rates of the whole.[70] Excluding the association–induction hypothesis as the sole means for creating solute asymmetries, however, was less tidy. Experimental refutation was prevented since contrary evidence was dismissed by fiat as experimental disruption, or it was circumvented by parameters that were adjusted as needed to save the hypothesis. Instead, resolution came through developing rival pump hypotheses and recognizing implausibilities in Ling's formulation that conflicted with accepted principles of cell structure and protein chemistry.

Fairness

Both the Hokins and Ling were invited speakers at international symposia, and they published numerous reviews in standard journals as well as research reports. Ling received a Research Career Development Award from NIH, and his papers through the 1970s acknowledged grants from NIH, NSF, and the Office of Naval Research. Moreover, Ling supervised a number of graduate students from the University of Pennsylvania. Neither the Hokins nor Ling were excluded from audiences or resources. On the other hand, consideration of Ling's views, initially noted in reviews,[71] disappeared as interest and patience evaporated.

Scientific Summary

The phosphatidic acid cycle was abandoned as a proposed component of the Na^+/K^+ pump when it failed a crucial experimental test of kinetic competency. Nevertheless, the underlying observations—specific phospholipid turnover stimulated by neurotransmitters and hormones—were subsequently tied to a major intracellular signalling system. By contrast, the association–induction hypothesis was based on negligible positive data and formulated in such a way that refutation was prevented. It foundered with the development of rival theories and when its central implausibilities were not remedied.

NOTES TO CHAPTER 13

1. Biographical information is from L. E. Hokin (1987); L. E. Hokin and Hokin-Neaverson (1989); and interviews with L. E. and M. R. Hokin (1993).
2. L. E. Hokin (1951).
3. L. E. Hokin and Hokin-Neaverson (1989) recount learning of an abstract on acetylcholine-stimulated labeling by ^{32}P-P_i of "nucleoproteins," isolated by a particular extraction procedure (and their eventual demonstration that labeling was due instead to contaminating phospholipids).
4. Why measure phospholipids? L. E. Hokin and Hokin-Neaverson (1989) report suspicions that phospholipid breakdown products contaminated the "nucleoproteins"; M. R. Hokin (interview, 1993) remembers a tradition of examining everything and throwing nothing away.
5. M. R. Hokin and Hokin (1953). In these and subsequent experiments using tissue slices, the ^{32}P-phosphate was presumed to be incorporated into ATP intracellularly, with this labeled ATP then donating its radioactivity to the phospholipids.
6. M. R. Hokin and Hokin (1954).
7. L. E. Hokin and Hokin (1955b). Their demonstration was made possible by Dawson's (1954) development of methods for separating and identifying phospholipids (actually, for separating their characteristic products produced by hydrolysis).
8. M. R. Hokin and Hokin (1953); L. E. Hokin and Hokin (1955a, 1956); M. R. Hokin et al. (1958); L. E. Hokin and Hokin (1959a), p. 1389–1390.
9. M. R. Hokin and Hokin (1959).
10. L. E. Hokin and Hokin (1960b).
11. The third paper, noting effects of added acetylcholine and that glands are innervated by cholinergic nerves, is Fänge et al. (1958).
12. L. E. Hokin and Hokin (1959b).
13. Ibid., p. 1068.
14. L. E. Hokin and Hokin (1960a).

15. L. E. Hokin and Hokin (1961), p. 836.

16. For example, the Hokins were invited speakers at international conferences in Stockholm and Prague and had been invited to write several review articles.

17. Thesleff and Schmidt-Nielsen (1962).

18. Yoshida et al. (1961, 1962); Nicholls et al. (1962).

19. L. E. Hokin and Hokin (1963a); the conference was held in 1961.

20. As pointed out by Judah and Ahmed (1964).

21. Ibid.

22. L. E. Hokin et al. (1963).

23. M. R. Hokin and Hokin (1961), p. 1016.

24. L. E. Hokin et al. (1963), p. 496.

25. M. R. Hokin and Hokin (1961).

26. Järnefeld (1961). He also showed that the apparent slow rate was not due simply to a rapid loss of labeling.

27. M. R. Hokin and Hokin (1964a); L. E. Hokin and Hokin (1964). Kirschner and Barker (1964) also pointed out the problem of rates, and Glynn et al. (1965) noted that no turnover of phospholipids occurred in eel electric organ.

28. Durell et al. (1968).

29. Michell (1975). He was, however, proposing a Ca^{2+} influx from the extracellular medium.

30. Rasmussen (1970) assembled convincing evidence for Ca^{2+} being a second messenger, then the only presumed second messenger that was not a cyclic nucleotide.

31. Michell et al. (1981).

32. Brockerhoff and Ballou (1961).

33. Berridge (1983).

34. Streb et al. (1983).

35. Takai et al. (1979a).

36. Takai et al. (1979b).

37. Troshin (1966).

38. Ling and Gerard (1949b). Ling later interpreted these not as "membrane potentials" but as "phase-boundary potentials."

39. Ling (1952).

40. Ling (1960, 1962). Also at the Institute were George Eisenman, Paul Mueller, and Donald Rudin.

41. Ling (1960, 1962, 1964, 1965, 1969).

42. Harris (1954b). Previously, Hodgkin and Keynes (1953c) had shown similar mobilities in cuttlefish axons.

43. Ling (1969).

44. Ling (1962).

45. Ling (1969).

46. Ling (1965).

47. Ling (1964, p. 244; 1965, pp. 87, 92); Ling and Ochsenfeld (1966), p. 820. This "holistic" concept, however, did not prevent Ling from attempting explanations in terms of the physical properties of the parts.

48. Ling calculated outflux by first extrapolating the slow phase of ^{22}Na outflux, as in Levi and Ussing's experiments (Fig. 3.2), back to zero time to evaluate the initial ^{22}Na content within the muscle cells. He then divided the initial content by the time the muscle was soaked in ^{22}Na prior to measuring outflux. However, the calculated flux (change in content/time)—on which Ling based his conclusion of metabolic impossibility—was one to two orders of magnitude greater than the outflux calculable from the slow phase of ^{22}Na outflux itself. Since this slow phase was the basis of Ling's calculation of ^{22}Na content, a startling inconsistency resulted. (The

slow phase was also the basis of Levi and Ussing's and Harris's evaluation of outflux rates.) I am indebted to Jeffrey Freedman for pointing out to me this internal contradiction.

49. Ling's 1960, 1962, 1964, 1965, and 1969 reviews ignored the process, despite the paper by Keynes and Swan (1959).

50. Ling (1952), p. 774.

51. For example, Bechtel (1982), citing Ling, described how philosophical principles could help biochemists avoid the error of believing in membrane pumps.

52. Ling (1969), p. 27.

53. E.g., Cope (1967, 1969).

54. E.g., Hinke et al. (1973). Hinke, however, did not subscribe to the association-induction hypothesis.

55. Berendsen and Edzes (1973).

56. Cope in Berendsen and Edzes (1973), p. 482.

57. Hinke et al. (1973).

58. Ling in Hinke et al. (1973), p. 295.

59. Palmer and Gulati (1977).

60. Ling (1977).

61. Gulati and Palmer (1977).

62. Ling and Negendank (1980).

63. Hilden and Hokin (1976) vs. Hilden and Hokin (1975).

64. Goldin (1977).

65. Hilden and Hokin (1975); Goldin (1977).

66. Ling and Ochsenfeld (1973).

67. Compare the studies on how ouabain affects the Na^+/K^+-ATPase (see Chapter 11).

68. In the published discussion following his 1952 paper, Ling's attention was called to the opportunity afforded by such cells.

69. Parker (1977).

70. M. R. Hokin and Hokin (1964a) did suggest a way to save their hypothesis: a tiny fraction, not detectable against the total labeling, could be turning over rapidly enough. But then the bulk of the labeling would be functionally irrelevant. Neither they nor anyone else pursued this implausible possibility.

71. For example, L. E. Hokin and Hokin (1963b); Albers (1967).

chapter 14

USING THE TRANSMEMBRANE CATION GRADIENTS: TRANSPORTERS AND CHANNELS

T HE Na^+/K^+ pump is a standard component of cells in higher animals, and there it consumes an estimated 10 to 60% of a cell's energy production.[1] This fraction is so huge that scientists as different as Conway and Ling considered pumps implausible on this basis alone. Such judgments assumed that pumps primarily offset wasteful leakiness. An alternative interpretation emerged from studies cited here: cation gradients are stores of potential energy and are built and maintained to power other cellular functions, notably transport and signalling.

Na$^+$-COUPLED TRANSPORT SYSTEMS

As Ling warned, "membrane theories" require separate pumps for the multitudes of solutes kept at nonequilibrium levels. Ling failed to imagine, however, that energy spent on the Na^+ gradient can be recaptured to drive linked transport systems for other solutes. Here I outline the origins and development of a few early formulations, skipping over efforts at deciphering the transport process (chiefly through kinetic analyses) and stopping before molecular mechanisms were defined; indeed, no mechanisms were understood at the molecular level by the mid-1990s.

But first a review of some terms may be useful. By the 1950s distinctions had been drawn between "simple diffusion"—the passage of solutes described by Fick's law—and "facilitated diffusion."[2] This latter process became associated with models of "carrier-mediated transport,"[3] having "saturable kinetics" (as if solutes bound to carriers for transport) with selectivity for solutes and competition between structurally similar ones (as if discrimination by carriers were imperfect). Carrier-mediated

Fig. 14.1 Robert K. Crane. (Photograph courtesy of Robert Crane.)

transport could be "passive," equilibrating to dissipate electrochemical gradients, or "active," accumulating solutes against their gradients.

Epithelial Sugar Transport

In 1950 common explanations for glucose transport across the intestinal epithelium invoked intracellular phosphorylation.[4] In such models, glucose diffused across the "luminal" (or "mucosal" or "outer") membrane into the epithelial cells; there it was rapidly phosphorylated, maintaining the gradient of free glucose from outside to inside while forming higher concentrations of free *plus* phosphorylated glucose inside than free glucose outside. Then a phosphatase at the opposite ("serosal") membrane of the epithelial cells converted phosphorylated glucose to locally high concentrations of free glucose, which diffused out into the bloodstream. Cellular synthesis of glucose-1- and glucose-6-phosphate was well established, but in 1956 Robert Crane (Fig. 14.1) in St. Louis showed that glucose modified chemically to prevent phosphorylation on carbons 1 and 6 was accumulated by intestinal tissue nonetheless.[5]

 If the glucose was not accumulated by phosphorylation, then how? E. Riklis and Quastel provided a vital clue in 1958.[6] They were concerned with ways salts affected intestinal absorption (these were issues of contemporary interest), and, using isolated intestinal segments, they showed that sugar transport from media bathing the luminal surface to media bathing the serosal surface (1) increased when the K^+ concentration of the bathing media was raised from 6 to 16 mM; (2) ceased when Na^+ was removed; and (3) required Mg^{2+} for K^+ to stimulate. Most of that paper dealt with the K^+ effect, which they localized to the serosal surface, but they did not pursue these issues. In 1960 T. Z. Csáky, a veteran explorer of sugar transport who was then on sabbatical leave in Ussing's department, confirmed the need for Na^+ and localized it to the luminal media.[7]

Meanwhile, Crane had localized active sugar transport to the luminal cell surface.[8] He incubated intestinal strips with sugar, dissected out the cell layers, and measured their sugar contents. The highest concentrations were within cells lining the lumen, as would be expected if active transport across the cells' luminal membranes were followed by passive diffusion from them—across their serosal membranes—to the bloodstream. Since his studies indicated that sugars were not themselves altered during accumulation, Crane considered alternative driving forces at the luminal membrane.[9] In a review published in 1960 Crane proposed a "convection" model whereby flows of Na^+ and water across luminal membranes, driven by the Na^+ pump, created compensatory inward flows carrying glucose through selectively permeable regions.[10]

By August 1960, at a major meeting in Prague, Crane revised his scheme significantly.[11] From experiments showing that levels of sugar uptake were proportional to Na^+ concentrations in the luminal media and that cardiotonic steroids blocked sugar transport, Crane now depicted a carrier with sites for *both* glucose and Na^+ (Fig. 14.2*A,B*): glucose-Na^+-carrier complexes crossed the luminal membrane, with glucose and Na^+ then dissociating in the cytoplasm; a pump extruded Na^+, maintaining the Na^+ gradient while the glucose gradient grew. Energy from the Na^+ gradient built a glucose gradient.

Consistent with this model was Schultz and Ralph Zalusky's finding that adding sugar to the luminal media increased the shortcircuit current across the epithelium (this was equated to a Na^+ current from luminal to serosal media).[12] Moreover, phlorizin, a competitive inhibitor of sugar transport,[13] blocked the increased shortcircuit current.[14]

Csáky, however, interpreted the role of Na^+ quite differently. In 1960 he noted that "the sugar pump can function only if the sodium pump is also functioning simultaneously"; moreover, in 1961 he expanded the principle, arguing that "all active transport mechanisms are driven by ATP and . . . a 'pump ATP-ase' is involved" in all active transport systems, adding that "all nonelectrolyte pumps seem to be inhibited by the absence of sodium ions."[15] Csáky considered that Na^+ acted *intra*cellularly: "it is not the carrier which requires sodium . . . but the part of the transport system which is responsible for the conversion of chemical energy into pumping energy, [specifically] derived from . . . ATP."[16] Csáky was thus generalizing from a $(Na^+ + K^+)$–stimulated ATPase to a $(Na^+ + sugar)$–stimulated ATPase to a coupled "pump adenosine triphosphatase [that] may be involved in all active

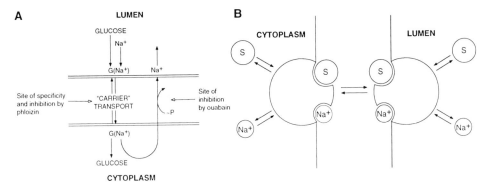

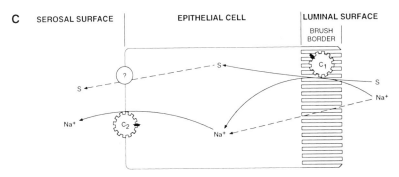

Fig. 14.2 Models for coupled Na$^+$-sugar cotransport. *A:* Relationship between carrier-mediated sugar transport and pump-driven ($\sim$P) Na$^+$ outflux. [Redrawn from Crane, 1983, Fig. 4, which was redrawn from Crane et al. (1961), Fig. 3. Reprinted with kind permission of Elsevier Science-NL, Amsterdam, The Netherlands.] *B:* A mobile carrier for Na$^+$ and glucose. [Redrawn from Crane (1965), Fig. 1. Reprinted by permission of the Federation of American Societies for Experimental Biology.] *C:* Relationship between a Na$^+$-sugar co-transporter at the luminal brush border membrane and a Na$^+$ pump at the serosal membrane. [From Schultz and Zalusky, 1964b), Fig. 7. Reprinted by permission of the Rockefeller University Press.]

transport process."[17] He thus envisioned a common energy-converting Na$^+$-ATPase coupled to various carriers, each specific for certain solutes. Accordingly, inhibition of sugar transport by cardiotonic steroids he universalized to those inhibitors being "general pump poison[s]."[18]

In 1962 Crane reported that inward Na$^+$ gradients drove sugar accumulation under anaerobic conditions, although he did not measure actual ATP levels.[19] A better response to Csáky's proposal was Crane's demonstration that reversed Na$^+$ gradients drove sugar *outflux.*[20] He first loaded the cells of an intestinal preparation—whose aerobic metabolism was blocked—with Na$^+$ and a nonmetabolizable sugar.[21] Then he incubated that tissue in the absence or presence of an outward Na$^+$ gradient. The reversed gradient drove sugar outflux against its concentration gradient (Table 14.1).[22] The following year, 1965, Csáky reported that ouabain inhibited intestinal

Table 14.1 Sugar Outflux Driven by an Outward Na⁺ Gradient*

Incubation Time	Na⁺ Concentration of Medium (mM)	Final Sugar Concentration (mM)		
		in Medium	in Tissue	Ratio
0	120	1.52	2.06	1.35
5 min	120	1.51	1.72	1.14
5 min	0	1.63	0.76	0.47

*Intestinal epithelial cells were first incubated for 5 min at 37°, under nitrogen, with an uncoupler of oxidative phosphorylation (dinitrocresol), 120 mM Na⁺, and 1.6 mM 6-deoxyglucose. Next they were incubated for 5 min longer, either in the same medium or in the absence of Na⁺ (substituting an organic cation). The content of 6-deoxyglucose in cells and in media was measured either before the second incubation or after it. (Earlier experiments showed that by the beginning of the second incubation, the Na⁺ concentration within the cells ranged from 77 to 116 mM.) [From Crane, 1964, Table 1. Reprinted by permission of Academic Press.]

sugar transport only from the serosal surface; as he acknowledged, this made direct inhibition of luminal sugar pumps unlikely.[23] Schultz and Zalusky had earlier described ouabain inhibiting shortcircuit currents from the serosal surface[24] and portrayed the two pumps at opposite poles of epithelial cells (Fig. 14.2C). And in 1971 Schultz and Richard Rose interpreted measurements of intracellular and transepithelial potentials as an electrogenic Na⁺ influx coupled to sugar (and to amino acid) transport.[25]

Simplified systems, however, could clarify the issues more decisively. In 1973 Ulrich Hopfer, while visiting Kurt Isselbacher in Boston, described glucose transport into luminal membrane vesicles.[26] They isolated from rat intestine "brushborder membranes" (so called because their finger-like microvilli project into the intestinal lumen) and showed with rightside-out vesicles formed from these membranes that (1) vesicles accumulated glucose by transport not by binding;[27] (2) other sugars competed for transport; (3) Na⁺ gradients drove transport (there was no ATP or other energy source present); and (4) phlorizin blocked both uptake and release.[28] The following year Hopfer, who was now in Zurich, reported electrogenic transport of sugars into vesicles.[29] An electrical potential across the vesicle membrane (inside negative) drove the entry of uncharged sugars, and they interpreted these observations as glucose-Na⁺-transporter complexes being moved by the electrical field (potentials were created and eliminated by using lipid-permeable ions and lipid-soluble ion carriers, "ionophores").

In 1976 Crane solubilized proteins from intestinal brushborder membranes with detergents and reconstituted the fractions in phospholipid vesicles to effect Na⁺-dependent sugar transport.[30] And in 1981 Giorgio Semenza in Zurich labeled an intestinal brushborder membrane fraction with a reactive derivative of phlorizin.[31] The criteria for identification included Na⁺-dependent labeling and protection by glucose against such labeling. Still, other proteins might be necessary for the functional system, and Semenza cautiously refered to a "component" of the transporter.

The molecular weight by SDS-PAGE was 72 kDa, and in 1983 Semenza used antibodies to select a 72 kDa fraction that appeared homogeneous by SDS-PAGE.[32] Ernest Wright in Los Angeles labeled a 75 kDa fraction using different reagents.[33] (By now images of lipid-soluble mobile carriers were replaced by those of transmembrane proteins cyclically exposing sites extracellularly and intracellularly.)

Attempts to characterize the functional proteins, however, were frustrated by the proteins' instability and extremely low abundance in the membranes. So Wright adapted an alternative approach, "expression cloning."[34] He took mRNA from intestinal cells, which encodes all the proteins that the cells synthesize, and injected it into oocytes of *Xenopus laevis*. The appearance of Na^+-dependent glucose transport in the oocytes indicated that mRNA for the transporter was expressed. By testing various fractions of the mRNA, Wright found one—representing 3% of the total mRNA—that produced such transport. He next synthesized cDNA segments from this mRNA fraction and inserted the cDNA molecules randomly into plasmids, which incorporated randomly into bacteria. To find which of the resulting 2,000 clonal colonies contained transporter cDNA, Wright first synthesized mRNA from the plasmid cDNA of each colony by transcription *in vitro*. He then injected each mRNA preparation into an oocyte, testing for expression of the proper mRNA by measuring any Na^+-dependent sugar transport by the oocyte.[35] With the appropriate clonal colony identified, he then sequenced its plasmid cDNA, deducing a protein of 73 kDa with an estimated 11 transmembrane segments.

Other Na⁺-linked Systems

Two brief examples will illustrate additional interests, each converging on Na^+-dependent transport: (1) accumulation of amino acids, and (2) extrusion of Ca^{2+}.

(1) Epithelial transport of amino acids in kidney and intestine was recognized well before 1950. Unlike active sugar transport, however, active amino acid transport occurs in most animal cells, and in 1952 Halvor Christensen and Thomas Riggs in Boston described a convenient system for studying cellular accumulation using "Ehrlich ascites cells," mouse tumor cells.[36] After incubating these cells for 1 to 2 hours with 2 to 10 mM glycine, they found an 8.2-fold accumulation of this amino acid which was sensitive to metabolic inhibitors such as cyanide and to cations. They also found that replacing extracellular K^+ with Na^+ lowered the concentration ratio for glycine (concentration in cells:concentration in medium) to 5.7, and that reducing extracellular Na^+ to 26 mM lowered the ratio to 3.5. Reducing intracellular K^+ to 108 mM, by incubating at 5° for 4 hours, lowered the ratio to 2.2, whereas reducing K^+ to 52 mM, by such incubation in the absence of extracellular K^+, lowered the ratio similarly, to 1.9 (intracellular Na^+ was not measured but was assumed to rise as K^+ fell). Moreover, glycine accumulation was accompanied by a fall in intracellular K^+ and a rise in intracellular Na^+. But unlike Riklis and Quastel, Christensen and Riggs did not measure uptake in the absence of extracellular Na^+. Like Riklis and Quastel, they focused on K^+, suggesting "a common step . . . by which K^+ and the amino acid are transferred" and noting that "the energy for amino acid concentration [could be]

drawn from the store of potential energy inherent in the asymmetric distribution of potassium."[37]

Six years later Christensen and Riggs, now in Ann Arbor, pointed out that Na^+ moved reciprocally with K^+ during amino acid accumulation and that "the excess of Na influx could occur in the form of a complex between the carrier, glycine, and sodium."[38] But from comparing *net* K^+ loss to Na^+ gain they concluded that "K^+ efflux is more likely to be the significant factor."[39] Christensen's interest lay in distinguishing the complex variety of amino acid transport systems, and he did not pursue the cation linkage.

In 1961, however, H. G. Hempling and D. Hare in New York measured ^{42}K fluxes during glycine accumulation and calculated that "more free energy is required . . . than is available from potassium efflux."[40] And in 1963 Erich Heinz in Frankfurt clearly demonstrated the link between glycine uptake by Ehrlich ascites cells and extracellular Na^+.[41] The following year George Vidaver in Bloomington described glycine transport into avian red blood cells associated with an inward Na^+ gradient and glycine outflux associated with an outward Na^+ gradient.[42] Csáky found that active transport of amino acids as well as sugars declined when Na^+ was absent,[43] and Schultz and Zalusky reported in 1965 that the shortcircuit current across the intestine increased with alanine transport.[44] That link was pursued by Schultz and Peter Curran, who published an influential review in 1970, "Coupled Transport of Sodium and Organic Solutes."[45] During Glynn's visit to Yale University he collaborated with Curran in showing that outward fluxes of alanine from the intestine could drive Na^+ outflux against its gradient.[46] Specific doubts about the electrochemical potential being quantitatively adequate to drive transport were addressed by Alan Eddy in Manchester and by Heinz, who included contributions of the electrogenic Na^+/K^+-ATPase in calculating the driving potential.[47]

In 1984 Hermann Koepsell in Frankfurt dissolved brushborder membranes from kidney tubules and reconstituted a fraction in phospholipid vesicles that effected Na^+-dependent amino acid transport.[48] But, as with sugar transporters, further isolation and characterization was slow. Not until 1992 did Yoshikatsu Kanai and Matthias Hediger in Boston report a sequence—by expression cloning—for a kidney and intestine Na^+-dependent glutamate transporter, with a deduced molecular weight of 57 kDa and 10 transmembrane segments.[49]

(2) By the later 1960s roles for Ca^{2+} in contractile and excitable tissues were established and mechanisms for controlling intracellular Ca^{2+} levels sought. H. Reuter and associates in Bern described ^{45}Ca outflux from heart sensitive to extracellular Na^+ concentrations and ^{45}Ca influx sensitive to intracellular Na^+; they considered that these fluxes represented a Na^+/Ca^{2+} exchange system.[50] At the same time, Mordecai Blaustein, Hodgkin, and associates described ^{22}Na outflux from squid giant axons linked to ^{45}Ca influx and ^{45}Ca outflux that "may be coupled to sodium entry."[51] And in 1973 Reinaldo DiPolo in Caracas, using internally perfused giant axons, demonstrated that ATP was not necessary for Na^+-dependent Ca^{2+} outflux.[52]

(Subsequently, DiPolo showed that ATP-driven Ca^{2+} pumps were also present in giant axons.[53])

In 1979 John Reeves and John Sutko in Dallas and Barry Pitts in Houston independently described vesicles formed from cardiac plasma membranes that could exchange Ca^{2+} in the medium for Na^+ in the vesicle; Pitts calculated a $3:1$ $Na^+:Ca^{2+}$ stoichiometry.[54] In 1990 Kenneth Philipson determined the amino acid sequence for the cardiac Na^+/Ca^{2+} exchanger by cDNA techniques, deducing a 108 kDa protein with 12 transmembrane segments.[55]

Generalizations

Transmembrane ion gradients serving as energy sources is now a central tenet of bioenergetics (and a major concept celebrated in this book), so it is worth noting when this idea arose. Crane stated in his memoir that he could find no other "diagram . . . which clearly expresses the concept of ion-coupling" in the proceedings of the Prague meeting or in papers published prior to his representation (Fig. 14.2*A*); he also reported that at the end of his talk in Prague in August 1960 "Peter Mitchell cried out, 'You've got it' " and hurried to the table where he (Crane) "reexplained the model," noting that "the generalization was, on the instant, made."[56] Not until 1965, however, did Crane write that "the mechanism of Na^+-dependent transport . . . may be ubiquitous," citing transport of amino acids and bile salts as well as sugars.[57]

As we shall see in Chapter 17, Mitchell presented a related mechanism immediately after the Prague meeting, and in 1963 he expounded on "how an asymmetric distribution of electrochemical activity of one . . . solute such as . . . Na^+ could give rise to the flow of a structurally unrelated solute."[58] There he also used the terms "symports" for systems in which the solutes move in the same direction, and "antiports" for those in which solutes move in opposite directions; alternative terms are "co-transporters" and "exchangers," respectively.

Within a few years this concept was widely recognized. For example, in his 1967 review, Albers stated that the "store of potential energy . . . generated [by the Na^+/K^+ pump] may be available for other cellular work," citing Crane and Mitchell.[59] A related distinction that was soon introduced was between "primary active transport," which is directly coupled to metabolic energy (as in ATPases), and "secondary active transport," which is coupled to transmembrane gradients.[60]

Experimental generalizations, moreover, disclosed multitudes of Na^+-coupled transporters (Table 14.2). From examining amino acid sequences of roughly 100 systems then determined, Milton Saier in 1993 described—on the basis of sequence similarities—a "superfamily" of sodium-solute transporters consisting of 11 families.[61] And while the studies noted here were in progress, analogous endeavors disclosed H^+-coupled transport systems, driven by transmembrane H^+ gradients, in plasma membranes of bacteria, plants, and fungi; in membranes of chloroplasts and mitochondria; and in other membranes of higher animals as well.

Table 14.2 Some Transporters Driven by the Na^+ Gradient*

Symporters (co-transporters)	Antiporters (exchangers)
Na^+-glucose	Na^+/Ca^{2+}
Na^+-amino acid (numerous)	Na^+/H^+
Na^+-neurotransmitter (numerous)	Na^+/Mg^{2+}
Na^+-phosphate	
Na^+-Cl^-	
Na^+-carbonate-bicarbonate	
Na^+-monocarboxylic acid	
Na^+-dicarboxylic acid	
Na^+-cholate	
Na^+-nucleoside	
Na^+-inositol	

*Stoichiometries are not shown; in some cases other ions (K^+, Cl^-) may also participate.

SIGNALLING BY ION FLUXES THROUGH CHANNELS

The Hodgkin-Huxley model, with its voltage- and time-dependent opening and closing of cation-specific channels, admirably accounted for action potentials of excitable tissues (see Chapter 5). Here I will outline some further steps in characterizing "voltage-gated" Na^+ channels. But I wish also to emphasize the growing recognition that ion fluxes are signalling mechanisms for *all* cells by pointing to examples of channels for other ions, of channels whose opening is affected by chemicals binding to them ("ligand-gated"), and of channels in intracellular as well as plasma membranes.

Voltage-gated Na^+ Channels

First, three developments of the Hodgkin-Huxley program should be noted. (1) To reconcile differences between measured isotope fluxes and electrical conductances, Hodgkin and Keynes proposed in 1955 that ions did not cross independently but "move in single file through narrow channels," supporting their interpretation with mathematical analyses and a mechanical model.[62] (Their term "channel" replaced "pore" as the standard designation.) (2) Hodgkin and Huxley pointed out that voltage-dependence implied "movement of some [charged] component of the membrane" to alter permeability.[63] Twenty-one years later Clay Armstrong and Francisco Bezanilla in Rochester succeeded in measuring that "gating-current," representing "molecular rearrangements . . . attendant on the opening and closing of . . . ionic channels in response to changes in . . . membrane [electrical] field."[64] (3) Hodgkin-Huxley channels displayed not only conductivity and gating but also selectivity. An apparent problem for "sieve" theories was how a channel large enough to accommodate hydrated Na^+ could exclude the smaller hydrated K^+. In 1960 Mullins, recognizing the considerable energy required to strip waters of hydration from ions, proposed that ions passed through channels with a single complete layer of hydration—but only if the channel walls interacted optimally, fitting snugly to replace the outer layers of hydration; too large as well as too small a channel would exclude the hydrated ion.[65] Mullins proposed 7 Å diameters for Na^+ channels, but a decade later Bertil Hille in Seattle calculated a much smaller "selectivity filter," 3.1 × 5.1 Å, from measuring the permeation of graduated solutes.[66] Hille pictured a Na^+ associated with three oxygens bordering the filter and with three more oxygens from water molecules (beside, in front, and behind); nevertheless, he could fit K^+ in such apertures and concluded that a "completely steric theory . . . fails entirely."[67] So, citing George Eisenman's analysis of selective cation binding in glass electrodes,[68] Hille added electrostatic interactions between cations and binding sites in the channel: selectivity reflected the energy difference between cation–water and cation–binding site interactions (functions of charge, dipole moment, and closeness of approach). Finally, Peter Läuger in Constance and Hille soon included kinetic as well as thermodynamic aspects.[69]

Next should be cited three technical advances that not only made possible the realization of certain obvious goals, such as the isolation and identification of channels, but also revealed unexpected details and characteristics of these channels. Preeminently important was the patch clamp technique. Previously, *summed* channel openings and closings were recorded by impaling cells with microelectrodes, but activities of *single* channels could only be inferred from the composite.[70] Then between 1976 and 1981 Erwin Neher and Bert Sakmann in Göttingen perfected methods for recording instead from tiny areas of membranes: cleaned cell surfaces were sucked tight against a polished glass electrode, so that the membrane patch was electrically isolated (Fig. 14.3A).[71] Openings and closings of single channels within that patch, seen as abrupt transitions between two states, thus underlay the smooth curves described by Hodgkin and Huxley (Figs. 5.3, 14.3D). With this technique in various forms (Fig. 14.3A–C), channel properties—conductivities and opening and closing probabilities—could be recorded for individual *molecules*. This was a spectacular feat that justly deserved the Nobel prize awarded in 1991. Now far smaller cells could be studied than was possible with impaling microelectrodes, and ion channels were soon catalogued across the biological spectrum. Alternatively, activities of one or a few channels could be recorded by incorporating them, either as purified channels or as membrane fragments, into artificial planar bilayers (see Chapter 10). Electrical measurements across the bilayer could then show channel openings and closings as a function of experimental conditions.[72]

The third technical advance came with the use of radioactive isotopes as labels. Whereas transport ATPases can be followed through purification procedures by monitoring enzymatic activity, channels have no functional identifier once membrane intergrity is lost, and reconstitution in vesicles or bilayers at each step is impractical. On the other hand, species competition has spawned diverse toxins aimed at Na^+ channels, toxins that bind tightly and specifically and that, when

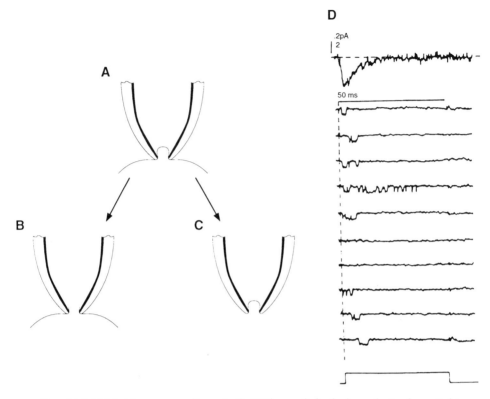

Fig. 14.3 Patch clamp recording. *A–C:* With a polished glass electrode, a tight seal is formed with a cleaned area of cell membrane (a gigaOhm seal), permitting "cell-attached" recording (*A*). With a pulse of suction the membrane beneath the electrode is broken, allowing direct access to the cytoplasm both for "whole-cell" recording and for altering cytoplasmic composition (*B*). By pulling the electrode away, a piece of membrane is detached, allowing recording from that piece as well as bathing the cytoplasmic surface of the piece with experimental media (*C*). [Redrawn from Hamill et al., 1981, Fig. 9. Reprinted by permission of Springer-Verlag.] *D:* Patch clamp recordings of isolated heart cells during a sustained depolarization from −140 to −50 mV (bottom trace) reveal openings (downward transitions) and closings (upward transitions) of Na$^+$ channels beneath the electrode. At the top is shown the summed current of the openings and closings from 76 depolarizations; compare with Fig. 5.3C. [Redrawn from Nilius (1989), Fig. 2. Reprinted by permission of the American Physiological Society.]

tagged with radioactive isotopes, provide useful labels for Na$^+$ channels. The first toxin applied extensively was tetrodotoxin, which is produced by puffer fish; a similar compound, saxitoxin, is produced by unicellular organisms.[73]

The expected course of purification, reconstitution, and sequencing can now be summarized. In the early 1980s William Agnew and Michael Raftery in Pasadena, William Catterall in Seattle, and their associates isolated 260–270 kDa tetrodotoxin-

and saxitoxin-binding proteins from eel electric organ and rat brain.[74] Soon afterward they, plus Robert Barchi in Philadelphia, reconstituted purified channel proteins in vesicles, demonstrating toxin-sensitive Na^+ fluxes[75] and voltage-dependent single-channel currents by patch electrodes.[76] In 1984 Numa's group deduced the amino acid sequence of the electric organ Na^+ channel using cDNA techniques: 1820 amino acids were present in four closely similar domains (Fig. 14.4A) with a calculated molecular weight of 208 kDa.[77] Two years later they expressed the specific mRNA in *Xenopus* oocytes to produce functional channels.[78]

From the amino acid sequence plausible models were soon devised. One such model (Fig. 14.4B) depicted a central pore that was flanked by transmembrane segments from each domain, with a chain of positively charged amino acids serving as the voltage sensor. This chain would move under applied voltage, pulling a gate to open the channel. A separate string of amino acids would then swing to block the channel (inactivation).[79]

Generalities and Particulars[80]

Sequence similarities revealed by cDNA techniques defined a family of voltage-gated cation channels, including several types of Na^+ channels, several types of Ca^{2+} channels of similar size, and several *dozen* types of K^+ channels (their sequences, however, were only a fourth as long, with functional channels formed from four subunits). On the other hand, voltage-gated Cl^- channels were structurally dissimilar. Yet another family was represented by ligand-gated channels. Numa's group had reported even earlier the sequence of an acetylcholine receptor having a channel that opened to allow passage of Na^+ and K^+ when acetylcholine binds. This structure contained five 50–58 kDa subunits of four types; by the later 1980s additional ligand-gated ion channels had been identified that were sensitive to different neurotransmitters but contained similar amino acid sequences.

Other channels of still different structure and at different sites were also recognized: gap junctions composed of six 30–50 kDa subunits in each plasma membrane, linking cells together; Ca^{2+} release channels of the endoplasmic reticulum, which were enormous tetramers of 565 kDa subunits; voltage-gated K^+ channels in protozoa, fungi, and plants; and even mechanically sensitive ion channels in bacteria. Also identified were channels for the passive fluxes of solutes and water that are essential to such processes as epithelial transport and the regulation of cell volume.

At the same time, various interests in diverse cellular activities converged on signalling mediated through ion channels. The functions regulated by such signalling included secretion by endocrine cells, fertilization of oocytes, sensory response in receptor cells, proliferation of lymphocytes, metabolic control in mitochondria, and intracellular coordination through second messenger systems. Although the relative ease of distinguishing channels by their electrical responses outpaced the understanding of particular physiological roles, established instances compel our appreciation of ion gradients exploited to transmit information.

A.

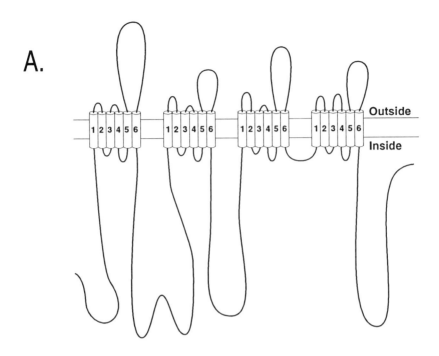

B.

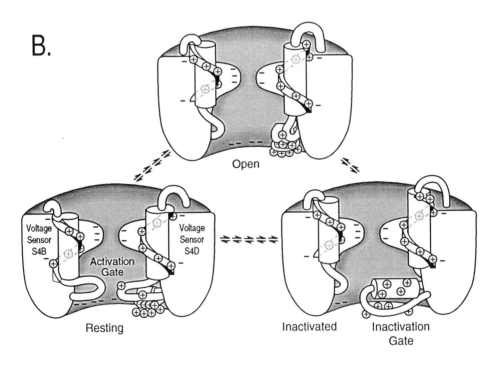

COMMENTARY

Places and People

Since Crane's formulation ranks so high among the accomplishments recorded here, even brief comments about his life may be instructive.[81] Robert Kellog Crane was born in New Jersey in 1919, the great nephew of Stephen Crane. Initially he planned to follow his father in civil engineering, but while a student at Washington College he became enamored with science and contemplated a career in biomedical research. Financial stringencies and war intervened; after brief employment by the chemical industry (working on explosives) and college teaching in Missouri, he entered the Navy, serving as a deck officer on a destroyer in the Pacific. Endowed by the GI Bill, he entered Harvard University after the war. There he studied carbon dioxide incorporation into animal tissues with E. G. Ball and obtained a Ph.D. in biochemistry in 1950. After a postdoctoral year in Boston with Lipmann, Crane moved to Carl Cori's biochemistry department at Washington University, where he rose from instructor to associate professor. There he worked on carbohydrate metabolism, including studies on hexokinase, using a spectrum of glucose analogs, and on carrier-mediated (passive) sugar transport into Ehrlich ascites cells. His venture into studying intestinal active transport was motivated by general curiosity about mechanisms, by possessing that spectrum of analogs, and by his insight into how the analogs might be used profitably. In 1961 he moved to Chicago Medical School as professor and chairman of biochemistry, and in 1966 to the College of Medicine and Dentistry of New Jersey as professor and chairman of physiology. He retired in 1986.

Experimental Design

Crane noted two approaches to devising experiments: answering specific questions and testing hypotheses, or "collecting data 'just to see what happens.' "[82] Although he found the former more congenial, he admitted that the latter "when done by an astute observer . . . is frequently the faster route to those things you want to know

---

Fig. 14.4 Structure of Na^+ channels. *A:* The polypeptide chain of the Na^+ channel is shown diagrammatically to emphasize the four similar stretches of six transmembrane segments, linked by extramembrane loops. [From Catterall, 1988, Fig. 3. Reprinted with permission, ©1988 American Association for the Advancement of Science.] *B:* A functional model of the Na^+ channel shows the selectivity filter, the positively charged voltage sensor that responds to changes in the electric field to open the activation gate, and the inactivation gate that then closes the channel. The four chains participating in channel opening and the one chain in inactivation correspond admirably with the m^3h dependence of the Hodgkin-Huxley equation. The multiple arrows linking the three states specified by Hodgkin and Huxley acknowledge the existence of additional states recognized subsequently. [Redrawn from Guy (1988), Fig. 4. Reprinted by permission of Academic Press.]

and often . . . the only route to those . . . you did not expect."[83] Induction—generalizing from observation—is vigorously criticized by many philosophers for many reasons. "Seeing what happens" certainly depends on background theories and organizing principles (although these may not be controversial).[84] Still, the utility, at least for biological research, of just "collecting data" has been under-appreciated by many commentators.

A different aspect of experimental design concerns the necessity of examining one variable at a time while holding all others constant—for example, varying Na^+ concentration while keeping glucose, K^+, ATP, phosphate, pH, and temperature constant. Maintaining those constancies is often difficult (varying Na^+, for example, may change K^+, ATP, and phosphate levels), and detecting and controlling relevant variables is easier in simplified systems (such as membrane vesicles that contain glucose transporters but not the Na^+/K^+-ATPase). An opposing and extreme view Crane also noted, quoting a critic's conclusion that what "I had measured was not interpretable in biological terms because I had broken [apart] the cells."[85] (Of course, experimenters must be alert to artifacts arising from any manipulation, and they must strive to integrate particular findings into models of entire cells [cf. Fig. 14.2C].)

New Methods

Once new techniques are available, hosts of applications often become apparent. Nevertheless, such obviousness is irrelevant to the usefulness of the answers obtained, which is the ultimate criterion in doing research. Here, wholesale efforts at transporter and channel reconstitution, at discovering still more channels by patch clamp methods, and at sequencing everything by cDNA techniques may have proceeded without much creative imagination (although modifying techniques for particular situations can be challenging). But the resulting information often provided new insights into functional mechanisms and relationships and disclosed common characteristics. For example, using tetrodotoxin to block only Na^+ channels supported the Hodgkin-Huxley model with its separate Na^+ and K^+ channels; sequencing the Na^+ and K^+ channels, however, demonstrated their difference as chemical entities. And the ubiquitous occurrence of channels became apparent only when they were discovered wherever anyone looked with the new techniques. An essential component of biological research during this period was uncovering new entities and processes.

Scientific Summary

From initial studies on epithelial sugar transport came the recognition that many solutes are actively transported across cell membranes as complexes of solute, protein transporter, and driving cation. Such systems thus use the cation's potential energy, stored in its electrochemical gradient, to effect transport. Common mechanisms and structurally related families of transporters were then identified through the reductionist program.

Another use for asymmetric ion distributions, a use that had been established earlier, is signalling by neurons and other excitable cells. Common mechanisms and structurally related families of ion channels were identified for these also. But from examining other cell types came the awareness that essentially all cells have such ion channels, and that ion movements across cell membranes, down their gradients, are a universal means for inter- and intracellular communication.

Many cells thus use a "Na^+ economy" founded on the Na^+/K^+-ATPase that creates the electrochemical gradient. This economy then supports the cells' necessary chores (although gradients of other ions can be exploited as well).

NOTES TO CHAPTER 14

1. For example, Edelman (1976); Erecinska and Silver (1994); Schramm et al. (1994).
2. Danielli (1954).
3. Wilbrandt and Rosenberg (1961).
4. Cf. Crane and Krane (1956).
5. Ibid.
6. Riklis and Quastel (1958).
7. Csáky and Thale (1960).
8. McDougal et al. (1960); they suggested a link to Na^+ transport, citing Riklis and Quastel. Csáky and Fernald (1961) also localized transport to the luminal surface.
9. According to his memoir (Crane, 1983), he considered phosphorylation of glucose carriers unlikely since the cell's mitochondria were so far from the microvilli of the luminal surface.
10. Crane (1960).
11. Crane et al. (1961); the model was published as an appendix, and in his 1983 memoir Crane noted that he devised it in Prague shortly before his talk. The experiments were described in detail in Bihler and Crane (1962) and Bihler et al. (1962).
12. Schultz and Zalusky (1963). Cf. Chapter 6.
13. Alvarado and Crane (1962).
14. Schultz and Zalusky (1964b).
15. Csáky and Zollicoffer (1960), p. 1058; Csáky et al (1961), p. 460; Csáky (1961), p. 1001. Csáky had by then returned to Chapel Hill.
16. Csáky (1963b), pp. 5, 6.
17. Ibid., p. 7.
18. Csáky (1963a), p. 162.
19. Bihler et al. (1962).
20. Crane (1964).
21. Nonmetabolizable sugars were used to prevent their providing cellular energy. In many of the studies cited here, nonmetabolizable sugars were used to avoid artifacts due to concomitant metabolism; one of the most widely used was 3-O-methylglucose, introduced by Csáky (1942).
22. Garrahan and Glynn had not yet reported reversal of the Na^+/K^+ pump driven by the ion gradients; an analogous reversal of a Na^+-glucose-ATPase could have been occurring in Crane's experiments.
23. Csáky and Hara (1965).
24. Schultz and Zalusky (1964a).
25. Rose and Schultz (1971).
26. Hopfer et al. (1973).

27. Criteria included uptake dependent on intracellular volume (adjusted experimentally by changing osmotic strengths of the media).

28. If uptake were due to binding, phlorizin should not block both release and uptake.

29. Murer and Hopfer (1974).

30. Crane et al. (1976).

31. Hosang et al. (1981).

32. Schmidt et al. (1983). Antibodies were prepared using purified brush border membranes as antigens, and fractionated to isolate antibodies blocking sugar transport.

33. Pearce and Wright (1984, 1985).

34. Hediger et al. (1987a,b).

35. For efficiency, various pooled fractions were screened initially.

36. Christensen and Riggs (1952); Christensen et al. (1952). Ehrlich ascites cells are transplanted into a mouse's peritoneal cavity where they multiply rapidly as free cells in the ascites fluid.

37. Christensen et al. (1952), p. 13.

38. Riggs et al. (1958), p. 1483.

39. Ibid.

40. Hempling and Hare (1961).

41. Kromphardt et al. (1963).

42. Vidaver (1964).

43. Csáky (1961).

44. Schultz and Zalusky (1965).

45. Schultz and Curran (1970).

46. Curran et al. (1970).

47. Philo and Eddy (1978); Heinz et al. (1980).

48. Koepsell et al. (1984).

49. Kanai and Hediger (1992). Hediger was the first author on Wright's expression cloning papers.

50. Reuter and Seitz (1968); Glitsch et al. (1970).

51. Baker et al. (1969); Blaustein and Hodgkin (1969), p. 498.

52. DiPolo (1973).

53. DiPolo (1978).

54. Reeves and Sutko (1979); Pitts (1979).

55. Nicoll et al. (1990).

56. Crane (1983), pp. 66, 67.

57. Crane et al. (1965), p. 474. Similar generalizations were made in his influential review (Crane, 1965). Csáky (1961) had proposed analogous transport systems for sugars, amino acids, and pyrimidines—but in terms of ATP-consuming pumps.

58. Mitchell (1963), p. 148. He also noted that "Crane et al. (1961) have hinted [that] the absorption of sugars . . . depend [sic] on the asymmetry created by the gradient of Na^+ across the cell membrane" (p. 158).

59. Albers (1967), p. 729.

60. Mitchell (1963) distinguished between "primary" and "secondary" transport.

61. Reizer et al. (1994).

62. Hodgkin and Keynes (1955b), p. 62. Their interpretation dealt with K^+ channels, and they did not find it necessary to propose similar characteristics for Na^+ channels.

63. Hodgkin and Huxley (1952d), p. 504.

64. Armstrong and Bezanilla (1973), p. 459. Their successful measurement relied on computer-averaging of multiple small signals and on blocking the overwhelming Na^+ channel current with tetrodotoxin.

65. Mullins (1960).

66. Hille (1971, 1972).

67. Hille (1972), p. 653. Hydrogen bonding proposed between organic solutes and the channel was a rationale for placing oxygens at the selectivity filter.

68. Eisenman (1962).

69. Läuger (1973); Hille (1975).

70. To resolve properties of individual "ion gates" from composite responses of neuromuscular acetylcholine receptors, Katz and Miledi (1972) applied statistical "noise analysis" in conjunction with their kinetic model.

71. Neher and Sakmann (1976); Hamill et al. (1981). An essential requirement was the high-resistance seal between electrode and membrane: a "gigaOhm seal."

72. Goodall et al. (1974); Miller and Racker (1976).

73. See Narahashi (1974).

74. Agnew et al. (1980); Hartshorne and Catterall (1981). Molecular weights were estimated by SDS-PAGE. In brain preparations a 38 kDa protein copurified; this was subsequently shown to be a subunit of brain but not electric organ Na^+ channels.

75. Weigele and Barchi (1982); Talvenheimo et al. (1982).

76. Rosenberg et al. (1984). The next year Catterall and associates reconstituted channels in planar bilayers (Hartshorne et al., 1985).

77. Noda et al. (1984).

78. Noda et al. (1986). They expressed mRNA of rat brain Na^+ channels whose sequence they also determined.

79. Guy (1988).

80. For surveys see Jan and Jan (1989); Hille (1992); and Peracchia (1994).

81. Biographical information is from Crane (1983).

82. Crane (1983), p. 51.

83. Ibid.

84. Criticisms range from complaints that "data" are accumulated always in the context of some enabling theory to claims that scientists only test hypotheses. Dependence on theory is a critical concern when alternative background theories lead to differently recognized data; in many examples cited here, this seems not to be the case. As for testing hypotheses, any search can be construed this way; for example, when exploring cation effects, this search can be interpreted as testing the hypothesis "cations affect glucose transport" and then measuring all conceivable effects that each cation might have to find which one(s) and how. That, however, is a far cry from testing more discrete hypotheses, such as "an inward gradient of Na^+ can alone drive the intracellular accumulation of glucose."

85. Crane (1983), p. 48.

chapter 15

CONTEMPORARY EVENTS: 1966 – 1985

D URING the decades covered in the four preceding chapters, biomedical sciences advanced expansively, but only selected events touching on the major themes can be cited. Studies on oxidative phosphorylation will be noted in the following chapters.

MEMBRANE STRUCTURE

By the early 1970s membrane models had evolved that differed from their predecessors significantly (Fig. 15.1). These incorporated conclusions from sundry approaches, and they developed concomitantly with changing views of transporters and channels; indeed, models now projected plausible avenues by which polar solutes could cross. Some strands of evidence and arguments were:

1. Reevaluation[1] of interfacial tensions between phospholipid head groups and water countered Danielli's rationale for interposing protein sheets (see Chapter 2).

2. Studies emphasizing hydrophobic interactions in protein structure and assembly[2] also argued against Danielli's and Robertson's extended protein conformation, which would require that hydrophobic side chains be exposed to polar lipid head groups and/or aqueous media.

3. Estimates of α-helical content of membranes suggested to S. J. Singer and to Donald Wallach in 1966—in the context of the above two arguments—that helical peptide chains penetrated membranes, inserted among the nonpolar lipid tails to maximize hydrophobic interactions (Fig. 15.1A; with the flanking sheets of protein discarded, Singer drew the lipid phase thicker).[3] Wallach recast Stein and Danielli's pore (Fig. 10.1B) as a "hydrophobic rod [having] a polar interior [through which] transport occurs."[4]

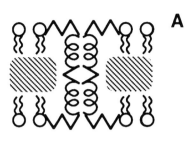

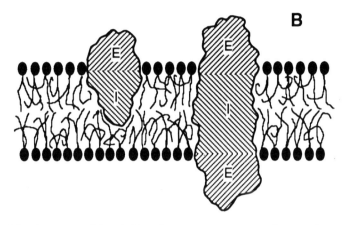

Fig. 15.1 Membrane models. *A:* Proteins penetrate across the membrane as α-helices; additional lipid is between the two phospholipid monolayers. [From Lenard and Singer (1966), Fig. 3. Reprinted by permission of S. J. Singer.] *B:* Globular proteins extend into and through the lipid bilayer; "E" refers to exterior and "I" to interior regions of the proteins, reflecting their surface hydrophilicity and hydrophobicity. [From Singer (1971), Fig. 13. Reprinted by permission of Academic Press.]

4. Freeze-fracture electron microscopy, moreover, provided images of proteins—globules about 85 Å in diameter—*within* membranes.[5] In this technique, frozen samples are mechanically fractured; the surfaces are then shadowed with metal films and these replicas examined by electron microscopy. A critical issue was where the cleavage occurred, and in 1970 Daniel Branton demonstrated convincingly that membranes fractured in a plane *between* the bilayer leaflets.[6] With this information, Singer modified his model in 1971 (Fig. 15.1B)[7] reviving the mosaic of protein and lipid that Höber and others had advocated in the 1930s (see Chapter 2). Singer also distinguished between "intrinsic" membrane proteins, embedded in the bilayer and resisting extraction, and more loosely associated "peripheral" ones, extractable by less drastic means (peripheral proteins could contribute to the observed membrane thickness, and he now returned to a simple bilayer).

5. Concurrently, studies on red blood cell membranes by Mark Bretscher and by Theodore Steck, Grant Fairbanks, and Wallach demonstrated that identifiable proteins were exposed at both surfaces, so that these proteins extended *across* the membrane.[8]

6. A year earlier, experiments by L. D. Frye and M. Edidin revealed an intermixing of surface proteins when two cells were fused: proteins in the lipid bilayer could diffuse laterally.[9] From such evidence, together with physical studies documenting lipid mobility, Singer and Garth Nicolson created in 1972 their "fluid mosaic" model of oriented, globular proteins diffusing in a fluid bilayer.[10] This model has endured (although more recent evidence indicates that diffusion of most membrane proteins is restricted[11]).

As for model membrane systems, planar bilayers were developed further during this period and were exploited for such purposes as examining single channel properties (see Chapter 14). Transformation of multilaminar vesicles into those surrounded by a single lipid bilayer was also achieved,[12] which permitted the reconstitution of membrane proteins into functional systems (see Chapter 11) as well as the examination of bilayer properties per se. Associated with these studies was the recognition and characterization of ionophores capable of promoting ion movements across lipid barriers. Two general types were distinguished. One, represented by the peptide antibiotic gramicidin, formed conducting channels spanning the membrane.[13] The other, represented by the donut-shaped antibiotic valinomycin, formed lipid-soluble complexes able to diffuse across lipid films, acting as mobile carriers.[14]

PROTEINS

Membrane proteins began to be identified and characterized at this time, and these successes were made possible by the development of detergent extraction and electrophoretic fractionation techniques. In 1971 Fairbanks, Steck, and Wallach separated red blood cell membrane proteins by SDS-PAGE into six major bands revealed by a conventional protein stain and three bands by a glycoprotein stain.[15] In 1975 Motowo Tomita and Vincent Marchesi sequenced the prominent glycoprotein, named "glycophorin," by conventional degradative techniques. Glycophorin had a molecular weight of 31 kDa with 131 amino acids, contained a single hydrophobic stretch of amino acids, and had sugar chains linked to 16 amino acids.[16] This was the first integral membrane protein sequenced. The most prevalent integral membrane protein of red blood cells, often referred to as Band 3 from its relative position by SDS-PAGE, was sequenced by Ron Kopito and Harvey Lodish in 1985 using cDNA techniques. This protein had a molecular weight of 103 kDa, with 929 amino acids and an estimated 12 transmembrane segments.[17]

Earlier, Marchesi extracted spectrin from red blood cell ghosts (it was named for its source) using dilute aqueous media.[18] This fibrous protein, a peripheral membrane protein by Singer's subsequent terminology, was later shown to attach to

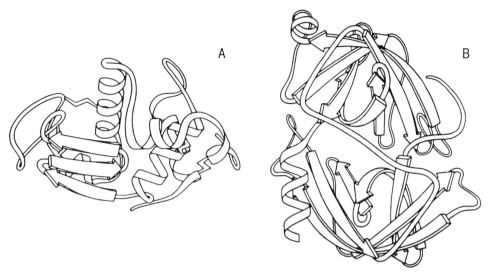

Fig. 15.2 Protein structure. Ribbon diagrams show how the peptide backbone twists; α-helices are represented as coils and β-sheets as arrows. *A:* lysozyme. *B:* Elastase (a serine protease akin to chymotrypsin). [From Richardson (1981), Figs. 74 and 68. Reprinted by permission of J. S. Richardson and of Academic Press.]

Band 3 through another peripheral protein, ankyrin.[19] Similar linkages between integral membrane proteins and the fibrous meshworks of filaments and tubules running through the cell—the mechanical scaffolding termed the "cytoskeleton"— became recognized as essential structures controlling cell motility as well as shape.

And as the preceding chapters illustrated, identification of membrane proteins with physiological and biochemical processes became a growth industry in this period.

For nonmembrane "soluble" proteins, determinations of sequences and three-dimensional structures progressed far more rapidly. The consequent information allowed the categorization of soluble proteins into functional families containing similar sequences ("motifs"). This information also promoted the recognition of structural domains and folding patterns. Assisting these analyses were systems for depicting the complex peptide convolutions by clarifying symbols (Fig. 15.2). Computerized matchings of sequences to those in growing data banks also facilitated the categorization. But predicting folding patterns from amino acid sequences remained an unachieved goal despite concentrated efforts.

On the other hand, a qualitative appreciation of enzyme reaction mechanisms was increasing. In 1967 David Blow and associates described the X-ray crystal structure of the proteolytic enzyme chymotrypsin, and from this they derived a reaction process firmly rooted in the mechanisms of organic chemistry (Fig. 15.3): "the catalytic site contains an . . . aspartic acid hydrogen-bonded to a histidine . . . hydrogen-bonded to a serine [so that] the serine oxygen [polarized] by the buried

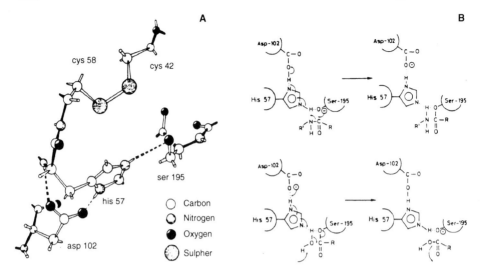

Fig. 15.3 Structure and reaction mechanism of chymotrypsin. *A:* Perspective sketch of amino acids at the enzyme active site. *B:* Proposed mechanism for protein cleavage, showing how three enzyme amino acids act on a protein (R'-NH-CO-R) to form an acylated enzyme and release one fragment, R'-NH$_2$ (top), with subsequent hydrolysis (deacylation) releasing HO-CO-R (bottom). [From Blow et al. (1969), Figs. 2 and 4. Reprinted by permission from *Nature*, © 1969 Macmillan Magazines Ltd.]

negative charge of the aspartic acid [would be] strongly nucleophilic."[20] Such interpretations established the expected form of mechanistic explanation.

TRANSPORTERS

Four examples of transporters—two from bacteria and two from mammals—can demonstrate the vigorous pursuit at that time of other transport systems.

Lac Permease

Lac permease is a H$^+$-lactose symporter in plasma membranes of *E. Coli.* First recognized as an inducible transporter of lactose, it is encoded by the *lac* operon, the focus of seminal determinations of bacterial gene regulation. In 1963 Mitchell proposed—in the context of his chemiosmotic hypothesis (see Chapter 17)—that lac permease functioned as a secondary active transporter driven by the electrochemical gradient of H$^+$ across plasma membranes.[21] Subsequent studies on intact bacteria and isolated membrane vesicles in the 1970s supported this view.[22] Initial attempts at purification were unsuccessful,[23] but in 1980 Benno Müller-Hill and associates deduced the amino acid sequence using cDNA methods: 46.5 kDa with 417 amino acids and 3 possible transmembrane segments.[24] And the following year T. H. Wilson, H. R. Kaback, and associates succeeded in purifying and reconstituting it in phospholipid vesicles.[25]

Bacteriorhodopsin

Bacteriorhodopsin is a light-driven H^+ pump in plasma membranes of *Halobacterium halobium* (halophilic bacteria growing optimally in 4 M NaCl; seawater is 0.6 M). Walther Stoeckenius had been studying membrane structure, chiefly by electron microscopy, when he turned to this organism in the late 1960s because its membrane reportedly had a subunit structure.[26] Although that claim proved erroneous, in his studies Stoeckenius separated by centrifugation a purple fraction that he and Dieter Oesterhelt showed to contain a single protein whose color was due to bound retinaldehyde (the same pigment as in the visual protein rhodopsin).[27] Two years later, in 1973, they described H^+ release driven by flashes of light, and they argued that *H. halobium* could use light energy to generate H^+ electrochemical gradients across its plasma membrane for metabolic use.[28] That same year Stoeckenius and Racker reconstituted the purple membrane in phospholipid vesicles to demonstrate light-driven H^+-pumping.[29] In 1975 a widely acclaimed paper by Richard Henderson and Nigel Unwin presented a structural model at 7 Å resolution, based on their electron microscopic studies and featuring seven transmembrane α-helices.[30] This model strongly supported arguments for intramembrane α-helices and established general features expected of integral proteins. It also satisfied expectations that ion transport progressed along transmembrane pathways between transmembrane protein strands. (By 1990 Henderson and associates increased the resolution to 3 Å in certain dimensions.[31]) The amino acid sequence of bacteriorhodopsin was published in 1979: 26 kDa with 248 amino acids.[32] This structural knowledge and bacteriorhodopsin's small size encouraged further study, abetted by another advantage: the light-driven reaction (involving conformational changes in the retinaldehyde and a related protonation-deprotonation of its linkage) could be readily followed as successive changes in optical absorption. By 1985 bacteriorhodopsin was the best understood membrane pump. (And by the mid-1990s models based on converging approaches depicted H^+ transfer from (*1*) aspartate-96, which is accessible to the exterior, to (*2*) the protein-bound rhodopsin, to (*3*) aspartate-85, which is accessible to the cytoplasm. The succession of transfers was facilitated by protein conformational changes altering the spatial relationships among these three sites, with the protein conformational changes being induced by light-driven isomerizations of rhodopsin.[33])

Glucose Uniporter

The characterization of secondary active transport systems for glucose in kidney and intestine was outlined in Chapter 14. Concurrently, parallel interests focused on passive, equilibrating glucose transport systems in ordinary cells (for technical reasons, red blood cells were experimental favorites). By 1966 such glucose transport was recognized as a facilitated diffusion process, often depicted by mobile carrier models with their derived kinetic properties reproducing observed fluxes. Not until 1977, however, did Peter Hinkle isolate the glucose transporter from red blood cells: a 55 kDa glycoprotein by SDS-PAGE that promoted glucose fluxes when reconstituted in phospholipid vesicles.[34] In 1985 Lodish and associates reported the amino acid sequence by cDNA methods: 46 kDa with 492 amino acids, one glycosylation site, and 12 transmembrane segments.[35] (Subsequently, a family of four GLUT transporters was identified in mammalian cells, which belongs to a huge superfamily of transporters that includes bacterial systems.)

Chloride/Bicarbonate Antiporter (Exchanger)

Since bicarbonate (HCO_3^-) is far more soluble in blood than carbon dioxide (CO_2), moving CO_2 from tissues to lungs is greatly expedited by converting it to HCO_3^-.) Carbonic anhydrase inside red blood cells catalyzes that hydration, and CO_2 readily diffuses across lipid barriers and can thus enter the cells. But HCO_3^- cannot diffuse across that barrier to exit from the red blood cells. Studies going back to the nineteenth century, however, demonstrated that red blood cells readily lost HCO_3^- while gaining Cl^-. Much progress was made toward characterizing that essential step, Cl^-/HCO_3^- exchange, during this period. In 1971 Philip Knauf and Aser Rothstein showed that disulfonic stilbenes specifically inhibited this exchange, and 3 years later Ioav Cabantchik and Rothstein reported that these chemicals bound to Band 3 of Fairbanks et al.[36] (Not only is the exchanger present in huge quantities—about a million molecules/cell, representing a fourth of the membrane protein—it is also extraordinarily fast, each exchanger moving 10^5 ions/sec.) By 1980 Cabantchik and associates reconstituted purified Band 3 in phospholipids to demonstrate anion exchange.[37] And, as noted above, Kopito and Lodish described the amino acid sequence from cDNA experiments in 1985.[38]

PROTEIN SYNTHESIS

Notable accomplishments in this area ranged from fusing details of protein synthesis into comprehensive models of ribosomal function to deciphering essential control mechanisms. This involved identifying repressors, promoters, and transcription factors that regulate the transcription of particular DNA sequences into RNA, and describing the subsequent processing of these transcripts into the mRNA that directs ribosomal protein synthesis.

For membrane proteins, further mechanisms must be employed for inserting the integral proteins into lipid bilayers.[39] By 1985 this process was frequently depicted in terms of "cotranslational insertion": proteins were synthesized by ribosomes bound to the endoplasmic reticulum, with the growing polypeptide chain threaded into the membrane sequentially or after first folding into hydrophobic domains that merged as a unit. (Attachment of ribosomes to the endoplasmic reticulum was directed by a "signal sequence" of the protein being synthesized, which interacted with receptors on the endoplasmic reticulum. This signal sequence was subsequently cleaved from the integral membrane protein.) Recognition that many membrane proteins were glycosylated was accompanied by descriptions of how sugars were attached initially when the proteins were in the endoplasmic reticulum, followed by extensive modification in the Golgi apparatus. But precise mechanisms for "targeting" proteins to the proper membranes (e.g., targeting the Na^+/K^+-ATPase to kidney tubule serosal membranes and not to luminal membranes) remained undiscovered.

Much mention has been made of sequencing by cDNA methods. Just beginning to be applied to these transporters were powerful new techniques for removing portions of a protein by altering the DNA specifying it (making deletions); for joining two proteins—or parts thereof—by splicing the relevant DNA (forming "chimeras"); and for altering individual amino acids within a protein by deleting the DNA sequences encoding them or substituting the DNA sequences encoding other amino acids (performing site-directed mutagenesis).

NOTES TO CHAPTER 15

1. See Haydon and Taylor (1963).
2. Perutz et al. (1965).
3. Lenard and Singer (1966); Wallach and Zahler (1966).
4. Wallach and Zahler (1966), p. 1558.
5. Branton (1966).
6. Pinto da Silva and Branton (1970). Such results also argued for lipid bilayer models and against protein-lipid subunits.
7. Singer (1971).
8. Bretscher (1971); Steck et al. (1971).
9. Frye and Edidin (1970). Such motility was soon confirmed in various cells.
10. Singer and Nicolson (1972).
11. See Jacobson et al. (1995).
12. Huang (1969).
13. See Veatch and Stryer (1977).
14. Ohnishi and Urry (1970).
15. Fairbanks et al. (1971). Proteins were stained using Coomassie blue and periodic acid-Schiff (PAS) reagent, respectively.
16. Tomita and Marchesi (1975). Glycophorin corresponded to the PAS-1 band of Fairbanks et al.
17. Kopito and Lodish (1985).
18. Marchesi and Steers (1968).
19. Bennett and Stenbuck (1979).
20. Matthews et al. (1967); Blow et al. (1969), p. 337. Although details of the "charge-relay" system for the serine proteases were debated, the proposal was an acknowledged triumph.
21. Mitchell (1963).
22. West (1970); Kaczorowski and Kaback (1979).

23. Fox and Kennedy (1965).

24. Büchel et al. (1980).

25. Newman et al. (1981).

26. Background information is from Stoeckenius (1994).

27. Oesterhelt and Stoeckenius (1971).

28. Oesterhelt and Stoeckenius (1973).

29. Racker and Stoeckenius (1974). The paper was written in 1973.

30. Henderson and Unwin (1975).

31. Henderson et al. (1990).

32. Ovchinnikov et al. (1979); Khorana et al. (1979).

33. Lanyi (1995).

34. Kasahara and Hinkle (1977).

35. Mueckler et al. (1985). They defined transporters of liver tumor cells, but later studies showed that red blood cells contained the same transporter.

36. Knauf and Rothstein (1971); Cabantchik and Rothstein (1974).

37. Cabantchik et al. (1980).

38. That sequence was for mouse Band 3; human Band 3 was not sequenced until 1989.

39. For an account of what was known of these topics by the end of this period see Darnell et al. (1986). For membrane proteins see also Wickner and Lodish (1985).

chapter 16

OXIDATIVE
PHOSPHORYLATION:
CHEMICAL-COUPLING
HYPOTHESIS

B Y 1953 mitochondria were known to catalyze oxidative phosphorylation: ATP synthesis that is coupled to a series of redox reactions over the respiratory chain (Chapters 2 and 7). With NADH—formed through oxidizing various intermediary metabolites—a yield of three ATP molecules was estimated for each oxygen atom reduced to water (P/O = 3). The respiratory chain, then known to include a sequence of cytochromes (b $\Rightarrow$ c_1 $\Rightarrow$ c $\Rightarrow$ a $\Rightarrow$ a_3), was coupled to ATP synthesis in such a way that redox reactions slowed drastically when ADP or P_i was absent. This tight "respiratory control" was lost in aged or damaged mitochondria. Respiration was also uncoupled from phosphorylation by diverse substances, including dinitrophenol. In addition, such uncouplers promoted ATP hydrolysis by mitochondria. The requisite enzymes and respiratory chain were associated with mitochondrial membranes; recent electron micrographs had revealed multiple infoldings of the inner mitochondrial membranes to form cristae, producing an extensive surface.

This and subsequent chapters describe efforts to develop these observations into an understanding of oxidative phosphorylation, delineating the mechanisms coupling oxidation to ATP synthesis. Initially, such studies represented a distinct scientific field; the next chapter recounts its unanticipated fusion with the field of membrane transport.[1]

SLATER'S FORMULATION

E. C. Slater (Fig. 16.1), from Keilin's group in Cambridge, published a highly influential review in 1953, in which he proposed a coupling mechanism that di-

Fig. 16.1 E. C. Slater. (Photograph courtesy of E. C. Slater.)

rected expectations for well over a decade.[2] This chemical-coupling hypothesis (Fig. 16.2A) illustrated how redox reactions between two successive members of the respiratory chain, A and B, trapped energy initially in a chemical complex $A{\sim}C$, which then donated its energy to the ultimate formation of ATP. C could be a group on A or an independent element; if it were the latter, C might be the same substance at each of the three coupling sites. Slater suggested that $A{\sim}C$ could also drive other reactions and serve as a general energy store within mitochondria. Independently, Lehninger proposed a similar scheme, which included further details of ATP synthesis (Fig. 16.2B).[3]

Slater's scheme closely paralleled the recently published mechanism for substrate-level phosphorylation by the glycolytic enzyme glyceraldehyde-3-phosphate dehydrogenase (Fig. 16.2C) of Efraim Racker (Fig. 16.3),[4] although Slater admitted that "there is no reason why similar reactions necessarily operate in respiratory chain phosphorylation."[5] It also satisfied the requirement that respiration could occur in the absence of ADP and P_i: hydrolysis of $A{\sim}C$ would allow oxidation to proceed unhindered by demands of phosphorylation. Uncouplers could act by stimulating this hydrolysis (Fig. 16.2D), which normally would be quite slow to ensure efficient formation of ATP. $A{\sim}C$ hydrolysis would also reverse the final steps, causing ATP hydrolysis; consequently, uncouplers would stimulate mitochondrial

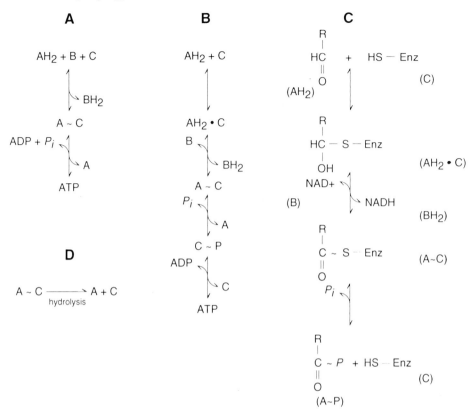

Fig. 16.2 Schemes for chemical intermediates. A/AH_2 and B/BH_2 are successive components of the respiratory chain in oxidized/reduced form. Scheme *A* is from Slater (1953), *B* from Lehninger (1954), and *C* from Racker and Krimsky (1952). In scheme *C*, HS-Enz represents the enzyme with its participating sulfhydryl group and R-CHO represents glyceraldehyde-3-phosphate (the forms corresponding to those in the preceding schemes are indicated in parentheses). Scheme *D* shows Slater's proposal for uncoupling phosphorylation from oxidation through hydrolysis of A~C [*A* and *D* are redrawn from Slater (1953). Reprinted by permission from *Nature*, ©1953, Macmillan Magazines Ltd. *B* is redrawn from Lehninger (1954). Reprinted by permission of John Wiley & Sons ©1954. *C* is redrawn from Racker and Krimsky (1952). Reprinted by permission of the American Society for Biochemistry and Molecular Biology.]

ATPase activity. The scheme thus followed an established precedent to accommodate known processes.

ATTEMPTS TO IDENTIFY INTERMEDIATES

Slater specified that identifying the proposed chemical intermediates was an "urgent task," and a host of investigators agreed.[6] Two efforts may represent these pursuits.

Fig. 16.3 Efraim Racker. (Photograph courtesy of V. P. Skulachev.)

Identifying A~C

In a series of papers published between 1960 and 1963, Gifford Pinchot, an accomplished biochemist in Baltimore, described the dissociation of a high-energy intermediate after incubating "sonicated" particles with NADH; the supernatant fraction from such incubations could then form ATP from ADP + P_i added during a subsequent incubation.[7] (The particles were obtained through ultrasonic disruption of the bacterium *Alcaligenes faecalis*. This book has ignored experiments with bacteria. Here I must acknowledge that studies on bacteria have been crucial to many aspects of physiology and biochemistry, and that many bacteria are capable of oxidative phosphorylation, with the respiratory chain and necessary enzymes embedded in the bacterial plasma membrane.)

Protein was released during the initial dissociating incubation with NADH, which Pinchot termed a "coupling enzyme." Thus, Slater's AH_2 (Fig. 16.2A) was NADH, C was the dissociable coupling enzyme, and A~C was NAD^+~enzyme that accepted P_i to become enzyme~P. Hydrolysis of enzyme~P was equivalent to uncoupling.

Experimental support included the following: (1) after undergoing dissociating incubations with labeled NADH, protein and labeled NAD^+ migrated together during electrophoretic separation; (2) net synthesis of ATP occurred with added ADP + P_i; (3) P_i was required (which argued against an adenylate kinase reaction simply synthesizing ATP from ADP alone); (4) added ATP was less effective than ADP (which argued against a mere P_i/ATP exchange reaction); and (5) the moles of NAD$^+$ released during the second incubation were roughly equivalent to the moles of ATP formed. But adding dinitrophenol during the subsequent incubation with ADP plus P_i did not inhibit ATP formation, as was expected from Slater's scheme.

Then in 1965 Gerda Pandit-Hovenkamp, who was working in Amsterdam where Slater was now professor of biochemistry, challenged Pinchot's findings and interpretations.[8] She reported, first, that material promoting P_i incorporation into nucleotides was released from the sonicated particles during anaerobic incubations at basic pH to the same extent without NADH as with it (Pinchot had reported that during incubation with NADH the incubation medium became basic, and oxygen was consumed). This observation discounted the involvement of NADH. Second, she reported that incorporation of P_i into ADP *preceded* incorporation into ATP (Fig. 16.4A), a sequence incompatible with P_i adding to preexisting ADP to form ATP. These observations suggested to her an alternative mechanism: a redox-energy independent P_i/ADP exchange (labeling ADP), followed by conversion of ADP to ATP by adenylate kinase (labeling ATP). She proposed that the exchange was catalyzed by polynucleotide phosphorylase, which catalyzes a reversible addition of nucleoside monophosphates (NMP) to polynucleotides (NMP$_n$), reversibly consuming nucleoside diphosphate (NDP) and releasing P_i:

$$ADP + NMP_n \Leftrightarrow NMP_{n+1} + P_i$$

Pandit-Hovenkamp, however, did not demonstrate polynucleoside phosphorylase activity in the fraction. Pinchot had acknowledged that adenylate kinase was present.

The next year Pinchot countered that argument.[9] He claimed that both Pandit-Hovenkamp's material released at basic pH without NADH and her extract catalyzing incorporation of P_i into ADP were functionally different from his,[10] and he reported his failures to demonstrate the presence of polynucleotide phosphorylase. But—significantly—he did not state whether his preparations labeled ADP more rapidly than ATP.

In a review article that year, Slater judged the evidence for Pinchot's proposal to be "inadequate."[11] Pinchot did not develop his arguments further, and interest waned.[12] Finally in 1972, Evangelos Moudriankis, who had worked in Pinchot's laboratory, confirmed Pandit-Hovenkamp's explanation: the total ^{32}P incorporated into nucleotides arose from ^{32}P-P_i/ADP exchange, which was due indeed to a demonstrated polynucleotide phosphorylase.[13]

Identifying C~P

An alternative and complementary approach was to search for phosphorylated intermediates by following the incorporation of radioactivity from incubations with la-

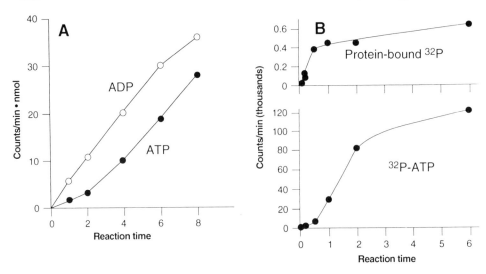

Fig. 16.4 Rates of labeling. In *A*, the supernatant fraction from *A. faecalis* particles, prepared by NADH treatment according to Pinchot's method, was incubated at 25° with ADP, ATP, and ^{32}P-phosphate. After the amount of time indicated (in minutes) the reaction was stopped by adding acid, and ADP and ATP were separated; the measured radioactivity in each is plotted as specific activity. [From Pandit-Hovenkamp, (1965), Fig. 1. Reprinted with kind permission of Elsevier Science-NL, Amsterdam, The Netherlands.) In *B*, rat liver mitochondria were first incubated at 0° with substrate for 10 minutes, and then ^{32}P-phosphate was added. After the additional minutes of incubation indicated, the reaction was stopped with acid and the endogenous ATP (lower panel) and protein-bound fractions (upper panel) were separated and their radioactivity measured. Note the different scales on the *Y*-axes. [Redrawn from Slater et al. (1964), Fig. 2. Reprinted by permission from *Nature*, © 1964, Macmillan Magazines Ltd.]

beled P_i. In the early 1960s, Boyer and associates used this straightforward approach to identify phosphorylated histidine in mitochondrial protein, proposing it as an intermediate of oxidative phosphorylation.

Boyer, then in Minneapolis, incubated mitochondria aerobically with substrate and ^{32}P-P_i in the absence of ADP (to prevent formation of ATP): mitochondrial protein was rapidly labeled.[14] Adding unlabeled P_i and ATP promptly decreased protein labeling, which indicated that dephosphorylation as well as phosphorylation was rapid. This would be expected of an intermediate in the rapid synthesis of ATP. The uncoupler dinitrophenol decreased labeling, although the concentrations required were far higher than those needed to block ATP formation.

Boyer next identified the phosphorylated group in the protein as a histidine, with a phosphoryl group bonded to a nitrogen of the histidine imidazole ring.[15] In addition, he reported that (1) ^{32}P-ATP labeled phosphohistidine prior to labeling the mitochondrial pool of P_i, and (2) ^{32}P-P_i labeled phosphohistidine prior to labeling the pool of mitochondrial ATP.[16] All this was consistent with the protein phospho-histidine being an intermediate between P_i and ATP:

$$P_i \Leftrightarrow \text{protein } P\text{-histidine} \Leftrightarrow \text{ATP}$$

Slater, while exploring the functions of potential intermediates, reexamined Boyer's proposal.[17] But when he measured labeling of mitochondrial protein with ^{32}P-P_i at $0°$ (so rates were slow enough to be determined accurately), he found that the rate of protein labeling was dramatically slower than the rate of ATP labeling (Fig. 16.2B). Slater concluded that this result "would seem to preclude phospho-histidine as an intermediate on the main pathway of oxidative phosphorylation."[18]

Boyer responded with measurements of his own, at $25°$ as well as $0°$: histidine labeling during incubations with ^{32}P-P_i rose more rapidly than did ATP labeling, and at the early time points it was greater when expressed in Boyer's graphs of the time course. But the ordinate in the graph was the *percentage* of ^{32}P-incorporation.[19] Slater replied that the proper plot was the *total* radioactivity incorporated at each time point: the total incorporation of ^{32}P into ATP was, even at the earliest time points, far greater than the total incorporation into phosphohistidine.[20]

Boyer continued to disagree with Slater's argument.[21] But by the time of Slater's reply the kinetics were irrelevant to identification. Boyer, now in Los Angeles, had in the meantime demonstrated that the phosphohistidine was a component of succinate thiokinase, an enzyme of the Krebs cycle involved in substrate-level phosphorylation.[22] So phosphohistidine was an intermediate in ATP synthesis, but in the synthesis of a relatively small fraction compared with that by oxidative phosphorylation. Still, Boyer had made a significant discovery, illuminating the reaction mechanism of a key enzyme of intermediary metabolism and demonstrating for the first time a phosphorylated imidazole in biological systems. But he had not identified an intermediate of oxidative phosphorylation.

OTHER DEVELOPMENTS: 1953–1965

Morphology

Early electron micrographs revealed two membranes surrounding mitochondria:[23] outer membranes separated mitochondria from cytoplasm; inner membranes enclosed the "matrix space" and were folded into cristae protruding into that space. Between these membranes lay the "intermembrane space" (Fig. 1.1).

Then in 1962, Humberto Fernández-Morán in Boston described globules 70–90 Å across lying atop the cristae.[24] These were most readily seen in negatively stained preparations where a short stalk was also revealed. Others soon confirmed the presence of such knobs on the matrix faces of inner mitochondrial membranes.[25] In 1964 Fernández-Morán, Green, and associates suggested that the globule, stalk, and "basepiece" (the associated region of the membrane) together contained the entire respiratory chain.[26] But that same year Britton Chance and Frederick Crane argued—from independent experiments in which fragmented membranes were separated into components—that the respiratory chain is contained within the membrane itself.[27] The next year, Racker, Chance, and associates demonstrated that the globules contained the coupling factor F_1 (see below). First, they prepared mitochondrial fragments capable of oxidative phosphorylation and ATP hydrolysis that contained membrane-attached knobs visible in electron micrographs. They then treated the fragments with trypsin and urea, which caused the globules to disappear and oxidative phosphorylation and ATPase activities to be lost, although oxidation over the entire respiratory chain was still demonstrable. Finally, they added purified F_1 to the stripped fragments, which restored both the knobs to the membranes and the ATPase activity.[28]

Inhibitors and Uncouplers

Inhibitors of specific steps in the respiratory chain (such as cyanide, antimycin A, and amobarbital) became important tools for establishing the sequence of redox reactions (see below). Uncouplers of oxidative phosphorylation were widely used for various purposes; a peculiar aspect, however, was that this common property was associated with quite diverse structures: uncouplers ranged from 6-carbon dinitrophenol to 15–amino acid gramicidin. A third class of reagents, inhibitors of oxidative phosphorylation, was specified by Lardy's description in 1958 of oligomycin.[29] It inhibited respiration in tightly coupled mitochondria, but the inhibition was relieved by adding uncouplers. Moreover, it inhibited mitochondrial ATPase activity. Lardy concluded that oligomycin "acts on an enzyme involved in phosphate fixation or . . . transfer."[30]

Respiratory Chain

During this period two new components of the respiratory chain were recognized by Green's associates: highly lipid-soluble ubiquinones and "nonheme-iron" proteins (containing iron not bound, as in cytochromes and hemoglobin, within heme groups).[31]

Placing the components in proper sequence was assisted by the distinctive optical spectra of NADH, flavoproteins, and individual cytochromes, and by the differences in spectra between oxidized and reduced states. This elucidation Keilin began and Chance continued elegantly in Philadelphia, combining innovative instrumentation and experimental design. Thus the steps at which various inhibitors acted were detectable by "crossover analysis":[32] upon adding an inhibitor during respiration, the redox states of the components upstream (beginning with NADH) from the site of inhibition became more reduced and those downstream (terminating in oxygen) more oxidized.

In like manner, the steps coupled to phosphorylation were detectable, in this case through depleting ADP in tightly coupled mitochondria. Upstream from the block respiratory chain components tied to ADP phosphorylation then became more reduced, while those downstream became more oxidized (compared with steady-state values when ADP was not limiting). In the absence of further inhibitors, the crossover point was between cytochromes c and a; in the presence also of low concentrations of azide, it lay between cytochromes b and c; and in the presence of additional azide, it lay between NADH and flavoprotein. These determinations agreed with the P/O ratios measured over the segments of the respiratory chain in experiments using specific inhibitors, electron donors, and electron acceptors to isolate those segments (e.g., the P/O ratio was nearly 1 when NADH was used as a substrate and oxidations beyond the flavoprotein were blocked by inhibitors).

Associated Reactions

ATP hydrolysis by mitochondria had been described earlier, as had its stimulation by dinitrophenol. As noted above, oligomycin inhibited ATP hydrolysis as well as ATP synthesis, in accord with one reaction being the reverse of the other.

Two exchange reactions described at this time also proved to be revealing subsequently. P_i/water exchange was measured as the loss of ^{18}O from labeled P_i, indicative of phosphorylation–dephosphorylation cycles.[33] P_i/ATP exchange was measured as the incorporation of ^{32}P from labeled P_i into ATP, indicative of ADP phosphorylation–ATP dephosphorylation cycles.[34] Both exchanges occurred anaerobically in the absence of substrate and were inhibited by dinitrophenol (but see Chapter 18).

Experiments at this time also showed that added ATP could drive the redox reactions of the respiratory chain in the reverse direction.[35] Moreover, Lars Ernster in Stockholm argued that "high-energy intermediates" could themselves, without first forming ATP, power mitochondrial processes,[36] thus endorsing Slater's proposal.

Fractionation

In the early 1960s, Youssef Hatefi and associates in Green's group successfully split the respiratory chain into four separable complexes that could be recombined to restore the complete redox system.[37] Complex I contained the NADH–ubiquinone reductase complex, including flavoprotein and nonheme-iron protein but no cytochromes. Complex III contained nonheme-iron protein plus cytochromes b and c_1. Complex IV contained the cytochrome oxidase complex, including cytochromes a and a_3. Complex II contained the succinate–ubiquinone reductase complex, including nonheme-iron protein; this complex represents an alternative entry to the respiratory chain beginning with succinate rather then NADH oxidation, bypassing the first coupling site for phosphorylation. (Cytochrome c, as Keilin demonstrated, is easily solubilized from mitochondria and is thus not represented in these complexes; to reconstitute the respiratory chain, cytochrome c must also be added.)

This fractionation grew from efforts to break the mitochondrial structure into fundamental components that could then be characterized individually. Mitochondrial fragments produced by ultrasonic irradiation, termed "electron transport particles" or "submitochondrial particles," were established at this time by Green's group[38] as a convenient, simplified system capable of coupling ATP synthesis to redox reactions over the respiratory chain.

Fractionating such preparations further, into components involved in coupling phosphorylation to oxidation, was an obvious next step. In 1960 Harvey Penefsky and Maynard Pullman in Racker's group in New York described a soluble protein extracted from disrupted submitochondrial particles that not only restored coupled phosphorylation (hence the name F_1, for the first coupling factor they isolated) but also had ATPase activity (hence the alternative name, F_1-ATPase).[39] Dinitrophenol stimulated ATPase activity, but surprisingly, oligomycin did not inhibit it. In 1963, however, Racker described a fraction that conferred oligomycin sensitivity to F_1, named F_o for its relationship to oligomycin.[40] The F_oF_1 complex thus had oligomycin-inhibitable ATPase activity.

Ion Transport and Swelling

Beginning in the early 1950s, several groups described accumulations of certain cations and anions—notably K^+, Ca^{2+}, and P_i—against apparent concentration gradients.[41] Accumulation required oxidizable substrates and was diminished by respiratory chain inhibitors and by dinitrophenol. Oligomycin did not inhibit such divalent cation accumulation, which supported proposals that high-energy intermediates could drive mitochondrial work without an intervening synthesis of ATP. In the absence of oxidizable substrates, however, added ATP could drive cation accumulation, and this process could then be inhibited by oligomycin.

Also published at this time were descriptions of mitochondrial swelling in isotonic media.[42] This swelling was promoted by particular agents, including Ca^{2+} and P_i, it required oxidizable substrates, and it was diminished by respiratory chain inhibitors and by dinitrophenol. On the other hand, adding ATP caused swollen mitochondria to shrink. Swelling was accompanied by gains in mitochondrial water content and shrinkage by losses. A popular model for ATP-induced shrinkage invoked mitochondrial contractile proteins to regulate organelle volume mechanically.[43]

COMMENTARY

Places and People

Again Cambridge was a wellspring. There Keilin established the cytochromes in a respiratory chain. There after World War II Slater came from Australia for doctoral research with Keilin. There during the war Chance came from the United States for his second Ph.D. as had Green before the war for his first. There after the war Mitchell completed his doctoral research (see Chapter 17). But this younger generation then moved on, Slater to Amsterdam, Chance and Green back to the United States, and Mitchell to Edinburgh.

Less expected, perhaps, was Madison, Wisconsin as a second source. There Lardy and Boyer entered in 1939 the graduate biochemistry program of the agriculture school, which was noted for its pioneering studies on nutrition by, among others, Carl Elvehjem (who had been a fellow in Cambridge).[44] There Lehninger too did his doctoral research in biochemistry, finishing a year sooner in 1942. Lardy alone stayed on, joining the Enzyme Institute of the University of Wisconsin, which had recruited Green as its first director in 1948. Green's development of sub-mitochondrial particles from bovine heart mitochondria as a standard preparation surely reflected also the presence in Madison of Oscar Mayer meatpackers.

During this period prominent groups also flourished in Stockholm (Ernster), Amsterdam (Slater), Philadelphia (Chance), Baltimore (Lehninger), Minneapolis (Boyer), and New York (Racker).

Equally diverse were the scientists. Racker was a refugee, arriving in the United States by way of Cardiff from Vienna, where he had received his M.D. in 1938 (after initially studying art). Slater was even more peripatetic, leaving Australia, where he had studied organic chemistry, to work for 9 years in Cambridge. His time there was interrupted by a brief stay with Ochoa in New York. He then migrated to Amsterdam (where he also served as the influential managing editor of *Biochimica et Biophysica Acta*) before retiring in England. Lardy published his first research, on vitamin D deficiency in cows, in the *Proceedings of the South Dakota Academy of Science*; his initial graduate school research was on preserving bull semen for artificial insemination. As his scientific career flourished, so too did his successes as a breeder of elegant Arabian horses. Chance, who was born to affluence, toiled assiduously and brilliantly, inventing instruments (such as dual-beam spectrophotometers), accumulating patents, perfecting techniques to solve essential problems, and winning an Olympic gold medal for yachting in 1952.

There was, however, no overlap between the major scientists whose work is cited in this chapter and those in preceding chapters who were engaged in traditional fields of membrane transport.

Methods

Enabling instruments newly applied in this period's research include oxygen electrodes, which provide continuous measurements of oxygen consumption (replacing cumbersome manometric devices); ion-specific electrodes, which provide continuous measurements of ion concentrations in the incubation media (such as K^+ electrodes for following mitochondrial fluxes); and dual-beam spectrophotometers, which monitor absorption at two wavelengths (one for following absorption peaks of reactant or product, the other for assessing confounding turbidity changes from mitochondrial swelling, shrinking, or aggregation). Advances in electron microscopy were rapidly followed by the identification of mitochondrial membranes and subsequently the knobs on inner membranes. Mitochondrial fragmentation progressed through the use of naturally occurring detergents (such as digitonin and the bile salts cholate and deoxycholate) and through physical methods using ultrasound and mechanical disruptors. Significantly, the resulting fragments were termed "particles"; their topography was unknown and apparently of little concern.

Artifacts

The revelations of microscopy may seem treacherous terrain, with potential artifacts lurking at every misstep. Certainly, biochemists were skeptical of Fernández-Morán's globules. Were they merely creations of negative staining? Reassurance came from independent approaches that detached, isolated, and restored the knobs, providing chemical and functional identities. But just how these real entities would appear without fixation and staining remained undetermined.

Microscopists are not, of course, alone in creating artifacts. A prominent proposal of this era was Green's "structural protein," which he claimed represented 35% of the mitochondrial membrane and provided stability and form without catalytic prowess.[45] (Identifying this protein was also a key impetus to Green's reformulation of general membrane structure in terms of protein-lipid subunits, at a time when the Danielli and Robertson models were losing favor.) Later studies,[46] however, revealed that "structural protein" was created by isolation procedures that denatured F_1.

Resolutions

Not all disputes are resolved so explicitly. In many cases, proposals lapse and controversies cease without such gratifying tidiness. Some scientists are loath to condemn their colleagues' views so bluntly, some recognize the marginal return in rebutting what few believe, some are too busy promulgating alternatives. Relying on published reports may then be misleading, for the research community also recognizes from unofficial comments as well as behavior patterns when claims are surrendered without an explicit *mea culpa*. Here Pinchot's counterargument was followed by long silence, efforts to demonstrate a mitochondrial contractile protein were abandoned, and once-extensive reports of ADP/ATP exchange activity[47] faded away—all without formal retractions by their advocates.

Scientific Summary

During this time further analyses revealed new components of the respiratory chain and established its reaction sequences; in addition, fractionation techniques showed that subunits of the membrane contain specific redox complexes that could be reconstituted into the overall chain. Other fractionation procedures allowed separation of an ATPase that, when reconstituted with stripped membranes, restored knob-like structures akin to those visible by electron microscopy of negatively stained mitochondria. Plausible arguments related this ATPase activity to a reversal of the steps synthesizing ATP. But how the redox reactions of the respiratory chain could drive ATP synthesis remained conjectural. In 1953 Slater proposed a chemical-coupling hypothesis, based on the precedent of substrate-level phosphorylation. Attempts to identify chemical intermediates, however, were unsuccessful.

NOTES TO CHAPTER 16

1. For histories see Ernster and Schatz (1981); Lehninger (1964); Nichols (1963); Slater (1966, 1981).

2. Slater (1953). An earlier proposal (Lipmann, 1946) depicted phosphate interacting before oxidation of AH_2 and would not allow respiration in the absence of phosphate.

3. Lehninger (1954). Chance and Williams (1956) subsequently proposed a more complex scheme that included an additional (unidentified) intermediate.

4. Racker and Krimsky (1952). "Substrate-level" phosphorylation refers to phosphorylation occurring with substrate oxidation (in this case oxidation of a glycolytic intermediate), rather than with oxidations along the respiratory chain.

5. Slater (1953), p. 978.

6. Ibid. Chance et al. (1967) tabulated 16 proposed intermediates; in light of the numbers of scientists then at work on the problem, that was a multitude.

7. Pinchot (1960, 1963); Pinchot and Hormanski (1962).

8. Pandit-Hovenkamp (1965). Earlier she had worked as a fellow in Pinchot's laboratory.

9. Pinchot and Salmon (1966).

10. According to Pinchot and Salmon, the difference lay in the fraction of "P_i-dischargeable ^{32}P incorporation," which was operationally defined as the loss of ^{32}P incorporation occurring after an intervening incubation with P_i, and interpreted as P_i interacting during the intervening incubation with the high-energy intermediate to dissipate its energy store.

11. Slater (1966), p. 384.

12. Omitting citations by Pinchot, there were 20 citations to Pinchot's first three papers on this topic in 1965, but the number declined to 4, 3, 2, 3, and 2 citations in subsequent years.

13. Adolfsen and Moudrianakis (1972).

14. Suelter et al. (1961).

15. Boyer et al. (1962).

16. Peter et al. (1963).

17. Slater et al. (1964).

18. Ibid., p. 783.

19. Bieber et al. (1964).

20. Slater and Kemp (1964).

21. Lindberg et al. (1965). Slater argued that a precursor must initially be present in greater quantity than its product. Boyer's argument concerned the labeling of precursor and product pools: with "an exchange sufficiently rapid so that the specific activity of the phosphohistidine is close to . . . that of the P_i . . . the participation of phosphohistidine would only demand that its rate of approach to maximal labelling equal or exceed that of ATP" (Bieber et al., 1964, p. 1317).

22. R. A. Mitchell et al. (1964).

23. Sjöstrand (1953); Palade (1953).

24. Fernández-Morán (1962).

25. See, for example, Parsons (1963); Stasny and Crane (1964).

26. Fernández-Morán et al. (1964).

27. Chance et al. (1964); Stasny and Crane (1964).

28. Racker et al. (1965). A neater argument would include a demonstration that F_1 restored phosphorylation also; however, such restoration, as Racker et al. noted, required additional coupling factors also removed during the stripping.

29. Lardy et al. (1958).

30. Ibid., pp. 588–589.

31. Crane et al. (1957); Beinert and Sands (1960).

32. Chance and Williams (1956).

33. Cohn (1953); Boyer et al. (1956). ^{18}O is a nonradioactive heavy isotope of the common ^{16}O, separable by mass spectroscopy. Water participates in the process of dephosphorylation, and ^{18}O in phosphate can then be transferred to water, thus disappearing from phosphate.

34. Boyer et al. (1956).

35. Klingenberg and Schollmeyer (1960); Chance (1961).

36. Lee et al. (1964).

37. Hatefi et al. (1962).

38. Blair et al. (1963).

39. Pullman et al. (1960); Penefsky et al. (1960).

40. Racker (1963). The F_o fraction was prepared from mitochondrial membranes first digested with the proteolytic enzyme trypsin and then extracted with urea; the resultant particulate fraction had no ATPase activity itself.

41. See, for example, Bartley and Davies (1954); Brierley et al. (1963); Slater and Cleland (1953); Spector (1953); Vasington and Murphy (1962).

42. Chappell and Greville (1958); Hunter and Ford (1955); Lehninger (1964); Raaflaub (1953).

43. See Lehninger (1964). At this time contractile mechanisms for electrolyte extrusion from cells were proposed by Kleinzeller and Knotková (1964).

44. Elvehjem became chairman in 1944; in 1958 he became president of the university.

45. Green and Perdue (1966).

46. Schatz and Saltzgaber (1971).

47. See Lehninger (1964).

chapter 17

OXIDATIVE PHOSPHORYLATION: CHEMIOSMOTIC COUPLING HYPOTHESIS

F AILURE to identify chemical intermediates was aggravating but not decisive. Absence of proof is not proof of absence, and in 1960 Lehninger reassured his colleagues that "it has not seemed necessary to postulate mechanisms other than those involving the classical common intermediate to explain available data."[1] Despite that counsel, Peter Mitchell (Fig. 17.1) the next year proffered his own tally of current shortcomings and invoked a radically different mechanism.[2]

MITCHELL'S 1961 FORMULATION

Mitchell's initial proposal for chemiosmotic coupling depicted three types of components inserted in—and oriented crucially across—an impermeable, insulating membrane.[3]

ATPases reversible to synthesize ATP (Fig. 17.2A). The enzyme active center was inaccessible to water but accessible to H^+ from one surface and to OH^- from the other—a vital orientation. For ATP synthesis from ADP + P_i (which requires removing water), H^+ exited to one surface, where the H^+ concentration was low, and OH^- to the other, where its concentration was low. The gradient for these ions thus drove ATP synthesis, although there was no *transmembrane* flow of either H^+ or OH^- through the ATPase.

The H^+/OH^- electrochemical gradient Mitchell expressed as the composite of chemical ("ΔpH") and electrical ("$\Delta\Psi$") potentials; for that sum as an electrical potential, $E = Z\Delta pH + \Delta\Psi$ (Z includes the constants of the Nernst equation relating concentration gradients to electrical potentials).[4] Mitchell calculated that consuming

Fig. 17.1 Peter Mitchell. (Photograph © Godfrey Argent Studio, London.)

one H^+ to synthesize one ATP required $\Delta pH = 7$ *or* $\Delta\Psi = 420$ mV, or intermediate values of each (illustrated as $\Delta pH = 2$ *and* $\Delta\Psi = 300$ mV).

Redox pumps generating H^+/OH^- gradients (Fig. 17.2B). The respiratory chain was oriented so hydrogens crossed the membrane in one direction, exiting as H^+ at one surface, while the extracted electrons crossed in the opposite direction, ultimately uniting with oxygen and H^+ to form OH^-. Starting with oxidizable substrate SH_2, three H^+ were released. These consumed three OH^- extracted from ADP and P_i when three ATP were synthesized, with one oxygen reduced (P/O = 3).

Ion carriers in the membrane. These carriers exchanged cation for cation and anion for anion, as H^+ for K^+, and OH^- for Cl^-. Thus other ions could be distributed across the membrane asymmetrically, driven by the H^+/OH^- gradient: ΔpH could be transformed into the potential of these ionic asymmetries.

In addition, Mitchell proposed that compounds like dinitrophenol uncoupled oxidation from phosphorylation by dissipating membrane gradients, acting as lipid-soluble carriers of H^+ or OH^-.

(Mitchell explained his term "chemiosmotic coupling" as the linking of chemical reactions like ATP synthesis to "the spatially directed channeling of the diffusion of a chemical compound . . . along a pathway specified . . . by the physical organization of the system."[5] Earlier, he distinguished between chemical work, the

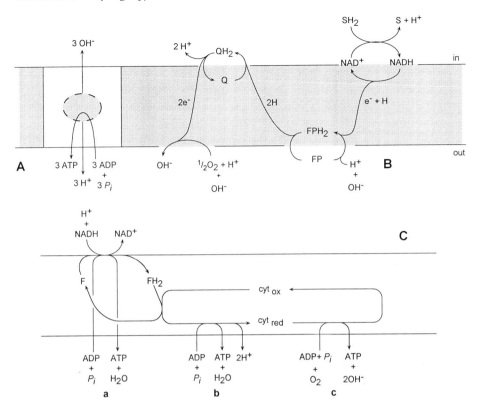

Fig. 17.2 Chemiosmotic hypothesis. *A*: The reversible ATPase is oriented so that H^+ and OH^- enter or leave from opposite membrane surfaces (for clarity, arrows are drawn to show only ATP synthesis). *B*: The respiratory chain also is oriented so that H^+ and OH^- are released at opposite surfaces, neutralizing the OH^- and H^+ produced in ATP synthesis. The redox chain begins with substrate (SH_2) oxidation by NAD^+ and concludes with oxygen reduction to OH^-. The designated in and out sides were acknowledged to be uncertain. [Redrawn from Mitchell (1961b), Figs. 1, 3. Reprinted by permission from *Nature*, © 1961 Macmillan Magazines Ltd.) *C*: Davies's scheme also featured a redox chain from NADH to oxygen with three sites of ATP synthesis (**a**, **b**, **c**), but only system **b** is associated with H^+ release. [Drawn from Davies and Krebs (1952), equations 3a, 4a, 5a. Reprinted by permission of the Biochemical Society.]

making and breaking of covalent bonds, and osmotic work, the creation of electrochemical gradients.[6] He eschewed the term "chemiosmosis"; his etymology was from ωσμος = push.)

Originality

Although analyzing creative processes and crediting priorities are not central objectives here, it seems worth recording some pathways to Mitchell's formulation, not least because accounts of this are contested.[7]

First is the case of reversible ATPases. Mitchell's studies on bacterial transport, which stemmed from his doctoral research on penicillin's mechanism of action, focused on phosphate and organic solutes. He interpreted such transport as "group-translocation": "metabolic energy is . . . converted to osmotic work by the *formation and opening of covalent links* between translocators in the membrane and the carried molecules *exactly as in enzyme-catalyzed group-transfer reactions.*"[8] As an example of group-transfer, Mitchell depicted glutamine synthesis, with phosphate transfer from ATP to enzyme and with glutamate transfer to enzyme before release as glutamine:

$$E + ATP \Rightarrow E\text{-}P + ADP$$

$$E\text{-}P + glutamate \Rightarrow E\text{-}glutamate + P_i$$

$$E\text{-}glutamate + ammonia \Rightarrow E + glutamine$$

Mitchell proposed that active transport of phosphate and glutamate-glutamine could occur if "enzyme molecules [were] oriented in a membrane . . . such . . . that certain components react with the active centre from one side . . . while . . . other components react . . . from the other."[9] Thus, "translocation" (active transport effected by "translocators") specifically involved making and breaking covalent bonds during "vectorial metabolism." Reactants and products *diffused* to and from enzyme active centers along pathways defined by the protein structure and its crucial orientation across the membrane. Mitchell contrasted this with mechanisms *forcing* solutes through membranes (cf. Fig. 8.10B).

In the preceding decade, Mitchell had claimed that his proposals also applied to oxidative phosphorylation.[10] But R. J. P. Williams in Oxford claimed priority for applying this mechanism to reversible ATPases.[11] Before Mitchell's 1961 proposal was submitted, Williams had concluded that the functional significance of multistep redox chains lay in separating in space the generation of H^+ and OH^-.[12] Such "dislocation" accomplished ATP synthesis "by the production of H^+ in one locality and its degradation in a second . . . by the reaction of OH^-."[13] But Williams was not referring to separation of H^+ and OH^- across membranes: "Our answer . . . removes the necessity of postulating semi-permeable membranes [for a] catalytic chain . . . could have all the functional significance normally ascribed to a semi-permeable membrane."[14] Williams's subsequent paper described generating H^+ and removing water in lipid phases, but this process was intramembrane and explicitly employed chemical intermediates like those Lehninger advocated.[15]

Second are proposals for redox pumps. These antedated Mitchell's formulation (enthusiasm was waning by this time, however; see Chapters 6 and 8), and Mitchell repeatedly cited Lundegårdh, Robertson, Conway, and Davies, noting that their models of redox systems were oriented in such a way that H^+ was produced at one surface and OH^- at the other.[16]

On the other hand, Davies recognized that gastric acid secretion required that more than four H^+ be formed for each O_2 consumed, so in 1950 he proposed a supplemental system driven by hydrolysis of energy-rich phosphate compounds.[17] Two years later he extended these concerns to redox chains associated with ATP synthesis (Fig. 17.2C).[18] Although the topography was not specified, Davies stated that ATP was formed without ion transport, the H^+ and OH^- generated neutralizing each other. He raised the possibility, however, of ATP produced by segments **a** or **c** being used to drive segment **b** backward, thereby transporting H^+ at the expense of ATP hydrolysis. Davies's formulation is rather murky, and as Fig. 17.2 shows, it is far from Mitchell's. But he did specify that energy stored in H^+ gradients could drive H^+-transporting ATPases backward to generate ATP. Davies, however, did not develop his idea, even though he claimed that it "leads directly to experiments."[19]

Another vision of energy storage in transmembrane gradients and its utilization was Crane's formulation coupling sugar transport to Na^+ flow down its gradient (see Chapter 14), which he presented at the 1960 Prague conference that Mitchell also attended. Despite Crane's recollection of Mitchell's enthusiasm for this mechanism, Mitchell did not cite Crane in his 1961 proposals, perhaps because Mitchell's model did not have H^+ flows through the ATPase and Crane's mechanism did not involve making and breaking covalent bonds. In his 1963 review, however, Mitchell did cite Crane's mechanism in the context of secondary transport.[20]

Third is the concept of ion accumulation by mitochondria. This topic was then attracting considerable interest (see Chapter 16), although exchanges for H^+ or OH^- were not a prime focus.[21]

Fourth is the question of mitochondrial membranes being impermeable to ions. This was indeed a new feature, as was Mitchell's proposal that uncouplers acted as lipophilic carriers to dissipate transmembrane ion gradients. (Williams depicted dinitrophenol interacting directly with chemical intermediates.[22]) Mitchell's representation of transmembrane enzymes, such as the ATPase, was also unorthodox: the reigning Danielli and Robertson membrane models featured proteins lying atop the lipid bilayer (see Chapters 2 and 10).[23] While the respiratory chain and related enzymes were then known to be associated with membranes, contemporaneous concepts were satisfied by enzymes on surfaces. Lehninger considered membranes a "necessity for structural organization . . . to minimize the path distance . . . and to maximize probability of interaction."[24]

Plausibility

Even if parts of Mitchell's formulation were clearly derivative, the whole was distinctly new. It also seemed in many specifics wildly implausible. Nerve and muscle, which were accepted paragons for exploiting membrane potentials, managed only 70–90 mV—not the several hundred that Mitchell calculated. Although stomach cells established a ΔpH of 5–6, such large gradients for ordinary mitochondria seemed similarly exorbitant. And at a time when plasma membranes were being recognized as readily permeable to cations, formulations of mitochondrial impermeability to H^+ seemed reactionary, especially in light of the known permeability to water (demonstrated by osmotic swelling). Moreover, requirements for a closed vesicular system clashed with studies of coupled phosphorylation using fragments described as "particles."[25] Williams also challenged the chemical plausibility of Mitchell's "field-driven" extraction of H^+ and OH^- from ADP + P_i.[26]

Development

Imagining novel mechanisms is no substitute for devising experiments that discriminate among alternatives. A virtue of Mitchell's proposal was that several features were approachable. Whether the proposal—in light of its perceived plausibility—merited such tests is another matter, for there seems little profit in refuting formulations the community judges unlikely. The challenge to test and explore then falls to its creator. Mitchell accepted that challenge (whereas Lundegård, Davies, and Williams did not develop their own conjectures). Thus Mitchell in 1961 described how adding uncouplers like dinitrophenol increased H^+ fluxes into mitochondria and bacteria "at rates equivalent to the rates of passage of . . . electrons through the [respiratory chain] during uncoupled . . . oxidation . . . [which is] consistent with the chemiosmotic coupling hypothesis."[27] (Mitchell termed this a "surprisingly elementary discovery."[28]) The next year Guy Grenville in Babraham announced his confirmation that uncouplers increased mitochondrial H^+ fluxes, and 5 years later Lehninger and associates described how dinitrophenol increased conductivity across artificial lipid bilayers, a result "qualitatively and quantitatively consistent with Mitchell's hypothesis."[29] And in 1962 Mitchell also reported H^+ outflux from mitochondria and bacteria during aerobic metabolism,[30] a significant confirmation (the direction of the flux was opposite that of Mitchell's initial guess; this was not a vital aspect, however). But at that point Mitchell's investigations were interrupted by illness, departure from Edinburgh, and the subsequent establishment of an independent research institute in Glynn House, near Bodmin, Cornwall.

In the meantime, Chappell interpreted gramicidin's ability to promote cation entry and to uncouple oxidative phosphorylation in terms of mitochondrial membranes normally impermeable to ions and in terms of H^+ ejection driven by redox reactions of the respiratory chain.[31] There were basic tenents of the chemiosmotic hypothesis.

In a 1962 review Lehninger commented that Mitchell's hypothesis had "a number of interesting features," citing proposals for how uncouplers act.[32] Slater's 1966 review described the hypothesis in detail, calling it "beautifully simple" and admitting that it "explains all the experimental findings . . . we have discussed."[33] And a prominent textbook also published in 1966 stated that "Mitchell's ideas have been most productive in focusing attention on some fundamental properties of membranes."[34] Nevertheless, these authors and most other investigators studying oxidative phosphorylation continued to favor chemical coupling hypotheses. Lehninger later admitted that the proposal "was not well received at first."[35]

MITCHELL'S 1966 REFORMULATION

Once he was resettled in Cornwall and joined by his collaborator from Edinburgh, Jennifer Moyle, Mitchell resumed experiments on oxidative phosphorylation. In 1965 Mitchell and Moyle described H^+ fluxes from mitochondria driven by the

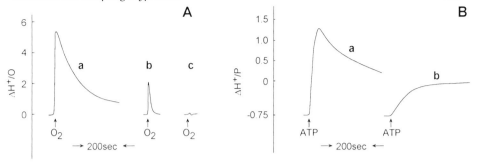

Fig. 17.3 Mitochondrial H^+ release. *A*: At the arrows, measured pulses of oxygen were added to anaerobic mitochondria in the presence of substrate, and H^+ release was measured with a pH electrode. Trace **a** shows the rise in H^+ release (expressed per oxygen atom added) followed by a fall as the oxygen was consumed and the gradient was dissipated. Extrapolation of the declining phase to the moment of addition, before the gradient was consumed by use, gives at least six H^+/O. Trace **b** shows a similar experiment but in the presence of dinitrophenol, making membranes leaky to H^+. Trace **c** shows mitochondria disrupted with a detergent. *B*: At the arrows, measured pulses of ATP were added to anaerobic mitochondria and H^+ release was measured identically (trace **a**, expressed per ATP molecule added). In trace **b** dinitrophenol and oligomycin were present, quantitating extraneous H^+ release (which sets the zero value for the vertical scale). Extrapolation in **a** to the moment of ATP addition gives at least two H^+/ATP. [From Mitchell and Moyle (1965), Figs. 1, 4. Reprinted by permission from *Nature*, © 1965 Macmillan Magazines Ltd.]

respiratory chain and by ATP hydrolysis (Fig. 17.3).[36] As acknowledged previously, fluxes were in the opposite direction from that in Mitchell's initial scheme, a difference that was unimportant mechanistically. Quite significant, however, was the different stoichiometry. This they now measured as (nearly) six H^+ from NADH to oxygen and (nearly) two H^+ from each ATP hydrolyzed. Consequently, the necessary gradient was cut in half, to $\Delta pH = 3.5$ *or* $\Delta\Psi = 210$ mV, or to intermediate values of each (such as $\Delta pH = 1$ *and* $\Delta\Psi = 150$ mV). Although relieved that the revised stoichiometry brought "potential differentials . . . within physiological limits,"[37] Mitchell was now forced to rewrite the underlying biochemistry.

Mitchell's reformulation, which was published as a book in 1966,[38] ranged over diverse issues, but a central concern was how the stoichiometry that was once so obviously one must necessarily now be two. For the reversible ATPase (Fig. 17.4*A*), this meant extracting not OH^- but O^{2-}, which was chemically less appealing. Mitchell's scheme pictured F_1 knobs atop membrane-embedded F_o. Two H^+ entering this structure helped form the anhydride X-I from two acids; next, the anhydride—a *chemical intermediate of the phosphorylation reaction*—was translocated, becoming energy-rich X~I where the H^+ concentration was low. X~I then drove ATP synthesis, with two H^+ released.

For the respiratory chain (Fig. 17.4*b*), two H^+ were released at each of three loops. This was straightforward, but the system of alternating H^+ release and e^-

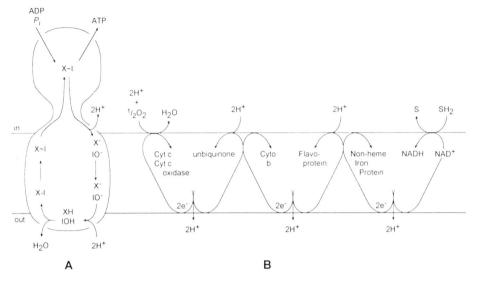

Fig. 17.4 Revised chemiosmotic scheme. *A*: The reversible ATPase is driven toward ATP synthesis by H^+ entering from the exterior and reacting with two hypothetical acids, X^- and IO^-, to form their anhydride, X-I. X-I is then translocated to a phase where it becomes "energy-rich" X~I, and this chemical intermediate then drives ATP synthesis. *B*: The respiratory chain is represented as three loops, composed of alternating hydrogen-transporting limbs (NADH, flavoprotein, ubiquinone) and e^- transporting limbs (non-heme iron protein, cytochrome b, cytochrome c–cytochrome oxidase), with two H^+ released at each loop. [Redrawn for Mitchell (1966), Figs. 7, 8, 11. Reprinted by permission of the Glynn Research Foundation Ltd.]

transfer seemed to some readers at odds with previous identifications of specific coupling sites and also with Hatefi's complexes (see Chapter 16). In particular, no H^+ were released by Complex IV (encompassing cytochrome c/cytochrome oxidase), despite prior evidence for a coupling site within that span. But Mitchell's scheme did depict the consumption of two H^+ to form water—and thus the loss of two positive charges from the matrix space—at that locus.

Overall, more plausible magnitudes for electrochemical gradients cost more problematic mechanisms. To some, Mitchell's facility in devising schemes seemed a tribute to his ingenuity rather than a revelation of biological reality.

DEVELOPMENT

Throughout the following decade, Mitchell's reformulation served often as a challenge rather than a guide. And while basic principles emerged triumphant, mechanistic details underwent drastic revisions. Chapter 18 will consider how models for ATP synthesis evolved. Here the course of some other issues will be noted.

Fig. 17.5 André T. Jagendorf. (Photograph courtesy of André Jagendorf.)

ATP Synthesis Driven by H^+ Gradients

Most significant for acceptance by Mitchell's peers was the early demonstration, clearcut and unequivocal, that H^+ gradients could indeed drive ATP synthesis. The key experiment, by André Jagendorf (Fig. 17.5) in Baltimore, involved photosynthetic phosphorylation. Photosynthetic processes include, as demonstrated in the 1950s, light-driven chains of redox reactions linked to ATP synthesis (then thought to be coupled by unidentified chemical intermediates). It is noteworthy that Mitchell explicitly embraced photosynthetic phosphorylation within his chemiosmotic hypothesis from its beginnings. (In green plants, the primary photosynthetic reactions occur in chlorophyll-containing chloroplasts. Like mitochondria, these organelles are bounded by double membranes, but the redox chain and ATP-synthesizing enzymes are in a third and innermost membrane system, the thylakoid membranes of the chloroplast grana.)

In 1963 Jagendorf described experiments in which chloroplasts illuminated in the absence of ADP and P_i formed energy-rich intermediates that could drive ATP synthesis when ADP and P_i were then added in the dark.[39] In 1964 Jagendorf, citing Mitchell, reported that illuminating isolated chloroplasts caused rapid losses of H^+ from the medium, losses that were blocked by uncouplers.[40] In 1965 he noted that intermediates were apparently formed in the dark when chloroplasts first incubated at pH 4.6 were then immersed in buffers at pH 8.6: Jagendorf concluded that the intermediate could be a "*trans*-membrane pH gradient."[41] So in 1966 he published the crucial "acid-bath" experiments.[42] He first equilibrated chloroplasts at 0° for 1 to 2 hours at pH 3.8 with inhibitors of the redox chain (blocking ATP generation by light); next he added ADP + P_i and simultaneously increased the medium pH to 8. ATP was synthesized. The only energy source was the H^+ gradient from chloroplast interior (acidified during the initial equilibration) to medium. The H^+ gradient was not simply forming some intermediate from the redox chain, for more molecules of ATP were synthesized than the number of redox chain molecules present. Moreover, the driving force for ATP synthesis was dissipated by uncouplers, and it decayed with time (measured using a lag between boosting medium pH and adding ADP + P_i) just as ATP synthesis by previously illuminated chloroplasts declined in the dark.

(The H^+ gradient for chloroplasts in Jagendorf's experiments is opposite that for mitochondria in Mitchell and Moyle's experiments. This topographical inversion, which was subsequently verified, is irrelevant to judging the mechanism's validity.)

That year both Racker and Slater confirmed Jagendorf's results.[43] H^+ gradients could drive ATP synthesis, and opponents of the chemiosmotic hypothesis retreated decisively. Chance now acknowledged the *reversible* formation of transmembrane gradients, although he allocated them to side reactions (Fig. 17.6).[44] Either way, Jagendorf's experiments established such gradients as critical elements, for even if they lay on side paths, the reversible arrows specified that forming these gradients would drive ATP synthesis whereas destroying the gradients by uncouplers or any leakiness would in turn dissipate energy stores.

On the other hand, analogous acid-bath experiments with mitochondria failed to drive ATP synthesis.[45] Mitchell, of course, provided reasons.[46] If chloroplast membranes were permeable to anions, they could utilize ΔpH to provide the H^+ influx without inducing inhibitory potentials, since passive anion fluxes would neutralize those potentials. But if mitochondrial membranes were impermeable to both cations and anions, any H^+ influx through the ATPase would be opposed by growing potentials, positive inside, arising from that influx. To demonstrate H^+-driven ATP synthesis in mitochondria, therefore, Mitchell and associates included in their media the K^+ ionophore valinomycin, since valinomycin by facilitating K^+ *out*flux would compensate electrically for H^+ *in*flux.[47] So first they equilibrated valinomycin-treated mitochondria in K^+-containing media at pH 8.8 to make mitochondrial interiors basic. Next they added enough acid to the media to create a pH differential of 4.6. ATP was synthesized.

Still more spectacular studies were published within a decade. In 1974 Racker and Stoekenius reconstituted bacteriorhodopsin (see Chapter 15) and mitochondrial ATPase in phospholipid vesicles.[48] These enzymes were oriented in the membranes so that light drove bacteriorhodopsin to pump H^+ into the vesicles, creating a gradient. H^+ outflux through the reversible ATPase then drove ATP synthesis. This study was elegantly direct; it also had the advantage that pH gradients were sustainable (and ATP synthesis continuable) as long as illumination continued (in mitochondria

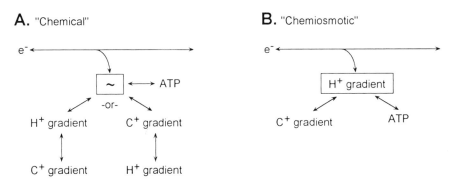

A. "Chemical"

B. "Chemiosmotic"

Fig. 17.6 Alternatives. *A*: The revised chemical-coupling scheme, after Jagendorf's experiment, showed the chemical intermediate formed by the respiratory chain ($\sim$) driving ATP synthesis and ion gradients: either primary H^+ transport or primary transport of another cation (C^+) then driving H^+ transport. All reactions were reversible. *B*: The chemiosmotic scheme showed the H^+ gradient as primary intermediate, driving ATP synthesis and cation transport. [Redrawn from Chance et al. (1967), Fig. 1. Reprinted by permission of the Federation of American Societies for Experimental Biology.]

the imposed gradients were lost in seconds). Racker and Stoekenius used partially purified ATPase preparations containing elements of the respiratory chain, but Yasuo Kagawa and associates in Tochigi-Ken, Japan remedied this difficulty the next year. They reconstituted bacteriorhodopsin and pure bacterial F_oF_1-ATPase in phospholipid vesicles:[49] H^+ gradients created by bacteriorhodopsin then had to be used directly by the ATPase. Mitchell's original distinction between his formulation and Slater's stipulated that no *chemical intermediate of the respiratory chain* was used as energy source for ATP synthesis. This was now demonstrated in the reconstituted model.

Also at that time, Peter Hinkle in Ithaca showed that ATP synthesis driven by imposed pH gradients was at least as fast as ATP synthesis driven by the respiratory chain.[50] Such "kinetic competence" demonstrated that H^+ fluxes could lie on the main path. Why then relegate them to a side path?

Primary and Secondary Cation Fluxes

Chance's accommodation included two alternatives: H^+ gradients driving other cation ("C^+") fluxes, or gradients of other cations driving H^+ fluxes (Fig. 17.6). Chance favored the latter, focusing on Ca^{2+} as the "other cation."[51] Berton Pressman in Chance's group focused on K^+. Indeed, Pressman published a scheme like Chance's the year before but with an *irreversible* arrow from energy-rich intermediate to K^+ transport.[52]

When Pressman discovered that valinomycin promoted energy-dependent K^+ influx, he initially considered that valinomycin somehow "triggered" a mitochondrial K^+ pump.[53] With demonstrations that valinomycin facilitated K^+ diffusion across hydrophobic barriers (such as artificial bilayers[54]), Pressman coined the term "ionophore" for such lipophilic ion carriers.[55] But that property was difficult to reconcile with valinomycin stimulating active K^+ transport. Pressman suggested that valinomycin allowed K^+ access to pumps buried within hydrophobic domains (an odd localization).

Mitchell, however, argued that the observed asymmetrical distributions of K^+ could not be achieved by K^+ pumps working while valinomycin kept membranes leaky to K^+, and, since valinomycin facilitated passive fluxes, that the asymmetrical K^+ distributions were instead secondary to electrochemical gradients established by primary H^+ transport.[56] Thus Pressman's scheme required membranes generally impermeable to K^+ but permeable to H^+, whereas Mitchell's scheme required membranes generally impermeable to H^+ but permeable (with the aid of valinomycin) to K^+.[57] Mitchell's titration experiments indicated that mitochondrial membranes were indeed only slowly permeable to H^+ (although during ATP synthesis fluxes through the reversible ATPase matched rates of respiration).[58] Subsequent studies revealed primary transporters for H^+ and secondary transporters for K^+ (K^+/H^+ exchangers[59]).

Membrane Potentials

Scepticism about membrane potentials continued even after their magnitude was cut in half in Mitchell's reformulation. An early assessment of ΔpH by Chance bolstered disbelief. From experiments using a standard pH indicator, the dye bromthymol blue, Chance reported in 1967 that after adding substrates "the magnitude [of the inferred pH change] is too small by a factor of . . . 10 [and the initial rate of H^+ transport] is over 1000-fold slower than the rate of oxidation" of respiratory chain components.[60] This seemed a straightforward refutation. The next year, however, Mitchell argued that bromthymol blue might not be a valid indicator of intramitochondrial pH: bromthymol blue is amphiphilic and charged, so it would partition at polar–nonpolar interfaces of membranes and redistribute as transmembrane electrical potentials changed.[61] Moreover, Mitchell reported experiments supporting his analysis. Two-thirds of the dye could be titrated immediately when substrates were absent, so only a third seemed to be inaccessible within mitochondrial interiors. By contrast, adding substrates made all the dye immediately titratable, which could be interpreted as the negatively charged dye migrating from interior to exterior interface as substrate oxidation created a negative interior (by H^+ extrusion).

Mitchell's arguments and experiments at the very least neutralized Chance's attempted refutation. Then in 1972 another indicator, 9-aminoacridine, was shown to distribute across chloroplast membranes according to imposed pH gradients[62] (just as it distributed across phospholipid vesicle membranes[63]), and in 1975 Hagai Rottenberg, working with Chance's group, reported that 9-aminoacridine fluorescence reflected the energy-dependent formation of substantial pH gradients across mitochondrial membranes.[64]

Demonstrations of transmembrane electrical potentials were achieved earlier. In 1970 Vladimir Skulachev and associates in Moscow described the distribution of lipophilic cations and anions across mitochondrial membranes in accord with Mitchell's hypothesis.[65] Subsequent studies using other indicators of $\Delta\Psi$ confirmed those observations and interpretations.[66]

(Another consequence of these studies was support for Mitchell's assessment[67] that while the driving force across chloroplast membranes is chiefly ΔpH, the driving force across mitochondrial membranes is chiefly $\Delta\Psi$.)

The critical issue, of course, was whether $\Delta pH + \Delta\Psi$ was sufficient to drive ATP synthesis. Mitchell calculated the sum available for ATP synthesis as 230 mV by measuring distributions of K^+ in the presence of valinomycin (K^+ diffusing passively in response to the electrochemical gradient, which is then calculable.).[68] But in 1966 Pressman reported that in his experiments 340 mV would be necessary (calculated from the free energy required for ATP-driven transport with the concentrations of ATP, ADP, and P_i present, and based on a H^+/ATP stoichiometry of two); similarly, Slater in 1973 calculated from experiments on ATP synthesis that 350 mV would be required.[69] If Pressman's and Slater's measurements—and the underlying assumptions—were correct, the discrepancy between measured and required potentials seemed a severe blow to Mitchell's hypothesis. Slater proposed that not all the energy from the respiratory chain was conserved in the H^+ gradient. Still, if Mitchell's measurements—and his underlying assumptions—were correct, at least the major fraction was stored in that gradient. And the ratio of Mitchell's to Pressman's and Slater's values suggested a revision that would make feasible the H^+ gradient as sole intermediate (see below).

Primary H^+ Transport

Additional support for chemiosmotic formulations came from studies in the mid-1970s by Racker, Hinkle, Kagawa, and associates demonstrating that individual, purified components, when reconstituted in phospholipid vesicles, each transported H^+: F_oF_1 (when driven by ATP hydrolysis) and Complexes I, III, and IV (when driven by redox reactions).[70]

Q Cycle

According to Mitchell's 1966 reformulation, ubiquinone was the hydrogen carrier in loop 3 (Fig. 17.4), so two H^+ needed to be extruded for each molecule of reduced ubiquinone added. Subsequent experiments, however, yielded a value of four.[71] One remedy was to reassign ubiquinone to loop 2: then two H^+ would be transported by that loop and another two by loop 3. But no hydrogen carrier for a revised loop 3 was obvious. So instead, Mitchell conceived the Q cycle,[72] a "fiendishly clever"[73] scheme merging loops 2 and 3 to achieve the requisite yield with only ubiquinone as the hydrogen carrier (Fig. 17.7A). Although begun[74] during a sleepless night in 1975, this formulation involved intricately balanced arguments to satisfy theoretical con-

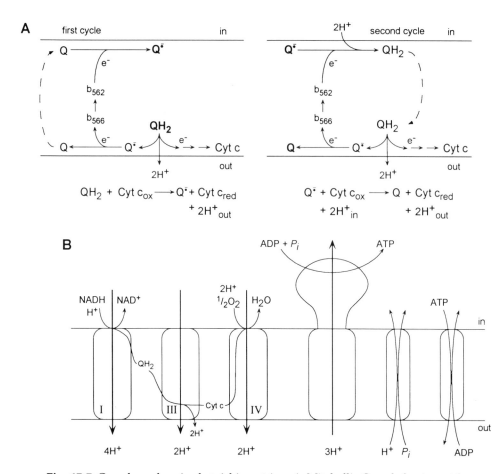

Fig. 17.7 Q cycle and revised stoichiometries. *A*: Mitchell's Q cycle begins with reduced ubiquinone (QH_2) that is oxidized to ubiquinone (Q) in two cycles, thereby transferring its two electrons (e^-) singly to oxidized cytochrome c (Cyt c_{ox}). In the first cycle QH_2 is oxidized to a semiquinone ($Q^{·-}$) and in the second, this semiquinone is oxidized to Q. The oxidations involve one electron transfer through two hemes (b_{566} and b_{562}) of cytochrome b. Q and QH_2, being lipid soluble and uncharged, can diffuse readily across the membrane (dashed lines), whereas the transmembrane potential restricts movement of charged $Q^{·-}$. Four H^+ are ejected, although only two H^+ disappear from the matrix space. [Redrawn from Mitchell, 1976, Scheme 24. Reprinted by permission of Academic Press.] *B*: Hinkle's allocation of H^+ extrusion and H^+ consumption results in P/O = 2.5 for NADH oxidation. Four H^+ are transported during NADH reduction of ubiquinone by complex I, two by the Q cycle of complex III, and two by cytochrome oxidase of complex IV. In addition, two H^+ disappear from the matrix space during water formation and two H^+ appear on the exterior as part of the Q cycle. Thus, 10 H^+ and 10 positive charges move from inside to out. For ATP synthesis, three H^+ are consumed, with a fourth required for P_i influx. Therefore, P/O = 2.5. [Redrawn from Hinkle et al. (1991), Fig. 4. Reprinted by permission, ©1991 American Chemical Society.]

straints (thermodynamic and spectroscopic) as well as complex experimental observations. Since then it has survived exhaustive scrutiny.[75]

Nonvesicular ATP Synthesis

As Greville pointed out, Mitchell's hypothesis would be "irrevocably refuted if respiratory chain . . . phosphorylation could be obtained with enzymes . . . in solution or [in] 'open vessicles' " where no transmembrane gradients were possible.[76] A number of reports claimed—briefly—to achieve just that.[77] For example, Edmond Hunter in St. Louis described ATP synthesis by a "soluble" redox system in 1980.[78] But the next year Racker responded that neither he nor several others could reproduce this results, and Edmund Bäuerlein subsequently attributed Hunter's results to contaminating adenylate kinase-like reactions.[79]

ACCEPTANCE AND FURTHER REVISIONS

After citing the potential for refutation, Greville concluded his 1969 review: "Assuming no such disaster for the Mitchell hypothesis, we may expect that acceptance would derive from gradually accumulating evidence in its favor."[80] That prediction was borne out. The 1977 *Annual Review of Biochemistry* published an unprecedented review of oxidative phosphorylation containing six commentaries.[81] Although not specified as a capitulation nor as a universal acquiescence to Mitchell's hypothesis, the joint review was understood as certifying Mitchell's triumph. Boyer focused on mechanisms of ATP synthesis (see Chapter 18); Chance discussed inadequacies of Mitchell's loops; Ernster concentrated on distinctions between loops and H^+ pumps; Slater expressed general agreement with Mitchell's principles but reserved the possibility for coupling modes more direct than H^+ gradients; whereas Racker bluntly admitted that "the basic [means for] the coupling of oxidation to phosphorylation has been solved in terms of the chemiosmotic formulation."[82] Mitchell circumspectly noted that the "chemiosmotic hypothesis is independent of the mechanisms coupling $[H^+]$ translocation reversibly to [ATP synthesis/hydrolysis]."[83]

The following year Mitchell was awarded the Nobel prize for chemistry. But significant revisions of the mechanisms and broadenings of the concepts followed in the next decade.

Stoichiometries

The possibility that more than two H^+ were transported at each coupling site had been raised by several studies, and in 1976 Lehninger mounted a forceful challenge.[84] He considered that Mitchell and Moyle's approach underestimated H^+ outflux for several reasons, but chiefly because concurrent P_i influx was accompanied by H^+ influx. Consequently, Lehninger and associates measured H^+ stoichiometries after washing mitochondria free of endogenous P_i and using an inhibitor of the H^+-P_i symporter. They then found an *average* of three to four H^+ ejected per coupling site from measurements over the respiratory chain spans of complexes I, III, and IV, and of III and IV.

These averages were supported by examinations of individual respiratory chain complexes. For Complex I, which oxidizes NADH to reduce ubiquinone, Giovanni Azzone in Padua and Mårten Wikström in Helsinki provided in the early 1980s convincing evidence that four H^+ were ejected.[85] For Complex III, Mitchell's Q cycle yielded four H^+, as noted above; however, if that sequence begins with reduced ubiquinine (QH_2, ubiquinol), the product of Complex I, then only two H^+ are removed from the matrix space (cf. Figs. 17.7A,B). For Complex IV, whose cytochrome oxidase oxidizes cytochrome c and reduces oxygen to water, Mitchell granted no role in H^+ transport (although in his scheme two H^+ were removed from the matrix space to form water). But in 1977, Wikström described H^+ ejection with a stoichiometry near four[86] (although that magnitude was debated for a decade). In 1985 Mitchell acknowledged H^+ transport by cytochrome oxidase. He then proposed a loop analogous to the Q cycle,[87] but this remained a minority view. (In 1995 the X-ray crystal structure of bacterial cytochrome oxidase was published, revealing a probable pathway for two H^+ transported from in to out, as well as the site at which two H^+ are consumed to form water.[88])

Meanwhile, Hinkle in 1983 reported a stoichiometry of nearly three H^+ transported for each ATP synthesized by inside-out submitochondrial particles.[89] By then Martin Klingenberg in Munich had demonstrated that H^+-P_i import into the matrix space consumed one H^+, while ATP export was associated with an electrogenic exchange for ADP.[90] Overall, each ATP synthesized and exported would thus consume four H^+. Lehninger reconciled this stoichiometry with P/O = 3 by depicting four H^+ ejected per coupling site (12 H^+ would then be ejected between NADH oxidation and oxygen reduction).[91] But in 1991, by scrupulously excluding sources of experimental error (such as oxygen adsorption to Teflon stirring bars), Hinkle calculated that P/O = 2.5, which corresponded to only 10 H^+ ejected (Fig. 17.7B).[92] (This nonintegral value is, of course, consistent with chemiosmotic formulations, where the unit of energy conservation is H^+ transport rather than ATP synthesis.) That revised stoichiometry also satisfied earlier concerns about energetic adequacies of the H^+ gradient. The cost to Mitchell's formulation was yet further rewritings of mechanisms.

Loops and Pumps

Mitchell's 1966 reformulation, including the later Q cycle, neatly described for each NADH oxidized two H^+ transported over each of three loops. The loops in these schemes displayed "direct coupling" in which the hydrogen being oxidized became the H^+ transported. Others argued for "indirect coupling" by pumps: redox reactions driving H^+ transport through induced protein conformational changes. By the late 1980s indirect coupling mechanisms (pumps) were favored for Complexes I and IV, whereas Q cycle loops were acknowledged for Complex III (the consumption of H^+ to form water by Complex IV can also be interpreted as a loop).

GENERALIZATION AND INTEGRATION[93]

Recognition of the Na^+/K^+ pump and Na^+-linked transport systems led to concepts of a "Na^+ economy" for mammalian cells: metabolic energy is convertible into transmembrane Na^+ gradients, which in turn can be tapped to power cellular work (see Chapters 8, 14). These early formulations by Crane and by Mitchell paralleled Mitchell's chemiosmotic coupling hypothesis and its "H^+ economy": redox energy is convertible into transmembrane H^+ gradients, which in turn may be tapped to power work by mitochondria, chloroplasts, and bacteria. Further broadening came with recognition that in certain bacteria the redox chain drives Na^+ rather than H^+ transport,[94] with the Na^+ gradient tapped by Na^+-consuming sytems to power cellular work. Morover, H^+ gradients serving as energy reservoirs are created across plasma membranes of fungi and plants by P-type H^+-ATPases (see Chapter 12) and also across vesicular membranes within mammalian cells by V-type H^+-ATPases (see Chapter 18).

Skulachev codified these processes as three laws of cell energetics.[95] (1) Cells transform energy from external sources—light and nutrients—into stores of cellular energy currencies. (2) Cells always possess two forms of energy currency—water-soluble energy-rich compounds and transmembrane gradients. (3) All cellular energy needs are satisfied by such interconvertible currencies. (One could also add that cellular signalling processes similarly rely on utilizing water-soluble compounds and transmembrane gradients, as in the stimulated phosphorylation of proteins by ATP and the stimulated fluxes of ions through gated channels.)

With this perception of integrated cellular functions came a scientific integration as well—the fusion of membrane transport, oxidative phosphorylation, and photosynthesis fields. When the *Journal of Biological Chemistry* began separating its table of contents by topics in 1981, one of the six sections was "Membranes and Bioenergetics."

COMMENTARY

Places and People

"[E]njoyment of family life, home building and creation of wealth and amenity, restoration of buildings of architectural and historical interest, music, thinking, understanding, inventing, making, sailing": this is how Peter Dennis Mitchell (1920–1992) described his recreations for *Who's Who*. This list reflects his scientific life as well. Mitchell entered Jesus College, Cambridge as an undergraduate in 1939 and emerged with a Ph.D. in biochemistry in 1950. His thesis research was supervised by Danielli, and Mitchell also acknowledged Keilin's deep influence. Mitchell did not depart for Edinburgh until 1955 (where he then directed the chemical biology unit of the Department of Zoology), so he overlapped with Slater, who remained an amicable although critical colleague. Serious illness and some aversion to academic life prompted Mitchell's departure for warmer Cornwall in 1963. The next two years he devoted chiefly to restoring a regency mansion on the verge of collapse, converting one wing to spartan laboratories and another to comfort and amenity for wife and family (including recreating plaster ceilings carved with gigantic images of medals that a previous occupant had garnered in the Napoleonic wars). At Glynn House Mitchell established the Glynn Research Institute, which was primed with family funds and supported by solicited contributions and research grants.

Slater invited Mitchell to explain the chemiosmotic hypothesis at a meeting on oxidative phosphorylation in Bari in 1965. In preparation for the meeting, Mitchell and Moyle (who also had received her Ph.D. in biochemistry from Cambridge) rapidly completed fundamental measurements of H^+ transport driven by oxygen pulses and ATP hydrolysis. This work laid the groundwork for the future, with Moyle ably supervising the laboratory and Mitchell thinking, understanding, inventing, making (and sailing). They were visited by numerous scientists—Glynn House is a short distance from a major stop on the London-Penzance rail line—including in the early years Slater, Jagendorf, Chappell, and Hinkle (who spent a postdoctoral year there from 1967 to 1968 after obtaining his Ph.D. with Racker). There Mitchell and his wife were entertaining and gracious hosts.

And in keeping with his vacation house in France was Mitchell's *savoir faire, joie de vivre,* and (of course) *amour propre.* Mitchell was also resolute in his convictions, assertive, and determined. His considerable charm was linked, however, with some stiff exclusions. Before publishing his 1961 paper in *Nature,* Mitchell initiated correspondence with Williams on chemical mechanisms,[96] yet Mitchell not only failed to acknowledge that assistance but avoided citing Williams's 1961 and 1962 papers for many years. Williams claimed that threats of libel lawsuits from Mitchell prevented Williams from publishing his own version of this history during Mitchell's lifetime.[97] Certainly, Mitchell's characterization of Crane's explicit formulation of 1960 ("As . . . Crane et al. . . . have hinted"[98]) and of Davies's proposal of 1962 ("suggestion . . . tentatively put forward"[99]) were not notably generous. Grumblings about Mitchell's failure to acknowledge work done by others have been frequent; however, that characteristic can be documented widely among the scientists featured in this book.

Animosity among the competitors there certainly was, to the extent that even an elementary textbook commented in 1970: "Since a number of strong, even arrogant personalities quickly became either proponents or opponents of the Mitchell hypothesis, the intensity of the response . . . obstructed reason even longer than usual."[100] Antipathies toward the chemisomotic hypothesis also erupted as specific irritations. Thus, Pressman responded to Mitchell's penchant for neologisms:[101] "It appears that one of the purposes for introducing so many new terms for old concepts is to place his scientific adversaries at a tactical disadvantage since they are directed to dispute with a vocabulary owned by Mitchell."[102] And David Wilson and associates, who were also from Chance's group, rejoiced that their claim of nonvesicular ATP synthesis (a claim that was short-lived, however) "removed [energy-coupling reactions] from the vitalistic realm of membrane biochemistry."[103]

Power

Since selecting cases to prove presuppositions is in vogue among some commentators, a selected example against their notion of power and politics resolving scientific disputes may be excusable. Mitchell's eccentric formulation was clearly his own and Mitchell's triumph was equally singular. How could that happen in light of the opposition from the powerful biochemical establishment, from scientists of greater personal wealth, from academics secure in prestigious institutions, and from opponents who controlled editorial boards, symposium speakers, and research funds? The astonishingly simple answer just might be that mitochondria indeed use transmembrane ion gradients as convertible energy stores, as increasing numbers of papers reported.

(Mitchell was, however, no solitary outsider. He attracted a circle of firm friends from all ranks, and his Glynn Research Foundation in 1991 boasted five patrons, all Nobel laureates.)

Fairness

Of course, acceptance of Mitchell's formulation would have been facilitated if obdurate egos in power had not felt threatened. Still, Mitchell was granted ample audiences; in the early years he was invited to speak at symposia in Italy and the United States in 1965, in Poland in 1966, and again in the United States in 1967 (where he was "American Society for Biological Chemistry Lecturer," sharing the platform with Chance, Jagendorf, Lardy, Lehninger, and Racker). Textbooks at that time described the chemiosmotic as well as chemical-coupling hypothesis. Particularly influential was Greville's even-handed review of 1969, which clarified formulations as well as results. Openness to new ideas is by no means perfectly attained by imperfect scientists, but self-interest coupled with self-doubt requires that circumspect scientists consider alternatives.

Strategies and Refutations

Proponents of chemical coupling needed (1) to identify an intermediate and (2) to rule out alternative hypotheses; for example, demonstrating the absence of electrochemical gradients would refute the chemiosmotic hypothesis. They failed in both.

For proponents of chemiosmotic coupling, refuting particular candidates for chemical intermediate left open the possibility that some other substance served that role. At best, the chemical-coupling hypothesis could be made unnecessary, prepped for Occam's razor. So their strategy was (1) to demonstrate the presence and sufficiency of electrochemical gradients and (2) to challenge attempted refutations. They succeeded at both, although with varying degrees of ease and elegance.

As an avowed disciple of Karl Popper, Mitchell wrote in 1960 that "[o]ne creates hypotheses in order to refute them," and in 1967 that "the rational approach to new hypotheses is to test them to destruction, if possible."[104] But Mitchell did not view these dicta as incompatible with fierce defenses of his hypothesis against attempted refutations. Complexities of experimental contingencies, validities of background assumptions, rigors of boundary conditions, and accumulated knowledge all vary from case to case, making idle any global pronouncements about the general practicality of refuting hypotheses in science. Thus, Mitchell circumvented refutation by reformulation (1961 to 1966 versions) and by reinterpreting results (bromthymol blue experiments), whereas Pinchot and Chance failed to salvage their proposals.

Scientific Summary

Mitchell's chemiosmotic coupling hypothesis proposed that for oxidative and photosynthetic phosphorylation, redox reactions drove H^+ transport across mitochondrial, bacterial, and chloroplast membranes. The electrochemical gradients so established then served as potential energy stores to drive ATP synthesis plus various transport systems. Accumulated evidence in favor of this hypothesis included demonstrations—using mitochondria, bacteria, chloroplasts, their vesicular fragments, and purified components reconstituted into phopholipid vesicles—that redox reactions created transmembrane electrochemical gradients, and that such gradients could power ATP synthesis as well as solute transport. Membrane transport thus represented a salient aspect of metabolic energy conservation and utilization. But the specific mechanisms for H^+ transport and its quantitative relationship to NADH oxidation and ATP synthesis, as originally proposed by Mitchell, were revised significantly in light of experimental explorations.

NOTES TO CHAPTER 17

1. Lehninger (1960), p. 954.
2. Historical accounts include Allchin (1991); Gilbert and Mulkay (1984); Prebble (1996); Robinson (1984, 1986b); Rowen (1986); Weber (1991); and Williams (1993).
3. Mitchell (1961b).
4. $Z = 2.303/F$; the notation $\Delta\Psi$ was introduced later.
5. Mitchell (1961b), p. 145.
6. Mitchell (1959).
7. Cf. Weber (1991), Williams (1993).
8. Mitchell (1959), p. 89. For active transport of inorganic ions, Mitchell proposed that organic carriers of such ions formed reversible covalent bonds with protein translocators.
9. Mitchell and Moyle (1958), p. 373.
10. See, for example, Mitchell (1957b).
11. Williams (1993).
12. Williams (1961).
13. Ibid., p. 16.
14. Ibid., p. 17.
15. Williams (1962).
16. See, for example, Mitchell (1963).
17. Davies and Ogston (1950).
18. Davies and Krebs (1952).
19. Ibid., p. 91.

20. Mitchell (1963).

21. Some noted H$^+$ release associated with cation uptake (e.g., Bartley and Amoore, 1958) but did not pursue the correlation at that time.

22. Williams (1962).

23. Danielli also drew models with transmembrane pores lined by single polypeptide layers (Fig. 10.1B) and with contractile strands traversing the bilayer (Fig. 8.7B).

24. Lehninger (1960), p. 952.

25. Moreover, Lehninger (1951) made mitochondria "leaky" in order to show NADH oxidation, and physical stresses can inactivate soluble as well as membrane-bound enzymes.

26. Williams (1982).

27. Mitchell (1961c); this study was extended in Mitchell and Moyle (1967a). The stimulation of F_1-ATPase by dinitrophenol (Pullman et al., 1961) would seem incompatible with Mitchell's mechanism, but I have found no published objections on this ground. A later explanation, which is irrelevant to uncoupling, is that F_1-ATPase is stimulated by certain anions (Kasahara and Penefsky, 1978).

28. Mitchell (1962), p. 31.

29. Greville in discussion to Mitchell (1963); Bielawski et al. (1966), p. 953.

30. Mitchell (1963). The symposium was in 1962.

31. Chappell and Crofts (1965).

32. Lehninger and Wadkins (1962), p. 51.

33. Slater (1966), p. 355.

34. Mahler and Cordes (1966), p. 609.

35. Lehninger (1967), p. 1334.

36. Mitchell and Moyle (1965). Such experiments were continued in Mitchell and Moyle (1967b, 1968).

37. Ibid. p. 151.

38. Mitchell (1966). A subsequent book (Mitchell, 1968) extended and generalized the theory.

39. Hind and Jagendorf (1963).

40. Neumann and Jagendorf (1964).

41. Hind and Jagendorf (1965), p. 3200. Jagendorf and Uribe (1967) acknowledged that Geoffrey Hind, then a postdoctoral fellow, called Jagendorf's attention to Mitchell's hypothesis.

42. Jagendorf and Uribe (1966). Hind and Jagendorf (1965) reported preliminary experiments showing ATP synthesis with an imposed pH gradient.

43. McCarty and Racker (1966); Tager et al. (1966). McCarty and Racker also showed that antibodies to the chloroplast ATPase blocked ATP formation, confirming that the proper ATP synthesizing component was indeed involved.

44. See, for example, Chance et al. (1967). Even Jagendorf (Jagendorf and Uribe, 1967) considered the gradient more likely to be on a side path.

45. Cited by Reid et al. (1966).

46. Mitchell (1966).

47. Reid et al. (1966).

48. Racker and Stoekenius (1974).

49. Yoshida et al. (1975).

50. Thayer and Hinkle (1975).

51. Chance (1965).

52. Moore and Pressman (1964).

53. Ibid., p. 566.

54. See, for example, Mueller and Rudin (1967).

55. Pressman et al. (1967).

56. Mitchell (1967).

57. As described by Greville (1969).

58. Mitchell (1961c, 1969); Mitchell and Moyle (1967a, 1969).

59. Martin et al. (1984).

60. Chance and Mela (1967), p. 830.

61. Mitchell et al. (1968).

62. Schuldinger et al. (1968).

63. Deamer et al. (1972).

64. Rottenberg and Lee (1975).

65. Grinius et al. (1970), and papers following.

66. See, for example, Laris et al. (1975).

67. Mitchell (1968).

68. Mitchell and Moyle (1969).

69. Cockrell et al. (1966); Slater et al. (1973). Such disparities continued to be reported; see, for example, Azzone et al. (1978).

70. Ragan and Hinkle (1975); Leung and Hinkle (1973); Hinkle et al. (1972); Sone et al. (1975).

71. Lawford and Garland (1973).

72. Mitchell (1976).

73. Harold (1986), p. 228.

74. Ibid.

75. See, for example, Trumpower (1990).

76. Greville (1969), p. 71.

77. See, for example, Painter and Hunter (1970); Wilson et al. (1972); Komai et al. (1976).

78. Painter and Hunter (1970).

79. Racker and Horstman (1972); Bäuerlein (1972).

80. Greville (1969), p. 71.

81. Boyer et al. (1977).

82. Ibid. p. 999.

83. Ibid. p. 1009.

84. Brand et al. (1976).

85. DiVirgilio and Azzone (1982); Wikström (1984).

86. Wikström (1977).

87. Mitchell et al. (1985).

88. Iwata et al. (1995).

89. Berry and Hinkle (1983).

90. Klingenberg (1980). At intramitochonrial pHs phosphate bears a single negative charge, so H^+-P_i is electrically neutral.

91. Alexandre et al. (1978).

92. Hinkle et al. (1991).

93. See Skulachev (1977, 1989); Harold (1986).

94. Tokuda and Unemoto (1982).

95. Skulachev (1992).

96. Williams (1993).

97. Ibid.

98. Mitchell (1963), p. 158.

99. Mitchell (1962), p. 35.

100. McGilvery (1970), p. 198.

101. Mitchell's terms included antiport and symport, osmotic barrier, periplasm, proticity, primary and secondary transport, and protonmotive force.

102. Pressman (1972), p. 598.

103. Wilson et al. (1972), p. 241.

104. Mitchell (1961a), p. 600; Mitchell (1967), p. 1374.

chapter 18

OXIDATIVE PHOSPHORYLATION: F_1, F_oF_1, AND ATP SYNTHASE

W HILE chemiosmotic principles garnered endorsements, however reluctant, Mitchell's mechanisms met persisting resistance, with skepticism bolstered by experimental challenges. This chapter traces the course to reformulating one of those mechanisms, that for the reversible ATPase. Although anaerobic bacteria may use this enzyme as a H^+ pump (creating electrochemical gradients that drive secondary active transport systems), oxidizing and photosynthetic bacteria as well as mitochondria and chloroplasts use the enzyme routinely to make ATP: as an ATP synthase. Identifications of this ATP synthase with the reversible F_oF_1-ATPase, cited in preceding chapters, were also linked to the working assumption, supported by accumulating evidence, that active sites catalyzing ATP hydrolysis by F_1 and by F_oF_1 were the active sites catalyzing ATP synthesis by F_oF_1.

CONFORMATIONAL MECHANISMS AND BINDING-CHANGE MODELS

The major rival to Mitchell's mechanism was developed by Paul Boyer (Fig. 18.1) in Los Angeles. His proposal had two roots: mitochondrial isotope exchange reactions first studied in the 1950s by Mildred Cohn and by Boyer (see Chapter 16), and a conformational coupling hypothesis proposed by Boyer in the next decade. The latter will be discussed next, for it directed Boyer's thoughts.

Conformational Coupling Hypotheses

In 1965 Boyer published a third coupling mechanism, with energy from redox reactions trapped as protein conformational changes that then drove the formation of "energy-rich" intermediates.[1] Boyer compared his proposal with models for the

Fig. 18.1 Paul D. Boyer. (Photograph courtesy of Paul Boyer.)

myosin ATPase, where energy from ATP hydrolysis drove muscle contraction through protein conformational changes, but he admitted that "no satisfactory methods are at hand for determining whether conformational change of protein is taking place in tightly coupled system."[2] Moreover, as Slater pointed out,[3] protein conformational changes were consistent not only with chemical intermediates but also with ion gradient intermediates. Indeed, protein conformational changes were then often invoked in models of membrane pumps (Fig. 8.10).

(Green developed more ambitious conformational hypotheses, featuring cooperative changes throughout the whole inner membrane.[4] These emerged from Green's models of membranes assembled from interlocking subunits (see Chapter 10) and from electron micrographs showing dramatically different membrane profiles that were correlated with distinct functional states. Establishing kinetic arguments so that morphological changes could be causal mechanisms was, however, difficult;[5] Green's formulations were not developed appreciably; interest in global models never flourished.)

Binding-change Mechanisms

In 1973 Boyer advanced a more modest scheme that focused on structural changes within the ATP synthase.[6] This was nonetheless a novel and far-reaching proposal,

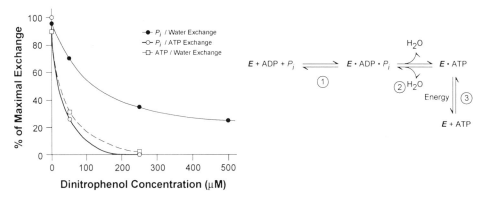

Fig. 18.2 Effects of uncouplers on mitochondrial isotope exchanges. Submitochondrial particles were incubated with 5 mM ATP, 0.5 mM ADP, 10 mM P_i, and the concentrations of dinitrophenol shown. P_i/water exchange was measured as the incorporation of ^{18}O from labeled water into P_i; ATP/water exchange as the incorporation of ^{18}O from labeled water into ATP; and P_i/ATP exchange as the incorporation of ^{32}P from labeled P_i into ATP. The scheme identifies the three reversible reactions involved in these exchanges. According to Boyer's 1973 reinterpretation, only reaction 3 requires added energy. [Redrawn from R. A. Mitchell et al. (1967), Fig. 4. Reprinted by permission of the American Society for Biochemistry and Molecular Biology.]

and one that sprang from reconsidering puzzling results reported in 1967. Boyer had then described ^{18}O exchange, catalyzed by submitochondrial particles, between labeled water and P_i (in the presence of ATP and ADP) that was markedly less sensitive to the uncoupler dinitrophenol (Fig. 18.2) than were the other exchanges, ^{18}O between labeled water and ATP and ^{32}P between labeled P_i and ATP.[7] To account for this disparity, he had suggested that P_i/water exchange represented an "independent . . . catalytic capacity."[8] In 1973 Boyer offered a startling reinterpretation: the P_i/water exchange in the presence of uncouplers reflected a reversible ATP synthesis but without release of that product—a synthesis achieved without adding energy. Boyer proposed that added energy (in the absence of uncouplers) instead drove ATP release. With this scheme (Fig. 18.2), reversals of reactions 1 and 2 represented P_i/water exchange, whereas reversals of reaction 3 (requiring added energy) allowed ATP release needed for P_i/ATP and ATP/water exchanges. (In the interim Boyer had established that ADP was necessary for exchange.[9])

To support this interpretation, Boyer now showed that other uncouplers inhibited similarly (sparing the P_i/water exchange was thus a general property, not peculiar to dinitrophenol), and that oligomycin blocked P_i/water exchange just as it did the other exchanges (so P_i/water exchange was probably linked to similar reactions of the ATP synthase, in contrast to his 1967 explanation). Moreover, Boyer also showed that in the presence of uncouplers tiny amounts of ^{32}P-ATP were formed from labeled P_i + ADP, rapidly attaining a steady level (interpretable as rapid synthesis of one ATP on the enzyme, with synthesis then ceasing because ATP could not dissociate without added energy).

But how could the initial synthesis of "energy-rich" ATP occur without added energy? Boyer suggested three possibilities: (1) pulling the reaction by tightly binding the product ATP; (2) decreasing the concentrations of product water; and/or (3) pushing the reaction through concentrating the reactants, ADP + P_i. From his data on P_i/water exchange Boyer focused on ATP binding, with ATP synthesis driven by the free energy from the tight binding. Conversely, releasing ATP required that energy be added to break the tight binding and restore the system to its initial state. Protein conformational changes, Boyer proposed, underlay the binding changes. (Slater, finding that nucleotides remained tightly bound to F_1 during purification, suggested independently in 1973 that "selective binding of ATP . . . could provide a method for the synthesis of ATP, by altering the position of the [reaction] equilibrium."[10] Moreover, the use of binding energy to drive enzymatic reactions was at this time being described for the myosin ATPase,[11] and the general case was being elaborated.[12])

In 1976 Boyer reported that adding dinitrophenol, which again equated to removing exogenous energy sources, decreased the affinity of the ATP synthase for ADP and P_i.[13] Conversely, adding energy should then increase the affinity for these reactants. So Boyer now amended his mechanism to specify that adding energy both increased the binding of reactants and decreased the binding of product. For this he drew an "alternate-site" model (Fig. 18.3A), with energy-driven binding of reactants on one subunit coincident with energy-driven release of product from the other. At the same time, enthusiasm for inter-subunit cooperativity was being fostered by progress in structural studies (see below). In 1979 Boyer gave his scheme the title "binding-change" mechanism to reflect the cyclical changes in affinity occurring through the reaction sequence.[14]

(Although P_i/water exchange was less sensitive to uncouplers than other exchanges [Fig. 18.2], some inhibition occurred. Boyer later explained this by identifying two exchange routes that were measured together in his 1967 and 1973 papers.[15] "Medium exchange" occurred when P_i was released into the medium. This was as sensitive to uncouplers as P_i/ATP and ATP/water exchanges, for P_i that had dissociated from the enzyme could not rebind in the absence of energy (presence of uncoupler), and thus labeling halted. "Intermediate exchange" occurred when P_i was formed during ATP hydrolysis but with that P_i still bound to the active site. This exchange was insensitive to uncouplers, for P_i bound to the active site could undergo a reversible loss of water, and thus labeling, when reversibly condensing with ADP to form ATP in the absence of added energy.)

While developing these conformational mechanisms, Boyer also found fault with Mitchell's mechanism. (1) Most significant was Boyer's contrast between Mitchell's directly-coupled mechanism, in which energy input, represented by two H^+ attacking P_i to extract O^{2-}, drove ATP *synthesis*, and Boyer's indirectly-coupled mechanism, in which energy input drove ADP + P_i *binding* and ATP *release*.[16] The

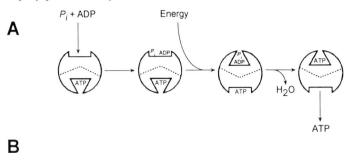

A

B

Fig. 18.3 Binding-change sequences for ATP synthases. *A*: Boyer's 1977 scheme for alternating-site catalysis shows ATP tightly bound at one site and ADP + P_i binding loosely to the other. Adding energy changes these affinities, so that the original ATP is released and ADP + P_i are bound tightly (ATP is bound even more tightly at that site, driving the reaction toward synthesis). [From Boyer et al. (1977), Fig. 1. Reprinted by permission from the *Annual Review of Biochemistry*, vol. 46, © 1977 Annual Reviews.] *B*: Cross's 1981 version, expanded to three sites, shows ATP bound tightly at one site and ADP + P_i bound loosely at a second. Another ADP + P_i bind to the open site. Adding energy drives affinity changes: ATP is discharged, leaving that site open. The initial ADP plus P_i are bound tightly, preparing them for conversion to the still more tightly bound ATP, and the second ADP + P_i are bound loosely. Thus, sites cycle through tight (T), loose (L), and open (O) configurations. [From Cross, 1984, Fig. 1. Reprinted by permission of the *Annual Review of Biochemistry*, vol. 50, © 1981 Annual Reviews.]

kinetics of P_i/water exchange, which indicated the step where energy was applied, strongly contradicted directly-coupled mechanisms,[17] a criticism that Mitchell did not counter. (2) Boyer as well as Williams challenged the likelihood of forming O^{2-} when ADP condensed with P_i,[18] criticisms Mitchell did not meet convincingly.[19] And (3) accumulating evidence revealing stoichiometries >2 H⁺/ATP also favored Boyer's mechanism, which had no intrinsic ratio for driving protein conformational changes. Mitchell's responses[20] were ingenious: he linked synthesis with an additional H⁺ to drive the import of ADP and P_i and the export ATP. But to most investigators this proposal was unconvincing.

On the other hand, Boyer went from advocating conformational *coupling* in 1973 to admitting that H⁺ gradients were the "likely" coupling intermediate in 1977 to bluntly accepting these gradients as the cause of the conformational changes in 1979.[21]

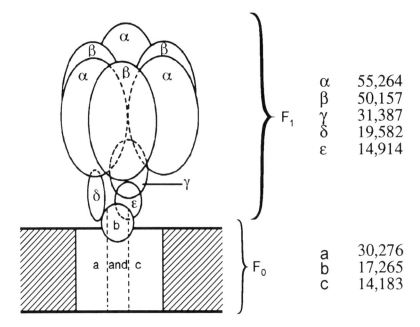

Fig. 18.4 Structure of ATP synthase from *E. coli*. F_1, composed of $\alpha_3\beta_3\gamma\delta\epsilon$ subunits, forms the knob and at least part of the stalk seen by electron microscopy. F_o, composed of $ab_2c_{10 \pm 2}$ subunits, forms the intramembrane base and perhaps part of the stalk. [From Walker et al (1982a), Fig. 1. Reprinted with permission from *Nature*, ©1982 Macmillan Magazines Ltd.] Molecular weights of the subunits, calculated from the amino acid sequences, are from Senior and Wise (1983).

STRUCTURE AND MECHANISM

Before we continue with this account of mechanistic models, concurrent studies on enzyme structure must be cited, for these studies shaped further formulations decisively.

Subunit Structure of F_1

In 1971 Alan Senior in Madison described five subunits of the mitochondrial enzyme. The stoichiometry, which was based on reported molecular weights for the complex and on the staining intensities of bands separated by SDS-PAGE, fitted $\alpha_3\beta_3\gamma\delta\epsilon$ (lettered from heaviest to lightest; see Fig. 18.4).[22] But 4 years later, Senior, now in Rochester, revised the stoichiometry to $\alpha_2\beta_x\gamma_2\delta_x\epsilon_2$ from counting protein sulfhydryl groups with a labeled sulfhydryl-binding reagent (x represented undetermined values).[23] Also supporting even numbers of subunits was Peter Pedersen's preliminary X-ray crystallographic study of 1978, which he interpreted as revealing two-fold symmetry.[24] Parallel studies with parallel uncertainties were reported for bacterial (*E. coli*) F_1: initially $\alpha_3\beta_3\gamma\delta\epsilon$, contradicted to $\alpha_2\beta_2\gamma_2\delta_{1-2}\epsilon_2$.[25]

Crucial to central issues like Boyer's alternating-site model of 1977, this stoichiometry was repeatedly reexamined. And in the early 1980s the weight of evidence turned to the earlier ratios. In 1979 William Allison in San Diego argued for $\alpha_3\beta_3$ after labeling the mitochondrial enzyme with a different reagent,[26] and in 1982 Robert Fillingame in Madison reported $\alpha_3\beta_3\gamma\delta\epsilon$ for the *E. coli* enzyme, on the basis of convincing experiments labeling the subunits *in vivo*.[27] Electron microscopic studies using antibodies to the α-subunit confirmed $\alpha_3\beta_3$ stoichiometries and showed α and β alternating in hexagonal arrays to form the bulk of the F_1 knobs.[28] And Richard Cross in Syracuse found six binding sites on the complex for a labeled analog of ATP: three sites that took up and lost the label rapidly, identified as catalytic sites, and three slowly exchangeable "non-catalytic" sites playing, perhaps, structural or regulatory roles.[29] Such results prompted reformulations, and models were expanded to show catalysis alternating among three sites (Fig. 18.3B).

The $\alpha_3\beta_3\gamma\delta\epsilon$ stoichiometry was further supported in the early 1980s by studies that established the amino acid sequences of each subunit. The sequence for the *E. coli* enzyme was determined independently by Masamitsu Futai in Okayama, Kaspar von Meyenburg in Copenhagen, and John Walker in Cambridge, by analyzing corresponding DNA sequences, and the sequence for the mitochondrial enzyme by Walker, using direct protein sequencing.[30] (Comparisons of bacterial and mammalian mitochondrial sequences showed that while each had five types of subunits and while α, β, and γ were similar, sequences of the smaller subunits of bacteria and mitochondria differed; mitochondria also had additional subunits.) Subsequent studies by Allison, Boyer, Cross, Pierre Vignais in Grenoble—and many others—assigned the substrate sites to interfaces between α-and β-subunits and identified numerous amino acids adjacent to the bound nucleotides.[31]

Subunit Structure of F_o

For the three subunits of F_o (subsequently named a, b, and c; see Fig. 18.4), Fillingame in 1982 reported a stoichiometry of $a_1b_2c_{10\pm1}$;[32] these were sequenced at the same time as the F_1 subunits. (As with F_1, mitochondrial F_o contained additional subunits.)

Earlier, Kagawa reconstituted purified bacterial F_o in phospholipid vesicles, dramatically increasing transmembrane H^+ conductance as if F_o were a H^+ carrier or channel.[33] Moreover, adding F_1 with F_o sharply reduced H^+ conductance as if F_1 controlled H^+ passage, blocking it directly or inducing changes in the F_o structure that obstructed fluxes. Adding dicyclohexylcarbodiimide (DCCD) also impeded H^+ conductance. (In 1966 Brian Beechey, at Shell Research, Ltd. in Sittingbourne, had found that DCCD, like oligomycin, inhibited mitochondrial oxidative phosphorylation.[34] This reagent proved quite valuable since bacterial oxidative phosphorylation is sensitive to DCCD although it is insensitive to oligomycin, and since DCCD, which reacts covalently with carboxyl groups, was later shown to label the c subunit of F_o—and a specific dicarboxylic acid in the peptide.[35])

Similar experiments using mitochondrial and *E. coli* F_o established it as a transmembrane H^+ channel whose fluxes were sensitive to F_1, DCCD, and (in mitochondria) oligomycin.[36] Those studies thus confirmed Mitchell's proposal that F_o conveyed H^+ across the membrane. They also undermined a last-ditch challenge to chemiosmotic principles (at least for mitochondria), claims that H^+ flowed not *across* mitochondrial membranes but *within* the membranes.

Aside from lingering antipathies toward Mitchell's formulations, rationales for such alternatives included: (1) apparent kinetic and thermodynamic shortcomings when calculations were based on transmembrane gradients, interpreted as the coupling H^+ not equilibrating with bulk aqueous phases but moving in "localized circuits" within membranes or at interfaces; and (2) abilities of alkaliphilic bacteria to synthesize ATP even in media above pH 10, where establishing H^+ gradients from out to in would seem exceedingly difficult.

In terms of the first rationale, positive evidence indicating localized circuits was minimal, whereas negative evidence against delocalized transmembrane gradients was challenged as artifactual.[37] Williams stipulated in 1982 that for localized circuits to be "accepted somebody somewhere *must* be able to show that vesicles are not necessary";[38] such evidence was not forthcoming. Furthermore, localized circuits between respiratory chain and ATP synthase imply integral coupling ratios (P/O ratios), whereas delocalized H^+ coupling allows nonintegral ratios, which had been established (see Chapter 17).

In terms of the second rationale, negative evidence against delocalized circuits was provocative, although the arguments were not insurmountable. Positive evidence included descriptions of altered F_o structures in alkaliphilic bacteria that *might* accept intramembrane H^+ flows directly from the respiratory chain.[39] Delineation of H^+ channels so oriented had not been achieved by the mid-1990s, however. And for mitochondria, no hint of branched channels in F_o—ones capable of accepting H^+ from the membrane interior as well as from the exterior phase—materialized either.

Mechanism and Structure

Meanwhile, in 1982 Penefsky, Cross, and Charles Grubmeyer reported fundamental measurements confirming the binding-change mechanism.[40] For the sequence

$$\boldsymbol{E} + \text{ATP} \Leftrightarrow \boldsymbol{E} \cdot ATP \Leftrightarrow \boldsymbol{E} \cdot ADP \cdot P_i \Leftrightarrow \boldsymbol{E} + \text{ADP} + P_i$$

ATP binding to isolated F_1 was favored over release by a factor of 10^{12}—an extraordinarily tight binding—whereas bound ATP was favored over bound ADP + P_i by a factor of merely 2; interconversion thus proceeded at negligible energy cost. In these experiments they added ATP at very low concentrations so that only one substrate site of F_1 would be occupied. Adding higher ATP concentrations, to fill the second and third substrate sites, increased the rate of product release from the first site a million-fold. The jump in velocity they explained as catalysis by one, two, and ultimately three active sites, with the dependence on ATP concentration reflecting the negative cooperativity between substrate sites.

Penefsky confirmed these values using intact F_oF_1 in submitochondrial particles.[41] He also showed that adding energy to the system—adding NADH to fuel the respiratory chain—drove the release of ATP previously formed at a single site.[42] And the rate of the product release was commensurate with observed rates of NADH-driven ATP synthesis. Such results complemented reports that modifying F_o (with oligomycin or DCCD) affected properties of the associated F_1,[43] that perturbing F_o induced structural alterations in F_1.

Binding Changed by Subunit Rotations

What might cause the sequential changes in affinity required by binding-change models? Boyer offered a strikingly novel answer in 1981:[44] If the $\alpha\beta$ ring rotated around the $\gamma\delta\epsilon$ core (or the core within the ring), then each $\alpha\beta$ pair could be distorted successively through the three reactive states of Fig. 18.3B. Although $\alpha_3\beta_3$ should have threefold symmetry, interactions with asymmetric $\gamma\delta\epsilon$ necessarily impart asymmetry to the whole, and thus rotation of the ring relative to the core could induce cyclical waves of structural deformations.

A few years later, Graeme Cox[45] in Canberra embellished that proposal, also noting parallels with bacterial flagellar models: certain bacteria propel themselves with rotating flagella, their rotary motors driven by transmembrane H^+ gradients[46] (for bacteria operating with "Na^+ economies," Na^+ fluxes drive flagellar rotation[47]). This analogy—and the vivid mechanical image of a cam-driven cycle of affinity changes—sustained enthusiasm during the 1980s (albeit without universal endorsement) despite a paucity of direct evidence.[48]

(In 1994 Walker and associates described the X-ray crystallographic structure of F_1 at 2.8 Å resolution.[49] With one site containing an ATP analog, the second containing ADP, and the third empty, each $\alpha\beta$ pair differed structurally, with deformations as large as 20 Å. Furthermore, these $\alpha\beta$ pairs encircled a central core containing the γ-subunit, which "look[ed] like a molecular bearing, lubricated by a hydrophobic surface."[50] But with the structure of F_o still unknown and the H^+ pathway uncertain, just how H^+ fluxes could turn γ within $\alpha_3\beta_3$ remained undefined. On the other hand, the structure did illuminate the mechanism of ATP hydrolysis: glutamate-188 of β activates water for an in-line nucleophilic attack on the terminal phosphate, while arginine-373 of α stabilizes the negative charge on the transition state phosphate. And with this structural information, Cross could design a dramatic experiment that strongly supported rotation:[51] (1) he genetically engineered a mutant F_1 susceptible to reversible inactivation through reversible cross-linking—by disulfide bond formation—of γ to *one* β; (2) after this cross-linking he labeled the two non-cross-linked βs by dissociation and reassembly; (3) he then reversed the cross-linking and allowed F_1 to catalyze ATP hydrolysis; and (4) he reformed the cross-links. These now joined γ to labeled as well as unlabeled βs (in different F_1 molecules): in at least some F_1 molecules γ must have moved during the catalytic cycles from its original β-partner and stopped next to a different β.)

Na$^+$-Driven ATP Synthase

The anaerobic bacterium *Propionigenium modestum* thrives by decarboxylating succinate. In 1984 Peter Dimroth in Munich proposed that Na$^+$ gradients in *P. modestum* established by membrane-bound decarboxylases drove membrane-bound reversible ATPases to synthesize ATP.[52] He then showed that (1) the ATPase was stimulated by Na$^+$; (2) it contained an F$_1$-like component (with α-, β-, γ-, δ-, ϵ-subunits) whose ATPase activity was not stimulated by Na$^+$; (3) it contained an F$_o$-like component (with a, b, c subunits) that, when added to its F$_1$, restored Na$^+$-activation; and (4) when reconstituted in phospholipid vesicles, this F$_o$F$_1$ coupled Na$^+$ transport to ATP hydrolysis.[53] Using *P. modestum* F$_o$ and *E. coli* F$_1$, Dimroth then constructed a hybrid F$_o$F$_1$ whose ATPase activity was activated by Na$^+$:[54] cation specificity followed from the F$_o$. Similar Na$^+$-F$_o$F$_1$ systems were soon identified in additional bacterial species. (Such functioning without H$^+$ also added yet another argument against models specifying direct participation of translocated H$^+$ in ATP synthesis.)

Families of Cation-Transporting ATPases

Whereas P-type cation-transporting ATPases contain one to three different subunits that are integral membrane proteins (see Chapter 12), F-type ATPases, which comprise ATP synthases of mitochondria, chloroplasts, and bacteria, contain eight or more different subunits separable into integral membrane (F$_o$) and peripheral membrane (F$_1$) proteins. And whereas reaction mechanisms of P-type ATPases involve enzyme phosphorylation, those of the F-type do not.[55] In their influential 1987 review in which these types were named, Pedersen and Carafoli distinguished a third type called "V" (for "vacuolar").[56] Five early examinations of V-type ATPases, cited below, will illustrate how another interest converged on cation-transporting ATPases.

Adrenal glands synthesize epinephrine for release into the bloodstream, but how do adrenal cells concentrate epinephrine in storage vesicles prior to its release? In 1962 Norman Kirshner described ATP-stimulated uptake of epinephrine by vesicles isolated from adrenal cells, uptake he interpreted as active transport.[57] Using vesicle membrane preparations, G. Taugner then showed that epinephrine influx correlated with ATP hydrolysis.[58] George Radda, observing that ATPase activity was enhanced and epinephrine uptake diminished by uncouplers of oxidative phosphorylation, suggested in 1975 that epinephrine "transport could . . . occur by an exchange or co-transport mechanism" driven by an ATPase "associated with the movement [of H$^+$] and a concomitant development of an electrochemical potential."[59] In 1979 David Apps and Gottfried Schatz reported likenesses between vesicle membrane ATPase and mitochondrial F$_1$, which included similar but not identical patterns of peptide bands separated by SDS-PAGE, reactions to antisera prepared against F$_1$, and responses to some—but not all—inhibitors.[60] And in 1984 Racker reconstituted in phospholipid vesicles the partially purified ATPase, demonstrating

ATP-driven H^+ transport.[61] These and hosts of further studies defined secondary active transport systems for epinephrine driven by H^+ electrochemical gradients established by H^+-ATPases.

Other investigations[62] revealed similar V-type ATPases not only in various vesicle membranes of microorganisms, plants, and animals (for example, in lysosomes, where acidification of the vesicle interior promotes degradative activity), but also in plasma membranes (for example, in kidney tubule cells, where acidification of urine is essential). And like some F-type ATPases, some V-type ATPases transport Na^+ preferentially. Further explorations also established that V-type ATPases were composed of "V_o" domains serving as H^+ channels (one peptide among the several with sequence similarities to c of F_o) as well as of "V_1" domains bearing sites for ATP hydrolysis (two peptides among the several with sequence similarities to α and β of F_1). Thus, V-type ATPases are close cousins to the F-type siblings, and the division of cation-transporting ATPases should be bipartite, P vs. FV.

COMMENTARY

Places and People

Although this chapter chronicles a second generation laboring across the globe—including Allison, Cross, Dimroth, Fillingame, Futai, Kagawa, Penefsky, Senior, and Walker—Boyer is the commanding presence. In graduate school at the University of Wisconsin he like Lardy was assigned to Paul Phillips, whose interest lay in large animal nutrition; he like Lardy recalled the inspiration of a conference on respiratory enzymes held in Madison in his third year (1941) when he was able to "listen to Hermann Kalckar describe P/O ratios greater than one and Fritz Lipmann mumble wisely about oxidative formation of ATP."[63] From a failed examination of vitamin E deficiency came secondary experiments on glycolysis that revealed stimulation by K^+: his report with Lardy and Phillips was the first account of monovalent cations stimulating an enzyme (see Chapter 7). Only a few of Boyer's accomplishments are noted here. Rather than recite others I will append two diverse observation by him.

> [M]any things can be cited as vital to . . . accomplishment. One is persistence, supported by an underlying optimism. Effort is difficult without optimism, and accomplishment is rare without effort.[64]

> The recognition of transmembrane electrochemical . . . gradients and of energy linked conformational changes may prove to be the two most fundamental contributions of the field of bioenergetics.[65]

Wholes and parts

Chapter 17 celebrated the triumph of Mitchell's chemiosmotic hypothesis, founded by contemplating *whole* mitochondria, chloroplasts, and bacteria, the membranes *and* the phases within and without. This chapter celebrates the triumph of Boyer's binding-change mechanism, which was established through dissecting organelles and organisms into parts—and into component proteins as well. Oddly, some philosophers find such reductionistic approaches unsavory,[66] and even some biologists are repelled. Thus, Ernst Mayr claimed that recognizing a joint as a hinge was sufficient insight into its function.[67] But surely the identification of F_oF_1 as a rotary enzyme adds a deeper comprehension of processes beyond mere indulgence in mechanical aesthetics. And through parallels with flagellar motors, we extend our appreciation of commonalities within the biological catalog.

This book should illustrate the requirement that understandings of functional wholes and of component parts must proceed together. Mitchell wrote in 1960: "We must strive to set up 'synthetic' or reconstituted membrane systems with which we can study directly . . . the processes."[68] That same year Lehninger, while advocating experiments on mitochondrial particles, wrote of "a biological necessity for structural organization of these [enzymes] in a . . . geometrically organized constellation in the membrane."[69]

Theory

At a conference in 1960 Mitchell apologized that "among biologists, the word theoretical has disreputable undertones."[70] In physics the beauty of theories commands allegiance, but in biology manifest complexities frustrate recognition of primary properties, and daunting diversities challenge inference and generalization. "Theory" is easily translated as "idle speculation." Imagined alternatives in the absence of experimental substantiation have had poor yields. Green's conformational hypothesis, embracing such trendy concepts as solitons, merely tarnished Green's reputation during his final years. Yet Mitchell's physically less sophisticated formulation succeeded. Insufficient information was available and mechanistic potentials were too numerous to deduce how oxidative phosphorylation must work. Why then did Mitchell's hypothesis triumph?

Mitchell's model sprang from the barest mechanistic necessities: membranes that are generally impermeable but that are endowed with makers and users of electrochemical gradients. While Slater focused on substrate-level phosphorylation, Mitchell recognized analogies with transport systems, applying his stark mechanism to a problem then foundering. But Mitchell's formulation prevailed not merely because he sensed the significance of mitochondrial membranes. Other coupling mechanisms were imaginable. It prevailed because the model when tested was verified. And it prevailed despite intimate aspects being erroneous. An appealing aspect of Mitchell's model was that H^+ gradients were formed directly by hydrogen carriers of the respiratory chain (which was confirmed only in part) and that transported H^+ reacted directly with $ADP + P_i$ to form ATP (which was clearly refuted). If Mitchell developed his model for such "wrong" reasons, experimental examinations demonstrated that it transcended its founding details.

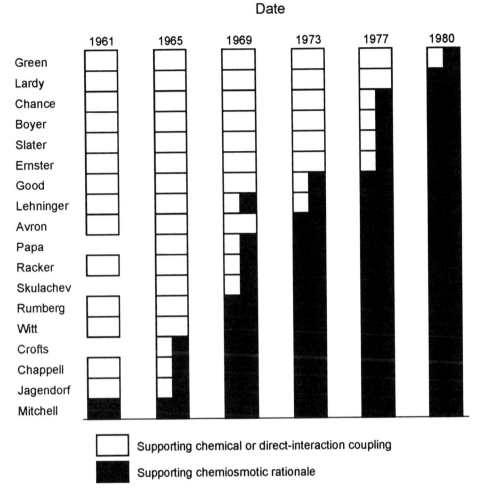

Fig. 18.5 Trends of support for the chemiosmotic hypothesis. Mitchell's assessment of support for or opposition to the "chemiosmotic rationale" by selected scientists is shown for various years. The earlier version that Gilbert and Mulkay discussed stopped at 1973 but included the same participants. [Redrawn from Mitchell (1981), Fig. 1.12. Reprinted by permission of Academic Press.]

Establishing Consensus

While exploring "discourse analysis," Nigel Gilbert and Michael Mulkay interviewed scientists active in oxidative and photophosphorylation research.[71] One focus was Mitchell's figure displaying the time course of conversion (Fig. 18.5). From diverse responses Gilbert and Mulkay concluded that criteria for recognizing a consensus were "contextually defined and contextually variable."[72] They identified essential concerns as (1) identifying members of the field; (2) attributing beliefs correctly; and (3) ensuring that perceptions of the issue were similar. Various scientists disagreed with Mitchell's figure and on all three aspects, implying that evidence of any true consensus evaporated in these disparate judgments.

But if one ventures beyond discourse analysis, the problem is less intractable. (1) Active participants can be easily identified, and Mitchell listed all major players of his generation (his figure includes those involved in photophosphorylation research, such as Mordhay Avron, whom I have skipped over). (2) While private beliefs may be unknowable, actions, printed comments, and reported experiments reveal public beliefs. (3) But how those beliefs were allocated into supporting or opposing categories certainly varied among the interviewees. For example, Gilbert and Mulkay paraphrase one pseudonymous scientist as believing himself to be wrongly categorized since he acknowledged only "an element of truth in the hypotheis" and "felt 'most of the details were nonsense'."[73] But historians willing to extract the scientific essence (membranes generally impermeable but endowed with makers and users of electrochemical gradients) from mechanistic details (which in all biological formulations undergo drastic revisions during development) can see a consensus forming very much like Mitchell's figure.

Different meanings attributed to "chemiosmotic hypothesis" also cloud assessments of how rational were those accepting or rejecting Mitchell's formulation. Again, if the bare principles cited above are the issue, then the crucial evidence was: (1) Mitchell's demonstration that uncouplers halted phosphorylation and promoted correspondingly rapid H^+ fluxes; (2) Mitchell's demonstration that respiratory chain redox reactions and ATP hydrolysis produced H^+ gradients; (3) Jagendorf's demonstration that H^+ gradients could drive ATP synthesis; and (4) Hinkel's demonstration that gradient–driven ATP synthesis proceeded rapidly enough to be the sole source. If rapid and reversible H^+ fluxes accompany redox reactions, then it seems capricious to exile them to redundant side reactions (Fig. 17.4). All this was apparent by 1975. Confounding arguments were Chance's early claim that redox reactions did not create electrochemical gradients, which was soon rebutted by Mitchell, Skulachev, and others, and Pressman's and Slater's claims of thermodynamic inadequacy. Here Mitchell's insistence on specific mechanisms misdirected attention, for criticisms based on thermodynamics were translatable into arguments about stoichiometry; these criticisms vanished if three rather than two H^+ drove ATP synthesis.

Nevertheless, major scientists responded holistically to particular issues of personal concern. Thus Boyer admitted that

> "For years I had been unable to accept the concept . . . that a transmembrane potential . . . might . . . transmit energy from oxidation to phosphorylation. This was because I could not see a satisfactory way to use [this] force to make ATP."[74]

This rationale echoes Fenn's opposition to sodium pumps on similar grounds four decades earlier (see Chapter 3). Using ATP hydrolysis to form cation gradients and using cation gradients to form ATP represent the same chemiosmotic challenge.

Scientific Summary

For ATP synthases of mitochondria, chloroplasts, and bacteria, Boyer's binding-change hypothesis described an indirect coupling to transmembrane electrochemical gradients. Catalysis proceeds sequentially at three active sites on the F_1 complex ($\alpha_3\beta_3\gamma\delta\epsilon$): at one site, tightly bound ATP is near equilibrium with ADP + P_i; at a second site, reactants ADP + P_i are bound loosely; and at a third site, product ATP is discharged. Steady-state synthesis results from cation fluxes through the F_o complex ($ab_2c_{10 \pm 2}$) driving sequential changes in binding affinity. The mechanism fitted isotope exchange experiments and direct measurements of equilibrium constants. Just how cation fluxes induce conformations having the requisite affinities remained unclear in the 1980s, but an attractive model invoked the asymmetric core of F_1 rotating relative to $\alpha_3\beta_3$, forcing the sequence of affinity changes.

NOTES TO CHAPTER 18

1. Boyer (1965).
2. Ibid., p. 1011.
3. Slater (1971).
4. Harris et al. (1968); Green and Baum (1970).
5. See, for example, Chance et al. (1969).
6. Boyer et al. (1973).
7. R. A. Mitchell et al. (1967).
8. Ibid., p. 1799. They also argued that ATP-driven reactions of mitochondria occur over routes separate from that for oxidative phosphorylation.
9. Jones and Boyer (1969).
10. Harris et al. (1973), p. 152. Later studies showed that these tightly bound nucleotides were not at catalytic sites, however.
11. Bagshaw and Trentham (1973).
12. Jencks (1975).
13. Kayalar et al. (1976).
14. Boyer (1979).
15. Rosing et al. (1977).
16. Boyer (1975).
17. This argument was developed in Boyer (1981a).
18. Boyer (1975); Williams (1975).
19. Mitchell (1975).
20. See, for example, Mitchell and Koppenol (1982).
21. Boyer et al. (1973); Kayalar et al. (1977), p. 2490; Boyer (1979).
22. Senior and Brooks (1971). Penefsky and Warner (1965) had shown that F_1 contains subunits, and MacLennan and Tzagaloff (1968) published photographs showing multiple bands after gel electrophoresis.
23. Senior (1975).
24. Amzel and Pedersen (1978).
25. Bragg and Hou (1975); Vogel and Steinhart (1976).
26. Esch and Allison (1979).
27. Foster and Fillingame (1982).
28. Lünsdorf et al. (1984).

29. Cross and Nalin (1982).

30. These results were described in a series of papers beginning with Kanazawa et al. (1981); Nielsen et al. (1981); Gay and Walker (1981); and Walker et al. (1985).

31. See, for example, Bullough and Allison (1986); Xue et al. (1987); Lunardi and Vignais (1982).

32. Foster and Fillingame (1982).

33. Okamoto et al. (1977).

34. Beechey et al. (1966).

35. See, for example, Stekhoven et al. (1972); Sebald et al. (1980).

36. See, for example, Glaser et al. (1980); Negrin et al. (1980).

37. Woelders et al. (1985).

38. Williams (1982), p. 2.

39. Ivey and Krulwich (1992).

40. Grubmeyer et al. (1982); Cross et al. (1982).

41. Penefsky (1985b).

42. Penefsky (1985c).

43. See, for example, Penefsky (1985a); Matsuno-Yagi et al. (1985).

44. Boyer and Kohlbrenner (1981).

45. Cox et al. (1984).

46. Silverman and Simon (1974); Larsen et al. (1974); Manson et al. (1977); Matsuura et al. (1977). Previously, speculations about the mechanism covered an imaginative range, including formulations by Ling (1969) in terms of his protein induction model, and by Mitchell (1972) in terms of electrophoresis of ions from the bacterium along the outside of a hollow flagellum, with a return flow through that tube.

47. Chernyak et al. (1983).

48. Cross and Duncan (1996) note two technical advances that furthered the idea of rotary mechanisms. (1) New mass spectroscopic techniques permitted analysis of the reaction pathways, demonstrating that the rate constants at each of the three catalytic sites was the same, i.e., the asymmetry of the enzyme did not cause asymmetries in the reaction pathways. (2) Methods for amino acid sequence determination permitted demonstrations that neither α nor β had a tripartite structure (which would have allowed each to react identically with γ, δ, and ϵ). Furthermore, several approaches provided evidence for cyclical changes in enzyme conformation; see, for example, Kandpal and Boyer (1987); Shapiro and McCarty (1988); and Gogol et al. (1990).

49. Abrahams et al. (1994).

50. Ibid., p. 628.

51. Duncan et al. (1995).

52. Hilpert et al. (1984).

53. Laubinger and Dimroth (1987, 1988).

54. Laubinger et al. (1990).

55. Absence of E-P intermediates in F-type ATPases is concluded from direct evidence against this (Webb et al., 1980) as well as the lack of evidence for it.

56. Pedersen and Carafoli (1987).

57. Kirshner (1962).

58. Taugner (1971).

59. Bashford et al. (1975), p. 155.

60. Apps and Schatz (1979). Cidon and Nelson (1983) criticized these experiments for their contamination with mitochondria. However, subsequent studies—showing similarities in amino acid sequences between the vesicular ATPase and mitochondrial ATPase—supported Apps and Schatz's conclusion.

61. Xie et al. (1984). Their starting material was "clathrin-coated vesicles," cytoplasmic vesicles formed from invaginations of the plasma membrane.

62. For general reviews see volume 122, *Journal of Experimental Biology* (1992).
63. Boyer (1981b).
64. Ibid., p. 239.
65. Ibid., p. 237.
66. See, for example, Kinkaid (1990).
67. Mayr (1982).
68. Mitchell (1961a), p. 598.
69. Lehninger (1960), p. 952.
70. Mitchell (1961d).
71. Gilbert and Mulkay (1984).
72. Ibid., p. 119.
73. Ibid., p. 117.
74. Boyer (1981b), p. 236.

chapter 19

CONCLUSIONS

THE half-century of scientific advances celebrated here began in the late 1930s
with images of cell membranes as lipoidal barriers, permeable only to those
polar solutes admitted through pores that select by size and charge; with mechanisms for maintaining cellular ion content rooted in absolute impermeabilities; and
with descriptions of intermediary metabolism limited chiefly to naming successive
products of catabolic pathways. By the late 1980s, these views had been drastically
revised. Membranes were recognized as dynamic structures containing hosts of
variously gated channels, specific transporters, and energy-consuming pumps. Cell
contents were thus actively selected by highly regulated systems. And usable energy
extracted during intermediary metabolism was stored in transmembrane electrochemical gradients as well as in "energy-rich" compounds like ATP. These reformulations reflect staggering proliferations of detail—of subcellular organization, of reaction sequences, of enzyme structures, and of molecular mechanisms. But these
reformulations also represent new generalities: families of transport ATPases,
modes of energy storage and interconversion, and systems for inter- and intracellular signalling.

Preceding chapters related those accomplishments. This afterview is concerned
with noting some characteristics of the endeavor, including the constraints on scientific theory choice and the rationality of conflict resolution.[1] But before venturing
further, three mingled meanings of "science" should be distinguished. (1) "Science"
is an accepted, although fluid, body of knowledge, the consensus codified in reviews
and textbooks (these also record uncertainties and unresolved controversies, which
are components of that knowledge also). (2) "Science" is an ideal approach to
attaining that knowledge, representing prescribed goals and means (although that
ideal has evolved and may continue to do so). (3) "Science" is the actual human
practice, the imperfect realization of (2).

Examples of (3) have been amply illustrated, but the diversity of individual
scientists and of their views should be emphasized again. Participating scientists
differed not only in their primary goals—from preserving blood and devising new

therapies to describing nerve conduction and understanding photosynthesis—but also in outlook and approach. Some professed a selfless quest for Knowledge. Others followed a calculated course of self-promotion. Some were certain of finding Truth. Others recognized that ultimately they would be found wrong. Some strove for unimpeachable data. Others scrambled to be first in print. Some bravely admitted error. Others grew increasingly recalcitrant. These scientists may thus seem to differ in their sense of what the aims and approaches of science ought to be. Persuasive arguments may be advanced that vindicate in the actual realm of scientific practice behaviors often criticized, including hasty publication, idiosyncratic interpretation, stubborn resistance to proclaimed refutations, and reward systems that resonate with a scientist's own self-interest.[2] Nevertheless, a common notion of proper scientific practice may be extrapolated from the deeds. That ideal of attaining knowledge through prescribed goals and means will be sketched below and its virtues extolled. And although this account is my interpretation, it is reinforced by general agreement, despite the diversity just acknowledged, with numerous scientists I know. But before pursuing that topic, two paragraphs about the historian's task are appropriate.

HISTORIES

Accounts are necessarily selective, and any historian's choices for inclusion and exclusion may be argued. I make no boast of infallibility in this coverage, but I do claim that the results and interpretations of the papers cited represent support for and challenge to the views of other scientists, affecting their thoughts and actions. The crucial issue is whether the evidence reported in the selected papers is sufficient to compel the changes in thoughts and actions that followed, and whether conflicting evidence not cited here could have compelled alternative responses. These should be the criteria of historical adequacy here.

History is written by the victors. Although this is frequently the case, that hoary cliché still ought not to imply that identifying who won and who lost in a given episode is necessarily problematic, or that reasons for the winning and the losing are undecipherable. By contrast, Steve Fuller proclaimed: "Whatever certainty we seem to have about whose side won in a particular scientific debate is more often the result of the winner's ability to suppress alternative accounts . . . than of some knock-down empirical demonstration."[3] How he quantitated that "more often" across all accounts is unclear, and contrary evidence—accounts by the losing side—have been noted here.[4] But the central issue is the *quality* of that "knock-down empirical evidence" which Fuller smugly disparages.

ASSUMPTIONS

Most biologists I know assume that there is a real world of organisms, cells, organelles, molecules, atoms, protons, electrons. But from acquaintance with neuroscience they recognize that perceptions are highly processed, and that innate computational devices shape our private representations which are then shared to create the communal view. (They also recognize that scientific instruments may confirm and extend our sensory channels.) If pressed, they acknowledge that scientific generalizations cannot be deduced from observing particular confirmations of hypotheses, and they may "justify" inductive inference by quasi-probabilistic accounts. Thus, scientific constructions are admittedly tentative and unprovable (by formal deductive logic).

So how can scientists believe they generate objective, reliable knowledge: formulations close and closer to the real world? If that real world contains parts and if those parts change and interact with some regularity, then the ideal scientific approach should, if any approach can, discover the parts, properties, and regular modes of interaction.[5] Philosophical nuances of antirealism—as well as the epistemological conundra of quantum mechanics—seem irrelevant to the concerns of most biologists.

If the real world is formless and inchoate, then there is nothing else to know. If the world is somewhere in between, searchers may report more than what exists; but that possibility would merely add another source of fallibility without suggesting alternative routes to knowing. Of course, the apparent world may be a dream (or fictional inputs to brains in vats), but still there seems to be no other game in town. (Nevertheless, even if the assumed regularities are actual, questions unanswerable through scientific approaches remain, including purpose and value and beauty and goodness.)

GOALS AND APPROACHES

The aim of science, according to a simplistic view, is to construct explanatory models representing entities of the (assumed) real world and their causal interactions. Here, concerns are linked to biology, where, as in other natural sciences, two realms are distinguishable: historical questions about origins and evolution, and functional questions about entities, parts, and interactions.

Questions addressed in this book fall in the latter category. And the answers are models replicating observed phenomena, thereby representing the (assumed) real entities and interactions and achieving "explanatory coherence" by assimilating results from multiple approaches. Such models are partial and restricted to the issues of interest—as for cells the organelles, for metabolic pathways the roster of enzymes and their catalytic capabilities, for enzymes the array of side chains at their active sites. The partial models are also hierarchical. Thus nerve action potentials can be described by models depicting the shape and magnitude of the voltage changes, the ion fluxes responsible for the voltage changes, and the conformational changes in membrane channels that permit the ion fluxes. Similarities are then specified in such terms as structure and identity, composition and function, action and mechanism, with reference to particular entities and processes established by experimental protocols. Generalizations are tied to the models and may be highly circumscribed and contextual.

The research presented here also reflects the guiding principle of explanatory reductionism,[6] which specifies that biological entities and processes are describable in terms of the entities and processes of chemistry. That principle does not denigrate the values of models expressed in other terms, but it does specify that no entities or processes inconsistent with the laws of chemistry are acceptable. It also sets a limit at which biologists can halt their explanatory regress. (Underlying that approach, of course, is the fundamental premise that biological processes *are* explainable in terms of the entities and processes of chemistry. From time to time that view is challenged, but no biological process has yet decisively failed such explanation.)

Construction of explanatory models then begins by searching for regularities, to identify entities and characterize their interactions. In practice this involves general strategies evident here: (1) isolation, separating to identify and examine, removing all other variables or holding them constant; (2) characterization, cataloging properties; (3) reconstitution, putting the system back together to ensure that the parts were not distorted by isolation and that the identified parts are sufficient; and (4) model building, fabricating an explanatory scheme. The range of approaches here included exploring and describing; inferring from accumulated evidence; and testing hypotheses. Frequently, the hypotheses were derived from general issues or from contemplating available data (as in the Hodgkin-Huxley model of the axon); occasionally they arose from complex theories (as did Mitchell's chemiosmotic mechanisms). Experimental strategies once developed may be adapted to broadly analogous quests. Precedent is a valuable guide to method as well as to hypothesis, but it is not guaranteed in either application.

Ideal approaches thus include: (1) describing accurately, using alternative, complementary methods; (2) controling variables to allow examination singly and in specified combination; (3) formulating the simplest explanatory model consistent with accepted principles and the observed phenomena;[7] and (4) testing all predictions of that model. Although these maxims appear commonsense, standards of practice have evolved to include such concerns as positive and negative controls, "blinded" observations, and statistical analyses. There are no mechanical algorithms for conducting scientific research, but there are aphorisms aplenty.

MEANS

The achievements related here depended crucially on developing new experimental approaches: for visualizing (electron microscopy, X-ray crystallography); for separating (centrifugation, chromatography, electrophoresis); for labeling (isotopes, antibodies); for measuring (spectroscopy, oxygen and ion-specific electrodes, voltage and patch-clamping); and for assessing (protein sequencing and sequence comparison, membrane protein reconstitution). The debt of new information to new techniques cannot be overemphasized, for scientific practice is opportunistic. Numerous unanswered questions remained at the end of the 1980s, clamoring for explication, but scientists can only examine what they have methods for examining.

Course

Thomas Kuhn, among others, stressed the role of problem-solving in the practice of science. He also distinguished between "normal science," the problem-solving that "articulates the paradigm," and "revolutionary science," the overthrowing of an old paradigm and the establishing of a new one.[8] As with many characterizations, the application of these distinctions is not so easy. Here I wish merely to note that much of the practice described in this book is identifiable as "normal science." Nevertheless, that course validated new experimental exemplars and inspired innovative explanatory models, for the branching issues and the revealed complexities fostered an evolutionary but explosive expansion. Answers to one question often revealed a dozen new problems to be solved. And in that process unsuspected relationships were uncovered (as in similarities between mechanisms for ATP synthesis and for driving flagellar motion), while previously plausible relationships were denied (as in proposed similarities between cation transport systems and contractile proteins).

CHARACTERISTICS OF SCIENTIFIC PRACTICE

Subjective and Objective Factors

Assessments of importance and novelty, intrusions of personal interests and ambition, responses to others' flexibility or intransigence—these all affect what experiments are done, what interpretations are stressed, and what aspects are noticed. Published reports are not mere repositories of data that readers may then download unthinkingly. But that trite observation does not imply that published reports contain neither verifiable results nor valid inferences that can be discovered by the cautious reader. There are many observations and measurements that all can agree about. Scientists' assumptions of a real world—and their personal experience that certain experiments repeatedly fail to satisfy their own expectations—also bolster belief in the objectivity of experiments honestly and competently performed. Such apparent constraints underlie not only scientists' claims about objective results but also their anticipations of rational choice between theories founded on such bases.

Choice of Projects and Choice of Theories

Another distinction that must be made is that between the choice an experimenter makes among guiding hypotheses when studying an unsettled issue, and the choice the scientific community makes, based on the accumulated evidence, to resolve an issue. The first choice may be difficult to justify when results are ambiguous and evidence incomplete. The scientist attempting to probe the issues often must follow one of several competing formulations, if only to design further experiments. Other factors may then intrude, including such "subjective" concerns as the likelihood of funding, the opportunity for patronage, the hope of personal glory. Rhetoric may persuade the ambivalent and power may recruit the insecure. On the other hand, in many of the cases described here, a time does come when the evidence available seems to be decisive despite all arguments, blandishments, and persuasions to the contrary. Individual perceptions of that moment may differ, in part through the time required for evaluating significance as well as for relinquishing cherished notions. But as we have seen, valid arguments do prevail.

(This reading of the past does not dismiss the likelihood that accepted evidence may be in error or that new evidence may overturn what once seemed assured. The concern is with choosing the better theory, not with establishing the best.)

Constraints

Two oft-cited contingencies could confound scientists' expectations of a real world constraining theory choice based on experiment. The first is that observations are theory dependent. Making any measurement depends on vast arrays of background assumptions and principles. But I find no instance here where rival theories led to different values for the same measurement. (Of course, the *significance* of some measurements was contested, with resolution then achieved through further [theory-dependent] studies.[9]) The second contingency is that theories are underdetermined by the evidence. This problem seems less dire than often professed.[10] Here, the range of alternatives available to contesting views has been tightly constrained by *shared beliefs*. Practical problems may impede the discrimination between recognized alternatives, and controversies may persist unresolved. But within the shared assumptions, new techniques can determine theories previously underdetermined. Obviously, not all alternatives are necessarily imagined by the rival camps, but rationality, not omniscience, is the issue.

Social Construction of Scientific Knowledge

The practice of science is never a solitary pursuit. Results are based on works of others, often in collaboration, and often in conflict with views of still others. One consequence is a diversity of approaches and analyses both broader and richer than anyone alone could muster. Another consequence is "negotiation" between rival factions through social ploys, from persuasion to coercion, clarification to distortion. In general, these disputes take the form of criticizing experiments and/or their interpretation. As such, they are, as described here, resolvable in most instances by further examination and closer analysis.

Ideal cooperative and competitive interactions may be fostered by benevolent generosity or hindered by entrenched power. These characteristics are not uniformly in abundance, and many scientists I know feel that their submitted papers and grant applications are not always reviewed fairly. But fairness is the standard publicly proclaimed.

The scientific community also establishes certain standards of practice through its apprenticeship system,[11] through reviews of submitted papers and grant applications, and through occassional edicts from professional societies. Traditional values, such as accepting $P < 0.05$ for statistical significance, become unquestioned norms.

Fallibility

Human error, ignorance, and misunderstanding accompany the actual practice of science and need no further belaboring here. Yet another acknowledgment of fallibility is the skepticism greeting unanticipated claims. Acceptance is then attained through careful confirmation, arguments against alternatives, and integration within the web of working knowledge. Since initial descriptions will later be shown to be inadequate, as further complexities are inevitably recognized and alternatives imagined, detractors may then label any founding hypothesis "wrong" despite the threads of verified consequences that persist.[12]

Trust and Skepticism

Between these opposing recourses scientists must pick their precarious paths, conscious of fallibility but aware that time will not permit their replicating all prior results. Active investigators must choose a guiding formulation among unestablished rivals, a choice based in part on trust—assessing claimants' past reliability—as well as on plausibilities, friendships, and hopes. Although previous experiments may not be replicated precisely,[13] the *consequences* of important findings are examined. Indeed, an experiment's significance springs from its providing new starting points for new investigations. (This reason alone should discourage attempts at fraud.[14])

RESOLVING CONFLICTS RATIONALLY

[S]cientists at the research front cannot settle disagreements through better experimentation, more knowledge, more advanced theories or clearer thinking.[15]

[T]here is no such thing as a unique critical experiment which can decisively adjudicate between one theory and another.[16]

[O]ne may hold onto any theory whatever in the face of any evidence.[17]

Few controversies . . . centre on such epistemic factors as accuracy and predicative capacities. The overwhelming majority turn on matters of goal orientation, social interests, and stubbornly held metaphysical beliefs.[18]

Undoubtedly, there are clarifications to be made about "settle," "decisively adjudicate," "hold onto," and "centre." But I will be happy to respond to any suggestions as to how:

1. Davson could maintain that red blood cells are impermeable to Na^+ and K^+ after studies by the Cohns and by Harris.

2. Conway could explain Na^+ transport across frog skin by redox pumps with Na^+/O_2 stoichiometries of four after studies by Zerahn and by Leaf and Renshaw.
3. The Hokins could defend their phosphatidic acid cycle against their own kinetic analysis.
4. Pinchot could rebut Pandit-Hovenkamp's measured rate of ADP labeling and Moudrianakis's finding polynucleotide phosphorylase activity.
5. Boyer could fault his reallocation of phosphohistidine to substrate-level rather than oxidative phosphorylation.
6. Chance could avoid capitulation to Jagendorf's demonstration that electro-chemical gradients could drive ATP synthesis.

Clearly, major reformulations followed reported results. If these results were not the sole motivation for each proselyte, analyses based on contemporary scientific standards reveal that the results were sufficient. Isotope studies by the Cohns and by Fenn's group and transport studies by Harris and by Danowski refuted the notion of membrane impermeability. These replicated arguments against red blood cell imper-meability surely made rational the abandonment of a previously dominant theory by 1942. On the other hand, refuting the chemical-coupling hypothesis, proving something does not exist, was strategically more difficult. And verifying the rival chemiosmotic hypothesis, which had complex predications, was protracted. Even so, by 1975, when H^+ transport by the respiratory chain was established and H^+ fluxes were shown to be kinetically adequate, acceptance of chemiosmotic principles could be deemed rational. At least in these cases, not only is rational choice demonstrable but the time when that rational choice should be made is determinable.[19] That all scientists did not accept these principles in unison (Fig. 18.5) reflects a separate issue, the imperfect practice of science.

Finally, three other issues affecting rational theory choice should be addressed.

Values

Here I will avoid issues about the epistemological or metaphysical consequences of being a "value." Instead, I will acknowledge Kuhn's list of scientific values,[20] but dismiss two of them as largely irrelevant to theory choice here: "simplicity" seems not to be a virtue advocated for discrimination,[21] and "fruitfulness," while a concern in selecting experimental paths, is not prominent in adjudicating between mature theories. "Scope" certainly was a cited concern but not one crucial to theory choice; moreover, estimations were arguable.[22] On the other hand, "accuracy"—how closely results meshed with predictions—and "consistency"—how well they agreed with accepted principles of science and with bodies of commonly accepted data—were obviously central concerns.

Accuracy is a straightforward notion, granted the assumption of a real world. The ability of explanatory models to match the phenomena (for theories to make accurate predictions) is an essential criterion in judging between rivals. But the cases described here did not display the hypothetical dilemma of theory A representing accurately phenomenon X but not Y and of theory B representing accurately Y but not X. I have found no contemporary arguments couched in these terms. The nearest approach is Racker's 1971 "scoreboard" comparing chemiosmotic and chemical-coupling hypotheses (Table 19.1).[23] Nevertheless, the three instances in which "chemical" was judged better did not represent generally accepted explanations at that time. (1) While Racker thought Ca^{2+} and K^+ transport was better explained by "chemical," Greville at that time argued for "chemiosmotic."[24] (2) Racker stressed that neither he nor others could confirm Hunter's newly published results.[25] (3) Although Racker felt that isotope exchange reactions (such as P_i/water exchange) were better explained by "chemical," he did not document this superiority (nor is it obvious to me that chemiosmotic formulations could not embrace these exchanges; Racker's rationale seems to have been that Mitchell ignored the issue). Instead, the arguments here were chiefly about whether the model described a particular phenomenon accurately (such as H^+/O ratios for redox pumps or reaction rates for hypothetical intermediates).

Consistency is a more complex issue, for that issue begs the question of consistency with what. Clearly beyond challenge were "established" principles of chemistry, such as the laws of thermodynamics, and of biology, such as general enzyme theory. Consistency with a mass of reported phenomena is more difficult to judge, since the validity of the phenoma may be challengable. Controversies can then be displaced from validity of theory to validity of phenomenon.

What I do not find, however, is an explicit competition between these two goals, of trading more consistent theories for more accurate ones, or vice versa.

Incommensurability

When paradigm shifts occur—as certainly happened in transitions from membrane impermeability to membrane pumps and from chemical to chemiosmotic coupling—

Table 19.1 "Scoreboard": Comparisons between Chemiosmotic and Chemical-Coupling Hypotheses.*

	Hypothesis	
	Chemiosmotic	Chemical
Role of membrane	+	−
Uncouplers and ionophores	+	−
H^+ translocation	+	−
Topography of respiratory chain	+	−
K^+ and Ca^{2+} transport	−	+
Hunter's experiments	−	+
Exchange reactions	−	+

*[From Racker and Horstman, 1972, Table 1. Reprinted by permission of Academic Press.]

the theories across that divide are claimed to be "incommensurable" since terms for entities and processes have different significances and references.[26] In the cases presented here, approaches and descriptions surely diverged. But experimental results were readily comprehensible from rival viewpoints; for example, the significance of Steinbach's experiment was clear to Conway, who fought hard against its validity on mutually comprehensible grounds, just as the significance of finding oxidative phosphorylation by nonvesicular mitochondrial fragments was clear to both camps. Both sides could recognize experiments that would refute one viewpoint; both sides could recognize experiments that would support one viewpoint.

Loss of Explanatory Power

Some investigators claim that when theories change, the replacing theories commonly do not explain all that the earlier ones did, so scientific knowledge is not cumulative.[27] (A standard example is the transition from Cartesian to Newtonian dynamics: the former explained why all planets circle the sun in the same direction, the latter did not.) But I find no loss of explanatory power in the major transitions recorded here, nor have I found contemporary protests that a supplanting theory left unanswered issues that were satisfactorily explained by a former theory.[28] Instead, the half-century of practice exhibited here resulted in the embellishment of older observations (for example, the modes of cation permeation through membranes) and their incorporation into broader issues (membrane protein structure, cellular communication, and metabolic control).

Justification

Elaborating the web of interrelated models is an easily identified goal of scientific practice. To what extent the proliferation savored here increases knowledge about the biological organisms may lie in the eye of the observer. Mayr, for example, was content with viewing the knee as a hinge. But I will specify one utilitarian realm benefiting from the investigations described in this book. Major therapeutic agents now include ion channel blockers for anesthesia (lidocaine, Xylocaine™), coronary insufficiency (nifedipine, Procardia™) and hypertension (diltiazem, Cardizem™); inhibitors of secondary active transport for depression (fluoxetine, Prozac™) and schizophrenia (reserpine, Serpasil™); and inhibitors of primary active transport for congestive heart failure (digoxin, Lanoxin™) and peptic ulcer (omeprazole, Prilosec™).

GENERALITY AND QUANTITATION

Here no claims are made about scientific practice beyond the cases illustrated. I invite other commentators to restrict their implications of generality that are based often, it seems,[29] on cases chosen to match their conclusions. Similarly, quantitations about scientific practice ("most," "few," "rarely") should be withheld until a random sampling of all scientific practice is examined.

RECOMMENDATION

Nevertheless, I will extract one general consideration for assisting scientific practice: openness. Scientists' fundamental reliance on experimental evidence can be realized only through open opportunities to conduct research and to communicate results freely and fairly. Despite some counterexamples, I think that the scientific practice described here was conducted accordingly. But in light of those counterexamples and of recognized human error, continuing concern is warranted. One approach to preserving diversity and maximizing innovation is through assuring pluralism: this means providing multiple, independent sources for supporting research and for disseminating results.

NOTES TO CHAPTER 19

1. Useful accounts include Cartwright (1983); Giere (1988); Kitcher (1993); and Laudan (1996).
2. Woodward and Goodstein (1996).
3. Fuller (1991), p. 362. He later repeated this statement (Fuller, 1993, pp. 9–10); however, the Bellman's stricture requires *three* tellings to make it true.
4. Admissions of error by Davson (1989) and by Hokin and Hokin-Neaverson (1989) have been noted. Another instance is Boyer (1981b), referring to his studies on phosphohistidine, in which be acknowledged bluntly "I was wrong" about what he had earlier characterized as "the most exciting [researches] of my professional career" (Boyer et al., 1963, p. 1080).
5. Salmon (1966), p. 53, dismisses such arguments as "impossibly vague." Still, I think this recognition displays the options available.
6. See, for example, Robinson (1986a, 1992).
7. A popular aphorism is Alfred North Whitehead's advice: "Seek simplicity, and then distrust it."
8. Kuhn (1970).
9. For example, Chance and Mitchell disputed what the changes in methylene blue adsorption signified; that conflict was resolved by clarifying studies on methylene blue and by using other dyes having different solubities (see Chapter 17).
10. Cf. Laudan (1996) for discussion.
11. For example, graduate students are indoctrinated in doubt through "journal clubs" where published papers are mercilessly castigated for faulty methods, absent controls, overlooked alternatives, and unjustified conclusions.
12. Such criticism was showered on Mitchell for the shortcomings of his initial formulations, whereas in the transport field, which had far more congenial scientists, no comparable condemnations were vented on the Albers-Post scheme for its initial shortcomings.
13. Precise replication may be required if entailed experiments fail. The difficulties encountered in replicating experiments have been advertized and some consequences of failing to replicate them described (Collins, 1985); however, no data about how often such

problems occur are available. A common response is that some experiments are easy to replicate, others are not.

14. The sole (recognized) instance of outright fraud relating to studies described in this book was the fabricated claim by George Webster of having found a chemical interemediate of oxidative phosphorylation (Webster et al., 1964; cf. Sanadi, 1965).

15. Collins and Pinch (1994), pp. 144, 145.

16. Morus (1993), p. 92: the review is repeating a conclusion from Marcello Pera's book, *The Ambiguous Frog.*

17. Willard V. O. Quine, quoted in Laudan (1996), p. 33.

18. Shortland (1988), p. 265; the review is drawing a conclusion from Engelhardt and Caplan (1987).

19. Contrast Kuhn (1970).

20. Kuhn (1977).

21. Of course, "simplicity" is invoked—and Occam's razor wielded—to dispose of extraneous bits.

22. Compare Conway's general advocacy of redox pumps (see Chapter 8) across the biological universe while excluding red blood cells.

23. Racker and Horstman (1972).

24. Greville (1969).

25. See Chapter 17.

26. Kuhn (1970).

27. Kuhn (1970); Laudan (1977).

28. Again, the nearest approach is Racker's "scoreboard."

29. See, for example, the bizarre selection of cases in Collins (1985); Collins and Pinch (1994); and Engelhardt and Caplan (1987).

appendix I

UNITS OF MEASUREMENT

Measurements here are based on the metric system, with fractions and multiples indicated by prefixes: 10^{-9}, nano (n); 10^{-6}, micro (μ); 10^{-3}, milli (m); 10^{-2}, centi (c); 10^3, kilo (k); 10^6, mega; 10^9, giga.

The unit of length is the meter (m); the Ångstrom (Å) is 10^{-10} m.

The unit of volume is the liter (L).

The unit of mass—commonly, if improperly, referred to as weight—is the gram (g).

Temperature in the text is listed as degrees Celsius (°C) but with the C omitted. In formulae the temperature T refers to the absolute scale, in kelvins (K).

Atomic and molecular masses (weights) are expressed as daltons (Da).

Quantities of molecules are expressed as gram-molecules, designated mole (mol), the grams of a molecular compound equal to its molecular mass. Thus, for glucose (molecular mass = 180 Da) one mole equals 180 g of glucose. A mole of any substance then contains a definite number of molecules, 6×10^{23}. For nonmolecular species—elements and their ions—the proper designation is gram-atom (g-atom), but—following common, convenient, but sloppy practice—this book uses mole for both.

Concentrations of molecules are expressed as moles per liter, designated molar (M). Thus, a one-liter solution containing 360 grams of glucose is a 2 M solution of glucose. For nonmolecular species, the proper designation is gram-atoms per liter, but again, following common, convenient, but sloppy practice, this book uses molar for both.

Acidity is expressed as pH, the negative logarithm of the hydrogen ion concentration. Thus, 1 μM H$^+$ (10^{-6} M) is equivalent to pH 6, 0.01 μM H$^+$ (10^{-8} M) to pH 8.

Electrical units are: for potential, volt (V); for charge, coulomb; for current, ampere; for resistance, ohm; and for capacitance, farad. Frequency is in herz (Hz).

Energy is expressed in calories (cal), and force in dynes.

appendix II

AMINO ACIDS AND PROTEINS

Amino acids have the general structure $X\text{-}CH(NH_3^+)-COO^-$, where X, the side chain, differs and provides characteristic properties. Common amino acids include

Acidic amino acids (with a second carboxyl): aspartate, glutamate

Basic amino acids: arginine, histidine, lysine

Neutral amino acids: alanine, asparagine, cysteine, glutamine, glycine, leucine, methionine, phenylalanine, proline, serine, threonine, tryptophan, tyrosine, valine.

Proteins are formed by amide ("peptide") bonds between the carboxyl of one amino acid and the amine of another. Amino acids in the sequence are numbered from the amine terminus; e.g., aspartate-257 would be the 257th amino acid from that end.

REFERENCES

Abercrombie, R.F., and P. DeWeer. Electric current generated by squid giant axon sodium pump: external K and internal ADP effects. *Am. J. Physiol.* 235:C63–C68, 1990.

Abood, L.G., and R.W. Gerard. Enzyme distribution in isolated particulates of rat peripheral nerve. *J. Cell. Comp. Physiol.* 43:379–392, 1954.

Abrahams, J.P., A.G.W. Leslie, R. Lutter, and J.E. Walker. Structure at 2.8 Å resolution of F₁-ATPase from bovine heart mitochondria. *Nature 370:621–628, 1994.*

Adair, G.S. Discussion. *Disc. Faraday Soc.* 21:285, 1956.

Addison, R. Primary structure of the *Neurospora* plasma membrane H⁺-ATPase deduced from the gene sequence. Homology to Na⁺/K⁺-, Ca²⁺-, and K⁺-ATPases. *J. Biol. Chem.* 261:14896–14901, 1986.

Adolfsen, R., and E.N. Moudrianakis. An exchange enzyme catalyzing incorporation of inorganic phosphate into adenosine diphosphate in *Alcaligenes faecalis. Arch. Biochem. Biophys.* 148:185–195, 1972.

Agnew, W.S., A.C. Moore, S.R. Levinson, and M.A. Raftery. Identification of a large molecular weight peptide associated with a tetrodotoxin binding protein from the electroplax of *Electrophorus electricus. Biochem. Biophys. Res. Commun.* 92:860–866, 1980.

Akera, T., F.S. Larsen, and T.M. Brody. The effect of ouabain on sodium-and potassium-activated adenosine triphosphatase from the hearts of several mammalian species. *J. Pharmacol. Exp. Ther.* 170:17–26, 1969.

Albers, R.W. Biochemical aspects of active transport. *Annu. Rev. Biochem.* 36:727–756, 1967.

Albers, R.W., S. Fahn, and G.J. Koval. The role of sodium ions in the activation of electrophorus electric organ adenosine triphosphatase. *Proc. Natl. Acad. Sci. USA* 50:474–481, 1963.

Albers, R.W., G.J. Koval, and G.J. Siegel. Studies on the interaction of ouabain and other cardio-active steroids with sodium-potassium-activated adenosine triphosphatase. *Mol. Pharmacol.* 4:324–336, 1968.

Aldridge, W.N. Adenosine triphosphatase in the microsomal fraction from rat brain. *Biochem. J.* 83:527–533, 1962.

Alexandre, A., B. Reynafarje, and A.L. Lehninger. Stoichiometry of vectorial H⁺ movements coupled to electron transport and to ATP syntheses in mitochondria. *Proc. Natl. Acad. Sci. USA* 75:5296–5300, 1978.

Allchin, D. *Resolving Disagreement in Science: The Ox-Phos Controversy, 1961–1977.* Ph.D. thesis, University of Chicago, 1991.

Allen, G., B.J. Trinnaman, and N.M. Green. The primary structure of the calcium-ion transporting adenosine triphosphatase protein of rabbit skeletal sarcoplasmic reticulum. *Biochem. J.* 187:591–616, 1980.

Altman, K.I. Some enzymologic aspects of the human erythrocyte. *Am. J. Med.* 27:936–951, 1959.

Alvarado, F., and R.K. Crane. Phlorizin as a competitive inhibitor of the active transport of sugars by hamster small intestine, *in vitro. Biochim. Biophys. Acta* 56:170–172, 1962.

Amzel, L.M., and P.L. Pedersen, Adenosine triphosphatase from rat liver mitochondria. *J. Biol. Chem.* 253:2067–2069, 1978.

Apps, D.K., and G. Schatz. An adenosine triphosphatase isolated from chromaffin-granule membranes is closely similar to F_1-adenosine triphosphatase of mitochondria. *Eur. J. Biochem.* 100:411–419, 1979.

Armstrong, C.M., and F. Bezanilla. Currents related to movement of the gating particles of the sodium channels. *Nature* 242:459–461, 1973.

Askari, A. (Na$^+$ + K$^+$)-ATPase: On the number of the ATP sites of the functional unit. *J. Bioenerg. Biomembr.* 19:359–374, 1987.

Askari, A., W.-H. Huang, and J.M. Antieau. Na$^+$, K$^+$-ATPase: Ligand-induced conformational transitions and alterations in subunit interactions evidenced by cross-linking studies. *Biochemistry* 19:1132–1140, 1980.

Avery, O.T., C.M. MacLeod, and M. McCarty. Studies on the chemical nature of the substance inducing transformation by a deoxyribonucleic acid fraction isolated from *Pneumococcus* type III. *J. Exp. Med.* 79:137–158, 1944.

Azzone, G.F., T. Pozzan, and S. Masari. Proton electrochemical gradient and phosphate potential in mitochondria. *Biochim. Biophys. Acta* 501:307–316, 1978.

Bader, H., R.L. Post, and G.H. Bond. Comparison of sources of a phosphorylated intermediate in transport ATPase. *Biochim. Biophys. Acta* 150:41–46, 1968.

Bader, H., R.L. Post, and D.H. Jean. Further characterization of a phosphorylated intermediate in (Na$^+$ + K$^+$)-dependent ATPase. *Biochim. Biophys. Acta* 143:229–238, 1967.

Bader, H., A.K. Sen, and R.L. Post. Isolation and characterization of a phosphorylated intermediate in the (Na$^+$ + K$^+$) system-dependent ATPase. *Biochim. Biophys. Acta* 118:106–115, 1966.

Bagshaw, C.R., and D.R. Trentham. The reversibility of adenosine triphosphate cleavage by myosin. *Biochem. J.* 133:323–328, 1973.

Baker, J.R. The cell-theory: A restatement, history, and critique. Part I. *Q. J. Microsc. Sci.* 89:103–125, 1948.

Baker, J.R. The cell-theory: A restatement, history, and critique. Part II. *Q. J. Microsc. Sci.* 90:87–108, 1949.

Baker, J.R. The cell-theory: A restatement, history, and critique. *Q. J. Microsc. Sci.* 93:157–190, 1952.

Baker, P.F., M.P. Blaustein, A.L. Hodgkin, and R.A. Steinhardt. The influence of calcium on sodium efflux in squid axons. *J. Physiol.* 200:431–458, 1969.

Ball, E.G., and J.F. Sadusk. A study of the estimation of sodium in blood serum. *J. Biol. Chem.* 113:661–674, 1936.

Bangham, A.D., M.M. Standish, and J.C. Watkins. Diffusion of univalent ions across the lamellae of swollen phospholipids. *J. Mol. Biol.* 14:238–252, 1965.

Bar, R.S., D.W. Deamer, and D.G. Cornwell. Surface area of human erythrocyte lipids: Reinvestigation of experiments on plasma membrane. *Science* 153:1010–1012, 1966.

Barlogie, B., W. Hasselbach, and M. Makinose. Activation of calcium efflux by ADP and inorganic phosphate. *FEBS Lett.* 12:267–268, 1971.

Bartley, W., and J.E. Amoore. The effects of manganese on the solute content of rat-liver mitochondria. *Biochem. J.* 69:348–360, 1958.

Bartley, W., and R.E. Davies. Active transport of ions by sub-cellular particles. *Biochem. J.* 57:37–49, 1954.

Bashford, C.L., R.P. Casey, G.K. Radda, and G.A. Ritchie. The effect of uncouplers on catecholamine incorporation by vesicles of chromafin granules. *Biochem. J.* 148:153–155, 1975.

Bastide, F., G. Meissner, S. Fleischer, and R.L. Post. Similarity of the active site of phosphorylation of the adenosine triphosphatase for transport of sodium and potassium ions in kidney to that for transport of calcium ions in the sarcoplasmic reticulum of muscle. *J. Biol. Chem.* 248:8385–8391, 1973.

Bäuerlein, E. A reinvestigation of Hunter's model system for oxidative phosphorylation. *Biochem. Biophys. Res. Commun.* 47:1088–1092, 1972.

Beadle, G.W., and E.L. Tatum. Genetic control of biochemical reactions in *Neurospora. Proc. Natl. Acad. Sci. USA* 27:499–506, 1941.

Bear, R.S., and F.O. Schmitt. Electrolytes in the axoplasm of the giant nerve fibers of the squid. *J. Cell. Comp. Physiol.* 14:205–215, 1939.

Beaugé, L.A., and I.M. Glynn. Sodium ions, acting at high-affinity extracellular sites, inhibit sodium-ATPase activity of the sodium pump by slowing dephosphorylation. *J. Physiol.* 289:17–31, 1979a.

Beaugé, L.A., and I.M. Glynn. Occlusion of K ions in the unphosphorylated sodium pump. *Nature* 280:510–512, 1979b.

Bechtel, W. Two common errors in explaining biological and psychological phenomena. *Phil. Sci.* 49:549–574, 1982.

Beechey, R.B., C.T. Holloway, I.G. Knight, and A.M. Roberton. Dicyclohexylcarbodiimide—an inhibitor of oxidative phosphorylation. *Biochem. Biophys. Res. Commun.* 23:75–80, 1966.

Beinert, H., and R.H. Sands. Studies on succinic and DPNH dehydrogenase preparations by paramagnetic resonance (EPR) spectroscopy. *Biochem. Biophys. Res. Commun.* 3:41–46, 1960.

Bendall, J.R. Effect of the "Marsh" factor on the shortening of muscle fibre models in the presence of adenosine triphosphate. *Nature* 170:1058–1060, 1952.

Bendall, J.R. Further observations on a factor (the "Marsh" factor) effecting relaxation of ATP-shortened muscle-fibre models, and the effect of Ca and Mg ions upon it. *J. Physiol.* 121:232–254, 1953.

Bennett, H.S., and K.R. Porter. An electron microscope study of sectioned breast muscle of the domestic fowl. *Am. J. Anat.* 93:61–89, 1953.

Bennett, V., and P.J. Stenbuck. The membrane attachment protein for spectrin is associated with band 3 in human erythrocyte membranes. *Nature* 280:468–473, 1979.

Berendsen, H.J.C., and H.T. Edzes. The observation and general interpretation of sodium magnetic resonance in biological material. *Ann. N.Y. Acad. Sci.* 204:459–485, 1973.

Bernal, J.D., and D. Crowfoot. X-ray photographs of crystalline pepsin. *Nature* 133:794–795, 1934.

Berridge, M.J. Rapid accumulation of inositol triphosphate reveals that agonists hydrolyze polyphosphoinositides instead of phosphatidylinositol. *Biochem. J.* 212:849–858, 1983.

Berry, E.A., and P.C. Hinkle. Measurement of the electrochemical proton gradient in sub-mitochondrial particles. *J. Biol. Chem.* 258:1474–1486, 1983.

Best, C.H., and N.B. Taylor. *The Physiological Basis of Medical Practice,* 2nd ed. Baltimore: Williams and Wilkins, 1939.

Bieber, L.L., O. Lindberg, J.J. Duffy, and P.D. Boyer. Rate and extent of labelling of bound phosphohistidine as related to its role in mitochondria. *Nature* 202:1316–1318, 1964.

Bielawski, J., T.E. Thompson, and A.L. Lehninger. The effect of 2,4-dinitrophenol on the electrical resistance of phospholipid bilayer membranes. *Biochem. Biophys. Res. Commun.* 24:948–954, 1966.

Bigay, J., P. Deterre, C. Pfister, and M. Chabre (1985). Fluoroaluminates activate transducin-GDP by mimicking the γ-phosphate of GTP in its binding site. *FEBS Lett.* 191:181–185, 1985.

Bihler, I., and R.K. Crane. Studies on the mechanism of intestinal absorption of sugars. V. *Biochim. Biophys. Acta* 59:78–93, 1962.

Bihler, I., K.A. Hawkins, and R.K. Crane. Studies on the mechanism of intestinal absorption of sugars. VI. *Biochim. Biophys. Acta* 59:94–102, 1962.

Blair, P.V., T. Oda, D.E. Green, and H. Fernández-Morán. Studies on the electron transfer system. *Biochemistry* 2:756–764, 1963.

Blake, C.C.F., D.F. Koenig, G.A. Mair, A.C.T. North, D.C. Phillips, and V.R. Sarma. Structure of hen egg-white lysozyme. *Nature* 206:757–761, 1965.

Blaustein, M.P., and A.L. Hodgkin. The effect of cyanide on the efflux of calcium from squid axons. *J. Physiol.* 200:497–527, 1969.

Blostein, R. Sodium pump-catalyzed sodium-sodium exchange associated with ATP hydrolysis. *J. Biol. Chem.* 258:7948–7953, 1983.

Blow, D.M., J.J. Birktoft, and B.S. Hartley. Role of a buried acid group in the mechanism of action of chymotrypsin. *Nature* 221:337–340, 1969.

Bodansky, M. *Introduction to Physiological Chemistry*, 4th ed. New York: Wiley, 1938.

Bonting, S.L., and L.L. Caravaggio. Studies on sodium-potassium-activated adenosinetriphosphatase. V. Correlation of enzyme activity with cation flux in six tissues. *Arch. Biochem. Biophys.* 101:37–46, 1963.

Bonting, S.L., K.A. Simon, and N.M. Hawkins. Studies on sodium-potassium-activated adenosine triphosphatase. I. Quantitative distribution in several tissues of the cat. *Arch. Biochem. Biophys.* 95:416–423, 1961.

Bordley, J., and A.M. Harvey. *Two Centuries of American Medicine*. Philadelphia: W.B. Saunders, 1976.

Bowman, B.J., S.E. Mainzer, K.E. Allen, and C.W. Slayman. Effects of inhibitors on the plasma membrane and mitochondrial adenosine triphosphatases of *Neurospora crassa*. *Biochim. Biophys. Acta* 512:13–28, 1978.

Boyer, P.D. Carboxyl activation as a possible common reaction in substrate-level and oxidative phosphorylation and in muscle contraction. *In* King, T.E., H.S. Mason, and M. Morrison (eds.) *Oxidases and Related Redox Systems*. New York: Wiley, 1965, pp. 994–1017.

Boyer, P.D. Energy transduction and proton translocation by adenosine triphosphatases. *FEBS Lett.* 50:91–94, 1975.

Boyer, P.D. The binding-change mechanism of ATP syntheses. *In* Lee, C.P., G. Schatz, and L. Ernster (eds.) *Membrane Bioenergetics*. Reading, MA: Addison-Wesley, 1979, pp. 461–479.

Boyer, P.D. On the mechanism of H^+-ATP synthase. *In* Skulachev, V.P., and P.C. Hinkle (eds.) *Chemiosmotic Proton Circuits in Biological Membranes*. London: Addison-Wesley, 1981a, pp. 395–406.

Boyer, P.D. An autobiographical sketch related to my efforts to understand oxidative phosphorylation. *In* Semenza, G. (ed.) *Of Oxygen, Fuels, and Living Matter*. Chichester: Wiley, 1981b, pp. 229–264.

Boyer, P.D., B. Chance, L. Ernster, P. Mitchell, E. Racker, and E.C. Slater. Oxidative phosphorylation and photophosphorylation. *Annu. Rev. Biochem.* 46:955–1026, 1977.

Boyer, P.D., R.L. Cross, and W. Momsen. A new concept for energy coupling to oxidative phosphorylation based on a molecular explanation of the oxygen exchange reactions. *Proc. Natl. Acad. Sci. USA* 70:2837–2839, 1973.

Boyer, P.D., M. DeLuca, K.E. Ebner, D.E. Hultquist, and J.B. Peter. Identification of a phosphohistidine in digests from a probable intermediate of oxidative phosphorylation. *J. Biol. Chem.* 237:PC3306–PC3308, 1962.

Boyer, P.D., D.E. Hultquist, J.B. Peter, G. Kreil, R.A. Mitchell, M. Deluca, J.W. Hinkson, L.G. Bulter, and R.W. Moyer. Role of the phosphorylated imidazole group in phosphorylation and energy transfer reactions. *Federation Proc.* 22:1080–1087, 1963.

Boyer, P.D., and W.E. Kohlbrenner. The present status of the binding-change mechanism and its relation to ATP formation by chloroplasts. *In* Selman, B.R., and S. Selman-Reimer (eds.) *Energy Coupling in Photosynthesis.* New York: Elsevier/North Holland, 1981, pp. 231–240.

Boyer, P.D., H.A. Lardy, and P.H. Phillips. The role of potassium in muscle phosphorylations. *J. Biol. Chem.* 146:673–682, 1942.

Boyer, P.D., H.A. Lardy, and P.H. Phillips. Further studies on the role of potassium and other ions in the phosphorylation of the adenylic system. *J. Biol. Chem.* 149:529–541, 1943.

Boyer, P.D., W.W. Luchsinger, and A.B. Falcone. O^{18} and P^{32} exchange reactions of mitochondria in relation to oxidative phosphorylation. *J. Biol. Chem.* 223:405–421, 1956.

Boyle, P.J., and E.J. Conway. Potassium accumulation in muscle and associated changes. *J. Physiol.* 100:1–63, 1941.

Boyle, P.J., E.J. Conway, F. Kane, and H.L. O'Reilly. Volume of interfibre spaces in frog muscle and the calculation of concentrations in the fibre water. *J. Physiol.* 99:401–414, 1941.

Bozler, E. Relaxation in extracted muscle fibers. *J. Gen. Physiol.* 38:149–159, 1954.

Bragg, P.D., and C. Hou. Subunit composition, function, and spatial arrangement in the Ca^{2+}- and Mg^{2+}-activated adenosine triphosphatases of *Escherichia coli* and *Salmonella typhimurium. Arch. Biochem. Biophys.* 167:311–321, 1975.

Brand, E., L.J. Saidel, W.H. Goldwater, B. Kassell, and F.J. Ryan. The empirical formula of β-lactoglobulin. *J. Am. Chem. Soc.* 67:1524–1532, 1945.

Brand, M.D., B. Reynafarje, and A.L. Lehninger. Stoichiometric relationship between energy-dependent proton ejection and electron transport in mitochondria. *J. Biol. Chem.* 73:437–441, 1976.

Branton, D. Fracture faces of frozen membranes. *Proc. Natl. Acad. Sci. USA* 55:1048–1056, 1966.

Brazier, M.A.B. The historical development of neurophysiology. *In* Field, J., H.W. Magoun, and V.E. Hall (eds.) *Handbook of Physiology. (Section 1), Neurophysiology.* Washington: American Physiological Society, 1959, pp. 1–24.

Bretscher, M.S. A major protein which spans the human erythrocyte membrane. *J. Mol. Biol.* 59:351–357, 1971.

Brierley, G., E. Murer, E. Bachmann, and D.E. Green. Studies on ion transport. *J. Biol. Chem.* 238:3482–3489, 1963.

Brockerhoff, H., and C.E. Ballou. The structure of the phosphoinositide complex of beef brain. *J. Biol. Chem.* 236:1907–1911, 1961.

Brotherus, J.R., L. Jacobsen, and P.L. Jørgensen. Soluble and enzymatically stable ($Na^+ + K^+$)-ATPase from mammalian kidney consisting predominantly of protomer $\alpha\beta$-units. *Biochim. Biophys. Acta* 731:290–303, 1983.

Brotherus, J.R., J.V. Møller, and P.L. Jørgensen. Soluble and active renal Na,K-ATPase with maximum protein molecular mass 170,000 ± 9000 daltons; formation of larger units by secondary aggregation. *Biochem. Biophys. Res. Commun.* 100:146–154, 1981.

Büchel, D.E., B. Gronenborn, and B. Müller-Hill. Sequence of the lactose permease gene. *Nature* 283:541–545, 1980.

Bullough, D.A., and W.S. Allison. Inactivation of the bovine heart mitochondrial F_1-ATPase by 5'-p-fluorosulfonylbenzoyl [^{3}H]-inosine is accompanied by modification of tyrosine 345 in a single β subunit. *J. Biol. Chem.* 261:14171–14177, 1986.

Burgen, A. S. V. The physiological ultrastructure of cell membranes. *Can. J. Biochem. Physiol.* 35:569–578, 1957.

Burton, A. C. The properties of the steady state compared to those of equilibrium as shown in characteristic biological behavior. *J. Cell. Comp. Physiol.* 14:327–349, 1939.

Cabantchik, Z. I., and A. Rothstein. Membrane proteins related to anion permeability of human red blood cells. *J. Membr. Biol.* 15:207–226, 1974.

Cabantchik, Z. I., D. J. Volsky, H. Ginsburg, and A. Loyter. Reconstitution of the erythrocyte anion transport system. *Ann. N.Y. Acad. Sci.* 341:444–454, 1980.

Caldwell, P. C., A. L. Hodgkin, R. D. Keynes, and T. I. Shaw. The effects of injecting "energy-rich" phosphate compounds on the active transport of ions in the giant axons of *Loligo.* *J. Physiol.* 152:561–590, 1960a.

Caldwell, P. C., A. L. Hodgkin, R. D. Keynes, and T. I. Shaw. Partial inhibition of the active transport of cations in the giant axons of *Loligo. J. Physiol.* 152:591–600, 1960b.

Caldwell, P. C., and R. D. Keynes. The utilization of phosphate bond energy for sodium extrusion from giant axons. *J. Physiol.* 137:12P–13P, 1957.

Caldwell, P. C., and R. D. Keynes. The effect of ouabain on the efflux of sodium from a squid giant axon. *J. Physiol.* 148:8P–9P, 1959.

Canfield, V. A., C. T. Okamoto, D. Chow, J. Dorfman, P. Gros, J. G. Forte, and R. Levenson. Cloning of the H,K-ATPase β subunit. *J. Biol. Chem.* 265:19878–19884, 1990.

Cantley, L. C., L. G. Cantley, and L. Josephson. A characterization of vanadate interactions with the (Na,K)-ATPase. *J. Biol. Chem.* 253:7361–7368, 1978.

Cantley, L. C., and L. Josephson. A slow interconversion between active and inactive states of the (Na-K)ATPase. *Biochemistry* 15:5280–5287, 1976.

Cantley, L. C., L. Josephson, R. Warner, M. Yanagisawa, C. Lechene, and G. Guidotti. Vanadate is a potent (Na,K)-ATPase inhibitor found in ATP derived from muscle. *J. Biol. Chem.* 252:7421–7423, 1977.

Carey, M. J., E. J. Conway, and R. P. Kernan. Secretion of sodium ions by the frog sartorius. *J. Physiol.* 148:51–82, 1959.

Caroni, P., and E. Carafoli. The Ca^{2+}-pumping ATPase of heart sarcolemma. *J. Biol. Chem.* 256:3263–3270, 1981.

Cartwright, N. *How the Laws of Physics Lie.* New York: Oxford University Press, 1983.

Carvalho, M. daG. C., D. G. deSouza, and L. deMeis. On a possible mechanism of energy conservation in sarcoplasmic reticulum membrane. *J. Biol. Chem.* 251:3629–3636, 1976.

Catterall, W. A. Structure and function of voltage-sensitive ion channels. *Science* 242:50–61, 1988.

Cavieres, J. D., and I. M. Glynn. Sodium-sodium exchange through the sodium pump: The roles of ATP and ADP. *J. Physiol.* 297:637–645, 1979.

Chance, B. Energy-linked cytochrome oxidation in mitochondria. *Nature* 189:719–725, 1961.

Chance, B. The energy-linked reaction of calcium with mitochondria. *J. Biol. Chem.* 240:2729–2748, 1965.

Chance, B., A. Azzi, I. Y. Lee, C.-P. Lee, and L. Mela. The nature of the respiratory chain. *FEBS Symp.* 17:233–273, 1969.

Chance, B., C.-P. Lee, and L. Mela. Control and conservation of energy in the cytochrome chain. *Federation Proc.* 26:1341–1354, 1967.

Chance, B., and L. Mela. Energy-linked changes of hydrogen ion concentration in sub-mitochondrial particles. *J. Biol. Chem.* 242:830–844, 1967.

Chance, B., D. F. Parsons, and G. R. Williams. Cytochrome content of mitochondria stripped of inner membrane structure. *Science* 143:136–139, 1964.

Chance, B., and G. R. Williams. The respiratory chain and oxidative phosphorylation. *Adv. Enzymol.* 17:65–134, 1956.

Chappell, J.B., and A.R. Crofts. Gramicidin and ion transport in isolated liver mitochondria. *Biochem. J.* 95:393–402, 1965.

Chappell, J.B., and G.D. Greville. Dependence of mitochondrial swelling on oxidizable substrates. *Nature* 182:813–814, 1958.

Charnock, J.S., and R.L. Post. Evidence of the mechanism of ouabain inhibition of cation activated adenosine triphosphatase. *Nature* 199:910–911, 1963.

Charnock, J.S., A.S. Rosenthal, and R.L. Post. Studies on the mechanism of cation transport. II. A phosphorylated intermediate in the cation stimulated enzymic hydrolysis of adenosine triphosphate. *Aust. J. Exp. Biol.* 41:675–686, 1963.

Chernyak, B.V., P.A. Dibrov, A.N. Glagolev. M.Y. Sherman, and V.P. Skulachev. A novel type of energetics in a marine alkali-tolerant bacterium. *FEBS Lett.* 164:38–42, 1983.

Chiesi, M., and G. Inesi. Adenosine 5'-triphosphate dependent fluxes of manganese and hydrogen ions in sarcoplasmic reticulum vesicles. *Biochemistry* 19:2912–2918, 1980.

Christensen, H.V., and T.R. Riggs. Concentrative uptake of amino acids by the Ehrlich mouse ascites carcinoma cell. *J. Biol. Chem.* 194:57–68, 1952.

Christensen, H.V., T.R. Riggs, H. Fischer, and L.M. Palatine. Amino acid concentration by a free cell neoplasm. *J. Biol. Chem.* 198:1–15, 1952.

Cidon, S., and N. Nelson. A novel ATPase in the chromaffin granule membrane. *J. Biol. Chem.* 258:2892–2898, 1983.

Clarkson, E.M., and M. Maizels. Distribution of phosphatases in human erythrocytes. *J. Physiol.* 116:112–128, 1952.

Cockrell, R.S., E.J. Harris, and B.C. Pressman. Energetics of potassium transport in mitochondria induced by valinomycin. *Biochemistry* 5:2326–2335, 1966.

Cohn, M. A study of oxidative phosphorylation with O^{18}-labeled inorganic phosphate. *J. Biol. Chem.* 201:735–750, 1953.

Cohn, W.E., and E.T. Cohn. Permeability of red corpuscles of the dog to sodium ion. *Proc. Soc. Exp. Biol. Med.* 41:445–449, 1939.

Cole, K.S. Dynamic electrical characteristics of the squid axon membrane. *Arch. Sci. Physiol.* 3:253–258, 1949.

Cole, K.S. Membrane excitation of the Hodgkin-Huxley axon. *Federation Proc.* 13:28, 1954.

Cole, K.S. Membrane excitation of the Hodgkin-Huxley axon. Preliminary corrections. *J. Appl. Physiol.* 12:129–130, 1958.

Cole, K.S. *Membranes, Ions, and Impulses.* Berkeley: University of California Press, 1968.

Cole, K.S., H.A. Antosiewicz, and P. Rabinowitz. Automatic computation of nerve excitation. *J. Soc. Industr. Appl. Math.* 3:153–172, 1955.

Cole, K.S., and H.J. Curtis. Electric impedance of the squid giant axon during activity. *J. Gen. Physiol.* 22:649–670, 1939.

Cole, K.S., and A.L. Hodgkin. Membrane and protoplasm resistance in the squid giant axon. *J. Gen. Physiol.* 22:671–687, 1939.

Collander, R. The permeability of plant protoplasts to non-electrolytes. *Trans. Faraday Soc.* 33:985–990, 1937.

Collins, H., and T. Pinch. *The Golem,* 2nd ed. Cambridge: Cambridge University Press, 1994.

Collins, H.M. *Changing Order.* London: Sage Publications, 1985.

Collins, H.M., and G. Cox. Recovering relativity: Did prophecy fail? *Soc. Stud. Sci.* 6:423–444, 1976.

Collings, J.H., A.S. Zot, W.J. Ball, L.K. Lane, and A. Schwartz. Tryptic digest of the α subunit of lamb kidney (Na^+ + K^+)-ATPase. *Biochim. Biophys. Acta* 742:358–365, 1983.

Conway, E.J. The physiological significance of inorganic levels in the internal medium of animals. *Biol. Rev.* 20:56–72, 1945.

Conway, E.J. Ionic permeability of skeletal muscle fibres. *Nature* 157:715–717, 1946.

Conway, E.J. Exchanges of K, Na and H ions between the cell and the environment. *Irish J. Med. Sci.* 6:593–609, 1947a.

Conway, E.J. Exchanges of K, Na and H ions between the cell and the environment. *Irish J. Med. Sci.* 6:654–680, 1947b.

Conway, E.J. The biological performance of osmotic work. A redox pump. *Science* 113:270–273, 1951.

Conway, E.J. A redox pump for the biological performance of osmotic work, and its relation to the kinetics of free ion diffusion across membranes. *Int. Rev. Cytol.* 2:419–445, 1953.

Conway, E.J. Some aspects of ion transport through membranes. *Symp. Soc. Exp. Biol.* 8:297–324, 1954.

Conway, E.J. Evidence for a redox pump in the active transport of cations. *Int. Rev. Cytol.* 4:377–396, 1955.

Conway, E.J. Nature and significance of concentration relations of potassium and sodium ions in skeletal muscle. *Physiol. Rev. 37:84–132, 1957a.*

Conway, E.J. Discussion. *In* Murphy, Q.R. (ed.) *Metabolic Aspects of Transport across Cell Membranes.* Madison: University of Wisconsin Press, 1957b, p. 194.

Conway, E.J. The redox pump theory and present evidence. *In* Coursaget, J. (ed.) *The Method of Isotopic Tracers Applied to the Study of Active Ion Transport.* London: Pergamon Press, 1959, pp. 1–27.

Conway, E.J. Principles underlying the exchanges of K and Na Ions across cell membranes. *J. Gen. Physiol.* 43 (suppl.) 17–41, 1960.

Conway, E.J., and P.J. Boyle. A mechanism for the concentrating of potassium by cells, with experimental verification for muscle. *Nature* 144:709–710, 1939.

Conway, E.J., and T.G. Brady. Source of the hydrogen ions in gastric juice. *Nature* 162:456–457, 1948.

Conway, E.J., O. Fitzgerald, and T.C. MacDougald. Potassium accumulation in the proximal convoluted tubules of the frog's kidney. *J. Gen. Physiol.* 29:305–334, 1946.

Conway, E.J., and D. Hingerty. Relations between potassium and sodium levels in mammalian muscle and blood plasma. *Biochem. J.* 42:372–376, 1948.

Conway, E.J., F. Kane, P. Boyle, and H. O'Reilly. Localization of electrolytes in muscle. *Nature* 144:752–753, 1939.

Conway, E.J., and E. O'Malley. Nature of the cation exchanges during short-period yeast fermentation. *Nature* 153:555–556, 1945.

Cope, F.W. NMR Evidence for complexing of Na^+ in muscle, kidney, and brain by actomyosin. *J. Gen. Physiol.* 50:1353–1375, 1967.

Cope, F.W. Nuclear magnetic resonance evidence using D_2O for structured water in muscle and brain. *Biophys. J.* 9:303–319, 1969.

Copenhaver, J.H., and H.A. Lardy. Oxidative phosphorylations: Pathways and yield in mitochondrial preparations. *J. Biol. Chem.* 195:225–238, 1952.

Cox, G.B., D.A. Jans, A.L. Femmel, F. Gibson, and L. Hatch. The mechanism of ATP synthase conformational change by rotation of the b-subunit. *Biochim. Biophys. Acta* 768:201–208, 1984.

Crane, E.E., and R.E. Davies. Chemical energy relations in gastric mucosa. *Biochem. J.* 43:xlii, 1948.

Crane, E.E., R.E. Davies, and N.M. Longmuir. Relations between hydrochloric acid secretion and electrical phenomena in frog gastric mucosa. *Biochem. J.* 43:321–342, 1948.

Crane, F.L., Y. Hatefi, R.L. Lester, and C. Widmer. Isolation of a quinone from beef heart mitochondria. *Biochim. Biophys. Acta* 25:220–221, 1957.

Crane, R.K. Intestinal absorption of sugars. *Physiol. Rev.* 40:789–825, 1960.

Crane, R.K. Uphill outflow of sugar from intestinal epithelial cells induced by reversal of the Na^+ gradient. *Biochem. Biophys. Res. Commun.* 17:481–485, 1964.

Crane, R.K. Na$^+$-dependent transport in the intestine and other animal tissues. *Federation Proc.* 24:1000–1006, 1965.

Crane, R.K. The road to ion-coupled membrane processes. *Comp. Biochem.* 35:43–69, 1983.

Crane, R.K., G. Forstner, and A. Eichholz. Studies on the mechanism of intestinal absorption of sugars. *Biochim. Biophys. Acta* 109:467–477, 1965.

Crane, R.K., and S.M. Krane. On the mechanism of the intestinal absorption or sugars. *Biochim. Biophys. Acta* 20:433–434, 1956.

Crane, R.K., P. Malathi, and H. Preiser. Reconstitution of specific Na$^+$-dependent D-glucose transport in liposomes by Triton X-100 extracted proteins from purified brush border membranes of hamster small intestine. *Biochem. Biophys. Res. Commun.* 71:1010–1016, 1976.

Crane, R.K., D. Miller, and I. Bihler. The restrictions on possible mechanisms of intestinal active transport of sugars. *In* Kleinzeller, A., and A. Kotyk (eds.) *Membrane Transport and Metabolism.* New York: Academic Press, 1961, pp. 439–449.

Cross, R.L. The mechanism and regulation of ATP synthesis by F$_1$-ATPases. *Annu. Rev. Biochem.* 50:681–714, 1981.

Cross, R.L., and T.M. Duncan. Subunit rotation in F$_o$F$_1$-ATP synthases as a means of coupling proton transport through F$_o$ to the binding changes in F$_1$. *J. Bioenerg. Biomembr.* 28:403–408, 1996.

Cross, R.L., C. Grubmeyer, and H.S. Penefsky. Mechanism of ATP hydrolysis by beef heart mitochondrial ATPase. *J. Biol. Chem.* 257:12101–12105, 1982.

Cross, R.L., and C.M. Nalin. Adenine nucleotide binding sites on beef heart F$_1$-ATPase. *J. Biol. Chem.* 257:2874–2881, 1982.

Csáky, T. Über die Rolle der Struktur des Gluoscmoleküls bei der Resorption aus dem Dünndarm. *Hoppe-Seylers Z. Physiol. Chem.* 277:47–57, 1942.

Csáky, T.Z. Significance of sodium ions in active intestinal transport of nonelectrolytes. *Am. J. Physiol.* 201:999–1001, 1961.

Csáky, T.Z. Effect of cardioactive steroids on the active transport of non-electrolytes. *Biochim. Biophys. Acta* 74:160–162, 1963a.

Csáky, T.Z. A possible link between active transport of electrolytes and nonelectrolytes. *Federation Proc.* 22:3–7, 1963b.

Csáky, T.Z., and G.W. Fernald. Localization of the "sugar pump" in the intestinal epithelium. *Nature* 191:709–710, 1961.

Csáky, T.Z., and Y. Hara. Inhibition of active intestinal transport by digitalis. *Am. J. Physiol.* 209:467–472, 1965.

Csáky, T.Z., H.G. Hartzog, and G.W. Fernald. Effect of digitalis on active intestinal sugar transport. *Am. J. Physiol.* 200:459–460, 1961.

Csáky, T.Z., and M. Thale. Effect of ionic environment on intestinal sugar transport. *J. Physiol.* 151:59–65, 1960.

Csáky, T.Z., and L. Zollicoffer. Ionic effect on intestinal transport of glucose in the rat. *Am. J. Physiol.* 198:1056–1058, 1960.

Curran, P.F., J.J. Hajjar, and I.M. Glynn. The sodium-alanine interaction in rabbit ileum. *J. Gen. Physiol.* 55:297–308, 1970.

Curtis, H.S., and K.S. Cole. Transverse electric impedance of the squid giant axon. *J. Gen. Physiol.* 21:757–765, 1938.

Curtis, H.S., and K.S. Cole. Membrane resting and action potentials from the squid giant axon. *J. Cell. Comp. Physiol.* 19:135–144, 1942.

Dame, J.B., and G.A. Scarborough. Identification of the hydrolytic moiety of the *Neurospora* plasma membrane H$^+$-ATPase and demonstration of a phosphoryl-enzyme intermediate in its catalytic mechanism. *Biochemistry* 19:2931–2937, 1980.

Dame, J.B., and G.A. Scarborough. Identification of the phosphorylated intermediate of the

Neurospora plasma membrane H⁺-ATPase as β-aspartyl phosphate. *J. Biol. Chem.* 256:10724–10730, 1981.

Danielli, J.F. Some properties of lipoid films in relation to the structures of the plasma membrane. *J. Cell. Comp. Physiol.* 7:393–408, 1936.

Danielli, J.F. Protein films at the oil-water interface. *Cold Spring Harb. Symp. Quant. Biol.* 6:190–193, 1938.

Danielli, J.F. Morphological and molecular aspects of active transport. *Symp. Soc. Exp. Biol.* 8:502–516, 1954.

Danielli, J.F. Structure of the cell surface. *Circulation* 26:1163–1166, 1962.

Danielli, J.F. Experiment, hypothesis and theory in the development of concepts of cell membrane structure 1930–1970. *In* Martonosi, A.N. (ed.) *Membranes and Transport.* New York: Plenum Press, 1982, Vol. 1, pp. 3–14.

Danielli, J.F., and H. Davson. A contribution to the theory of permeability of thin films. *J. Cell. Comp. Physiol.* 5:495–508, 1935.

Danielli, J.F., and E.N. Harvey. The tension at the surface of mackerel egg oil, with remarks on the nature of the cell surface. *J. Cell. Comp. Physiol.* 5:483–494, 1935.

Danowski, T. The transfer of potassium across the human blood cell membrane. *J. Biol. Chem.* 139:693–705, 1941.

Darnell, J., H. Lodish, and D. Baltimore. *Molecular Cell Biology.* New York: Scientific American Books, 1986.

Darrow, D.C. Tissue water and electrolyte. *Annu. Rev. Physiol.* 6:95–122, 1944.

Davies, R.E. Gastric hydrochloric acid production—the present position. *In* Murphy, Q.R. (ed.) *Metabolic Aspects of Transport Across Cell Membranes.* Madison: University of Wisconsin Press, 1957, pp. 277–293.

Davies, R.E., and H.A. Krebs. Biochemical aspects of the transport of ions by nervous tissue. *Biochem. Soc. Symp.* 8:77–92, 1952.

Davies, R.E., and A.G. Ogston. On the mechanism of secretion of ions by gastric mucosa and by other tissues. *Biochem. J.* 46:324–333, 1950.

Davson, H. The permeability of erythrocytes to cations. *Cold Spring Harb. Symp. Quant. Biol.* 8:255–268, 1940.

Davson, H. The effect of some metabolic poisons on the permeability of the rabbit erythrocyte to potassium. *J. Cell. Comp. Physiol.* 18:173–185, 1941.

Davson, H. *A Textbook of General Physiology.* New York: McGraw-Hill, 1951.

Davson, H. Biological membranes as selective barriers to diffusion of molecules. *In* Tosteson, D.C. (ed.) *Membrane Transport. People and Ideas.* Bethesda: American Physiological Society, 1989, pp. 15–49.

Davson, H., and J.F. Danielli. Studies on the permeability of erythrocytes. II. The alleged reversal of ionic permeability at alkaline reaction. *Biochem. J.* 30:316–320, 1936.

Davson, H., and J.F. Danielli. Studies on the permeability of erythrocytes. V. Factors in cation permeability. *Biochem. J.* 32:991–1001, 1938.

Davson, H., and J.F. Danielli. *The Permeability of Natural Membranes.* Cambridge: Cambridge University Press, 1943.

Davson, H., and J.M. Reiner. Ionic permeability: An enzyme-like factor concerned in the migration of sodium through the cat erythrocyte membrane. *J. Cell. Comp. Physiol.* 20:325–342, 1942.

Dawson, R.M.C. The measurement of ³²P labelling of individual kephalins and lecithin in a small sample of tissue. *Biochim. Biophys. Acta* 14:374–379, 1954.

Deamer, D.W., R.C. Prince, and A.R. Crofts. The response of fluorescent amines to pH gradients across liposome membranes. *Biochim. Biophys. Acta* 274:323–335, 1972.

Dean, R.B. Anaerobic uptake of potassium by frog muscle. *Proc. Soc. Exp. Biol. Med.* 45:817–819, 1940.

Dean, R.B. Theories of electrolyte equilibrium in muscle. *Biol. Symp.* 3:331–348, 1941.

Dean, R.B. Reminiscences on the sodium pump. *Trends Neurosci.* 10:451–454, 1987.

Dean, R.B., T.R. Noonan, L. Haege, and W.O. Fenn. Permeability of erythrocytes to radioactive potassium. *J. Gen. Physiol.* 24:353–365, 1940.

Degani, C., and P.D. Boyer. A borohydride reduction method for characterization of the acyl phosphate linkage in proteins and its application to sarcoplasmic reticulum adenosine triphosphatase. *J. Biol. Chem.* 248:8222–8226, 1973.

Degani, C., A.S. Dahms, and P.D. Boyer. Characterization of acyl phosphate in transport ATPases by a borohydride reduction method. *Ann. N.Y. Acad. Sci.* 242:77–79, 1974.

DeGowin, E.L., J.E. Harris, and E.D. Plass. Changes in human blood preserved for transfusion. *Proc. Soc. Exp. Biol. Med.* 40:126–128, 1939.

DeGowin, E.L., J.E. Harris, and E.D. Plass. Studies on preserved human blood. *JAMA* 114:855–857, 1940.

Deguchi, N., P.L. Jørgensen, and A.B. Maunsbach. Ultrastructure of the sodium pump. *J. Cell Biol.* 75:619–634, 1977.

Deisenhofer, J., O. Epp, K. Miki, R. Huber, and H. Michel. Structure of the protein subunits in the photosynthetic reaction centre of *Rhodopseudomonas viridis* at 3 Å resolution. *Nature* 318:618–624, 1985.

deMeis, L., and A.L. Vianna. Energy interconversion by the Ca^{2+}-dependent ATPase of the sarcoplasmic reticulum. *Annu. Rev. Biochem.* 48:275–292, 1979.

Deul, D.H., and H. McIlwain. Activation and inhibition of adenosine triphosphatases of subcellular particles from the brain. *J. Neurochem.* 8:246–256, 1961.

DeWeer, P., and D. Geduldig. Contribution of sodium pump to resting potential of squid giant axon. *Am. J. Physiol.* 235:C55–C62, 1978.

Dickerson, R.E. X-ray analysis and protein structure. *In* Neurath, H. (ed.) *The Proteins*, vol. 2. New York: Academic Press, 1964, pp. 603–778.

DiPolo, R. Calcium efflux from internally dialyzed squid giant axons. *J. Gen. Physiol.* 62:575–589, 1973.

DiPolo, R. Ca pump driven by ATP in squid axons. *Nature* 274:390–392, 1978.

DiVirgilio, F., and G.F. Azzone. Activation of site I redox-driven H^+ pump by exogenous quinones in intact mitochondria. *J. Biol. Chem.* 257:4106–4113, 1982.

Dixon, J.F., and L.E. Hokin. Studies on the characterization of the sodium-potassium transport adenosine triphosphatase. Purification and properties of the enzyme from the electric organ of *Electrophorus electricus*. *Arch Biochem. Biophys.* 163:749–758, 1974.

Draper, M.H., and S. Weidmann. Cardiac resting and action potentials recorded with an intracellular electrode. *J. Physiol.* 115:74–94, 1951.

Dubuisson, M. Chemistry of muscle. *Annu. Rev. Biochem.* 21:387–410, 1952.

Duncan, T.M., V.V. Bulygin, Y. Zhou, M.L. Hutcheon, and R.L. Cross. Rotation of subunits during catalysis by *Escherichia coli* F_1-ATPase. *Proc. Natl. Acad. Sci. USA* 92:10964–10968, 1995.

Dunham, E.T. Parallel decay of ATP and active cation fluxes in starved human erythrocytes. *Federation Proc.* 16:33, 1957a.

Dunham, E.T. Linkage of active cation transport to ATP utilization. *Physiologist* 1:23, 1957b.

Dunham, E.T., and I.M. Glynn. Adenosinetriphosphatase activity and ion movements. *J. Physiol.* 152:61P–62P, 1960.

Dunham, E.T., and I.M. Glynn. Adenosinetriphosphatase activity and the active movements of alkalai metal ions. *J. Physiol.* 156:274–293, 1961.

Durell, J., M.A. Sodd, and R.O. Friedel. Acetylcholine stimulation of phosphodiesteratic cleavage of guinea pig brain phosphoinositides. *Life Sci.* 7:363–368, 1968.

Dux, L., S. Pikula, S. Mullner, and A. Martonosi. Crystallization of Ca^{2+}-ATPase in detergent-solubilized sarcoplasmic reticulum. *J. Biol. Chem.* 262:6439–6442, 1987.

Ebashi, S. A granule-bound relaxation factor in skeletal muscle. *Arch, Biochem. Biophys.* 76:410–423, 1958.

Ebashi, S. Calcium binding and relaxation in the actomyosin system. *J. Biochem.* 48:150–151, 1960.

Ebashi, S. Calcium binding activity of vesicular relaxing factor. *J. Biochem.* 50:236–244, 1961.

Ebashi, S., and F. Lipmann. Adenosine triphosphate-linked concentration of calcium ions in a particulate fraction of rabbit muscle. *J. Cell Biol.* 14:389–400, 1962.

Edelman, I.S. Transition from the poikilotherm to the homeotherm. *Federation Proc.* 35:2180–2184, 1976.

Eisenman, A.J., L. Ott, P.K. Smith, and A.W. Winkler. A Study of the permeability of human erythrocytes to potassium, sodium, and inorganic phosphate by the use of radioactive isotopes. *J. Biol. Chem.* 135:165–173, 1940.

Eisenman, G. Cation selective glass electrodes and their mode of operation. *Biophys. J.* 2:259–323, 1962.

Emmelot, P., and C.J. Bos. Adenosine triphosphatase in the cell-membrane fraction from rat liver. *Biochim. Biophys. Acta* 58:374–375, 1962.

Emmelot, P., C.J. Bos, E.L. Benedetti, and P. Rümke. Studies on plasma membranes. *Biochim. Biophys. Acta* 90:126–145, 1964.

Engbaek, L., and T. Hoshiko. Electrical potential gradients through frog skin. *Acta Physiol. Scand.* 39:348–355, 1957.

Engelhardt, H.T., and A.L. Caplan, eds. *Scientific Controversies: Case Studies in the Resolution and Closure of Disputes in Science and Technology.* Cambridge: Cambridge University Press, 1987.

Engelhardt, W.A., and M.N. Ljubimowa. Myosine and adenosinetriphosphatase. *Nature* 144:668–669, 1939.

Epstein, C.J., R.F. Goldberger, and C.B. Anfinsen. The genetic control of tertiary protein structure: Studies with model systems. *Cold Spring Harb. Symp. Quant. Biol.* 28:439–449, 1963.

Epstein, W., and S. Schultz. Cation transport in *Escherichia coli.* V. Regulation of cation content. *J. Gen. Physiol.* 49:221–234, 1965.

Epstein, W., V. Whitelaw, and J. Hesse. A K^+ transport ATPase in *Escherichia coli. J. Biol. Chem.* 253:6666–6668, 1978.

Erecinska, M., and I.A. Silver. Ions and energy in mammalian brain. *Prog. Neurobiol.* 43:37–71, 1994.

Ernster, L. The merger of bioenergetics and molecular biology. *Biochem. Soc. Trans.* 22:253–265, 1994.

Ernster, L., and G. Schatz. Mitochondria: A historical review. *J. Cell Biol.* 91:227s–255s, 1981.

Esch, F.S., and W.S. Allison. On the subunit stoichiometry of the F_1-ATPase and the sites in it that react specifically with p-fluorosulfonylbenzoyl-5'-adenosine. *J. Biol. Chem.* 254:10740–10746, 1979.

Esmann, M., and J.C. Skou. Occlusion of Na^+ by the Na,K-ATPase in the presence of oligomycin. *Biochim. Biophys. Res. Commun.* 127:857–863, 1985.

Fahn, S., R.W. Albers, and G.J. Koval. *Electrophorus* adenosine triphosphatase: Sodium-activated exchange after N-ethyl maleimide treatment. *Science* 145:283–284, 1964.

Fahn, S., M.R. Hurley, G.J. Koval, and R.W. Albers. Sodium-potassium-activated adenosine triphosphatase of *Electrophorus* electric organ. II. Effects of N-ethylmaleimide and other sulfhydryl reagents. *J. Biol. Chem.* 241:1890–1895, 1966b.

Fahn, S., G.J. Koval, and R.W. Albers. Sodium-potassium-activated adenosine triphosphatase of *Electrophorus* electric organ. *J. Biol. Chem.* 241:1882–1889, 1966a.

Fairbanks, G., T.L. Steck, and D.F.H. Wallach. Electrophoretic analysis of the major polypeptides of the human erythrocyte membrane. *Biochemistry* 10:2606–2617, 1971.

Fänge, R., K. Schmidt-Nielsen, and M. Robinson. Control of secretion from avian salt gland. *Am. J. Physiol* 195:321–326, 1958.

Farley, R.A., C.M. Tran, C.T. Carilli, D. Hawke, and J.E. Shively. The amino acid sequence of a fluorescein-labeled peptide from the active site of (Na,K)-ATPase. *J. Biol. Chem.* 259:9532–9535, 1984.

Feng, T.P., and R.W. Gerard. Mechanism of nerve asphyxiation: With a note on the nerve sheath as a diffusion barrier. *Proc. Soc. Exp. Biol. Med.* 27:1073–1076, 1930.

Feng, T.P., and Y.M. Liu. The connective tissue sheath of the nerve as effective diffusion barrier. *J. Cell. Comp. Physiol.* 34:1–16, 1949.

Fenn, W.O. Electrolytes in muscle. *Physiol. Rev.* 16:450–487, 1936.

Fenn, W.O. The role of potassium in physiological processes. *Physiol. Rev.* 20:377–415, 1940.

Fenn, W.O. Introduction to muscle physiology. *Biol. Symp.* 3:1–8, 1941.

Fenn, W.O. Born fifty years too soon. *Annu. Rev. Physiol.* 24:1–10, 1962.

Fenn, W.O., and D.M. Cobb. Electrolyte changes in muscle during activity. *Am. J. Physiol.* 115:345–356, 1936.

Fenn, W.O., D.M. Cobb, A.H. Hegnauer, and B.S. Marsh. Electrolytes in nerve. *Am. J. Physiol.* 110:74–96, 1934.

Fenn, W.O., L.F. Haege, E. Sheridan, and J.B. Flick. The penetration of ammonia into frog muscle. *J. Gen. Physiol.* 28:53–77, 1944.

Fernández-Morán, H. Cell-membrane ultrastructure. *Circulation* 26:1039–1065, 1962.

Fernández-Morán, H., T. Oda, P.V. Blair, and D.E. Green. A macromolecular repeating unit of mitochondrial structure and function. *J. Cell Biol.* 22:63–100, 1964.

Findlay, M., W.L. Magee, and R.J. Rossiter. Incorporation of radioactive phosphate into lipids and pentosenucleic acid of cat brain slices. The effect of inorganic ions. *Biochem. J.* 58:236–243, 1954.

Finean, J.B., and J.D. Robertson. Lipids and the structure of myelin. *Br. Med. Bull.* 14:267–273, 1958.

Florkin, M. A history of biochemistry. *Comp. Biochem.* 30:1–343, 1972.

Flynn, F., and M. Maizels. Cation control in human erythrocytes. *J. Physiol.* 110:301–318, 1950.

Forbush, B., III. Rapid release of ^{42}K or ^{86}Rb from two distinct transport sites on the Na$^+$, K$^+$-pump in the presence of P$_i$ or vanadate. *J. Biol. Chem.* 262:11116–11127, 1987.

Forgac, M., and G. Chin. Na$^+$ transport by the (Na$^+$)-stimulated adenosine triphosphatase. *J. Biol. Chem.* 257:5652–5655, 1982.

Forte, J.G., P.H. Adams, and R.E. Davies. Acid secretion and phosphate metabolism in bullfrog gastric mucosa. *Biochim. Biophys. Acta* 104:25–38, 1965.

Forte, J.G., and R.E. Davies. Oxygen consumption and active transport of ions by isolated frog gastric mucosa. *Am. J. Physiol.* 204:812–816, 1963.

Forte, J.G., and R.E. Davies. Relation between hydrogen ion secretion and oxygen uptake by gastric mucosa. *Am. J. Physiol.* 206:218–222, 1964.

Forte, J.G., G.M. Forte, and P. Saltman. K$^+$-stimulated phosphatase of microsomes from gastric mucosa. *J. Cell. Physiol.* 69:293–304, 1967.

Foster, D.L., and R.H. Fillingame. Stoichiometry of subunits in the H$^+$-ATPase complex of *Escherichia coli. J. Biol. Chem.* 257:2009–2015, 1982.

Fox, C.F., and E.P. Kennedy. Specific labeling and partial purification of the M protein, a component of the β-galactoside transport system of *Escherichia coli. Proc. Natl. Acad. Sci. USA* 54:891–899, 1965.

Francis, W.L., and O. Gatty. The effect of iodoacetate on the electrical potential and on the oxygen uptake of frog skin. *J. Exp. Biol.* 15:132–142, 1938.

Frazier, H.S., and R.D. Keynes. The effect of metabolic inhibitors on the sodium fluxes in sodium-loaded frog sartorius muscle. *J. Physiol.* 148:362–378, 1959.

Friedenwald, J.S., and R.D. Stiehler. Circulation of the aqueous. *Arch. Ophthalmol.* 20:761–768, 1938.

Fruton, J.S. *Molecules and Life.* New York: Wiley-Interscience, 1972.

Frye, L.D., and M. Edidin. The rapid intermixing of cell surface antigens after formation of mouse-human heterokaryons. *J. Cell Sci.* 7:319–335, 1970.

Fuhrman, F.A. Inhibition of active sodium transport in the isolated frog skin. *Am. J. Physiol.* 171:266–278, 1952.

Fuhrman, F.A. Transport through biological membranes. *Annu. Rev. Physiol.* 21:19–48, 1959.

Fuhrman, F.A., and H.H. Ussing. A characteristic response of the isolated frog skin potential to neurohypophysial principles and its relation to the transport of sodium and water. *J. Cell. Comp. Physiol.* 38:109–130, 1951.

Fuller, S. Book review. *Am. Sci.* 79:361–363, 1991.

Fuller, S. *Philosophy of Science and its Discontents,* 2nd ed. New York: Guilford Press, 1993.

Ganser, A.L., and J.G. Forte. K⁺-stimulated ATPase in purified microsomes of bullfrog oxyntic cells. *Biochim. Biophys. Acta* 307:169–180, 1973.

Gárdos, G. Akkumulation der Kalium durch menschlichen Blutkörperchen. *Acta Physiol. Hung.* 6:191–199, 1954.

Garrahan, P.J., and I.M. Glynn. The behaviour of the sodium pump in red cells in the absence of external potassium. *J. Physiol.* 192:159–174, 1967a.

Garrahan, P.J., and I.M. Glynn. The sensitivity of the sodium pump to *external* sodium. *J. Physiol.* 192:175–188, 1967b.

Garrahan, P.J., and I.M. Glynn. Factors affecting the relative magnitudes of the sodium: potassium and sodium:sodium exchanges catalysed by the sodium pump. *J. Physiol.* 192:189–216, 1967c.

Garrahan, P.J., and I.M. Glynn. The stoicheiometry of the sodium pump. *J. Physiol.* 192:217–235, 1967d.

Garrahan, P.J., and I.M. Glynn. The incorporation of inorganic phosphate into adenosine triphosphate by reversal of the sodium pump. *J. Physiol.* 192:237–256, 1967e.

Gay, N.J., and J.E. Walker. The atp operon. *Nucl. Acids Res.* 9:2187–2194, 1981.

Giere, R.N. *Explaining Science.* Chicago: University of Chicago Press, 1988.

Gilbert, G.N., and M. Mulkay. *Opening Pandora's Box.* Cambridge: Cambridge University Press, 1984.

Gilbert, W. Towards a paradigm shift in biology. *Nature* 349:99, 1991.

Gill, T.J., and A.K. Solomon. Effects of ouabain on sodium flux in human red cells. *Nature* 183:1127–1128, 1959.

Glaser, E., B. Norling, and L. Ernster. Reconstitution of mitochondrial oligomycin and dicyclohexylcarbodiimide-sensitive ATPase. *Eur. J. Biochem.* 110:225–235, 1980.

Glitsch, H.G., H. Reuter, and H. Scholz. The effect of the internal sodium concentration on calcium fluxes in isolated guinea-pig auricles. *J. Physiol.* 209:25–43, 1970.

Glynn, I.M. Sodium and potassium movements in human red cells. *J. Physiol.* 134:278–310, 1956.

Glynn, I.M. The action of cardiac glycosides on sodium and potassium movements in human red cells. *J. Physiol.* 136:148–173, 1957a.

Glynn, I.M. The ionic permeability of the red cell membrane. *Prog. Biophys.* 8:242–307, 1957b.

Glynn, I.M. Sodium and potassium movements in nerve, muscle, and red cells. *Int. Rev. Cytol.* 8:449–480, 1959.

Glynn, I.M. Activation of adenosinetriphosphatase activity in a cell membrane by external potassium and internal sodium. *J. Physiol.* 160:18P–19P, 1962.

Glynn, I.M. The action of cardiac glycosides on ion movements. *Pharmacol. Rev.* 16:381–407, 1964.

Glynn, I.M. The Na^+, K^+-transporting adenosine triphosphatase. *In* Martonosi, A. (ed.) *The Enzymes of Biological Membranes.* New York: Plenum Press, 1985, Vol 3, pp. 35–114.

Glynn, I.M. The sodium pump. *J. R. Coll. Physicians Lond.* 23:39–49, 1989a.

Glynn, I.M. A recipe for hot ATP. *Curr. Contents,* number 33, p. 15, 1989b.

Glynn, I.M., and J.B. Chappell. A simple method for the preparation of ^{32}P-labelled adenosine triphosphate of high specific activity. *Biochem J.* 90:147–149, 1964.

Glynn, I.M., Y. Hara, and D.E. Richards. The occlusion of sodium ions within the mammalian sodium-potassium pump: Its role in sodium transport. *J. Physiol.* 351:531–547, 1984.

Glynn, I.M., and J.F. Hoffman. Nucleotide requirements for sodium-sodium exchange catalysed by the sodium pump in human red cells. *J. Physiol.* 218:239–256, 1971.

Glynn, I.M., J.F. Hoffman, and V.L. Lew. Some "partial reactions" of the sodium pump. *Phil. Trans. R. Soc. B* 262:91–102, 1971.

Glynn, I.M., and S.J.D. Karlish. ATP hydrolysis associated with an uncoupled sodium flux through the sodium pump: Evidence for allosteric effects of intracellular ATP and extracellular sodium. *J. Physiol.* 256:465–496, 1976.

Glynn, I.M., and V.L. Lew. Synthesis of adenosine triphosphate at the expense of downhill cation movements in intact human red cells. *J. Physiol.* 207:393–402, 1970.

Glynn, I.M., and D.E. Richards. Occlusion of rubidium ions by the sodium-potassium pump: Its implications for the mechanism of potassium transport. *J. Physiol.* 330:17–43, 1982.

Glynn, I.M., C.W. Slayman, J. Eichberg, and R.M.C. Dawson. The adenosine-triphosphatase system responsible for cation transport in electric organ: Exclusion of phospholipids as intermediates. *Biochem. J.* 94:692–699, 1965.

Goffeau, A., and C.W. Slayman. The proton-translocating ATPase of the fungal plasma membrane. *Biochim. Biophys. Acta* 639:197–223, 1981.

Gogol, E.P., E. Johnston, R. Aggeler, and R.A. Capaldi. Ligand-dependent structural variations in *Escherichia coli* F_1 ATPase revealed by cryoelectron microscopy. *Proc. Natl. Acad. Sci. USA* 87:9585–9589, 1990.

Goldacre, R.J. The folding and unfolding of protein molecules as a basis of osmotic work. *Int. Rev. Cytol.* 1:135–164, 1952.

Goldin, S.M. Active transport of sodium and potassium ions by the sodium and potassium ion-activated adenosine triphosphatase from renal medulla. *J. Biol. Chem.* 252:5630–5642, 1977.

Goldin, S.M., and S.W. Tong. Reconstitution of active transport catalyzed by the purified sodium and potassium ion-stimulated adenosine triphosphatase from canine renal medulla. *J. Biol. Chem.* 249:5907–5915, 1974.

Goldman, D.E. Potential, impedance, and rectification in membranes. *J. Gen. Physiol.* 27:37–60, 1943.

Goodall, M.C., R.J. Bradley, G. Saccomani, and W.O. Romine. Quantum conductance changes in lipid bilayer membranes associated with incorporation of acetylcholine receptors. *Nature* 250:68–70, 1974.

Gore, M.B.R. Adenosinetriphosphatase activity of brain. *Biochem. J.* 50:18–24, 1951.

Gorter, E., and F. Grendel. On bimolecular layers of lipoids on the chromocytes of the blood. *J. Exp. Med.* 41:439–443, 1925.

Graham, J., and R.W. Gerard. Membrane potentials and excitation of impaled single fibers. *J. Cell. Comp. Physiol.* 28:99–117, 1946.

Green, D.E., and H. Baum. *Energy and the Mitochondrion.* New York: Academic Press, 1970.

Green, D.E., and S. Fleischer. The role of lipids in mitochondrial electron transfer and oxidative phosphorylation. *Biochim. Biophys. Acta* 70:554–582, 1963.

Green, D.E., and J.F. Perdue. Correlation of mitochondrial structure and function. *Ann. N.Y. Acad. Sci.* 137:667–684, 1966.

Greenberg, D.M., W.W. Campbell, and M. Murayama. Studies in mineral metabolism with the aid of artificial radioactive isotopes. *J. Biol. Chem.* 136:35–46, 1940.

Greville, G.D. A scrutiny of Mitchell's chemiosmotic hypothesis of respiratory chain and photosynthetic phosphorylation. *Curr. Top. Bioenerg.* 3:1–78, 1969.

Griffiths, J.H.E., and B.G. Maegraith. Distribution of radioactive sodium after injection into the rabbit. *Nature* 143:159–160, 1939.

Grinius, L.L., A.A. Jasaitis, Y.P. Kadziauskas, E.A. Liberman, V.P. Skulachev, V.P. Topali, L.M. Tsofina, and M.A. Vladimirova. Conversion of biomembrane-produced energy into electric form. *Biochim, Biophys. Acta* 216:1–12, 1970.

Grubmeyer, C., R.L. Cross, and H.S. Penefsky. Mechanism of ATP hydrolysis by beef heart mitochondrial ATPase. *J. Biol. Chem.* 257:12092–12100, 1982.

Grundfest, H. Bioelectric potentials in the nervous system and in muscle. *Annu. Rev. Physiol.* 9:477–506, 1947.

Grundfest, H. General neurophysiology. *Prog. Neurol. Psychiatr.* 5:16–42, 1950.

Gulati, J., and L.G. Palmer. Potassium accumulation in frog muscle. *Science* 198:283–284, 1977.

Guy, H.R. A model relating the structure of the sodium channel to its function. *Curr. Top. Membr. Transpt.* 33:289–308, 1988.

Hager, K.M., S.M. Mandala, J.W. Davenport, D.W. Speicher, E.J. Benz, Jr., and C.W. Slayman. Amino acid sequence of the plasma membrane ATPase of *Neurospora crassa:* Deduction from genomic and cDNA sequences. *Proc. Natl. Acad. Sci. USA* 83:7693–7697, 1986.

Hahn, L., and G. Hevesy. Potassium exchange in the stimulated muscle. *Acta Physiol. Scand.* 2:51–63, 1941.

Hahn, L., and G. Hevesy. Rate of penetration of ions into erythrocytes. *Acta Physiol. Scand.* 3:193–223, 1942.

Hahn, L.A., G.C. Hevesy, and O.H. Rebbe. Do the potassium ions inside the muscle cells and blood corpuscles exchange with those present in the plasma? *Biochem. J.* 33:1549–1558, 1939.

Hakim, G., T. Itano, A.K. Verma, and J.T. Penniston, Purification of the Ca^{2+}- and Mg^{2+}-requiring ATPase from rat brain synaptic plasma membrane. *Biochem. J.* 207:225–231, 1982.

Hald, P.M., M. Tulin, T.S. Danowski, P.H. Lavietes, and J.P. Peters. The distribution of sodium and potassium in oxygenated human blood and their effects upon the movements of water between cells and plasma. *Am. J. Physiol.* 149:340–349, 1947.

Hall, K., G. Perez, D. Anderson, C. Gutierrez, K. Munson, S.J. Hersey, J.H. Kaplan, and G. Sachs. Location of the carbohydrates present in the HK-ATPase vesicles isolated from hog gastric mucosa. *Biochemistry* 29:701–706, 1990.

Hamill, O.P., A. Marty, E. Neher, B. Sakmann, and F.J. Sigworth. Improved patch-clamp techniques for high-resolution current recording from cells and cell-free membrane patches. *Pflügers Arch.* 391:85–100, 1981.

Harold, F.M. *The Vital Force.* New York: W.H. Freeman, 1986.

Harris, D.A., J. Rosing, R.J. van de Stadt, and E.C. Slater. Tight binding of adenine nucleotides to beef-heart mitochondrial ATPase. *Biochim. Biophys. Acta* 314:149–153, 1973.

Harris, E.J. The transfer of sodium and potassium between muscle and the surrounding medium. *Trans. Faraday Soc.* 46:872–882, 1950.

Harris, E.J. Linkage of sodium- and potassium-active transport in human erythrocytes. *Symp. Soc. Exp. Biol.* 8:228–241, 1954a.

Harris, E.J. Ionophoresis along frog muscle. *J. Physiol.* 124:248–253, 1954b.

Harris, E.J., and G.P. Burn. The transfer of sodium and potassium ions between muscle and the surrounding medium. *Trans. Faraday Soc.* 45:508–528, 1949.

Harris, E.J., and M. Maizels. The permeability of human erythrocytes to sodium. *J. Physiol.* 113:506–524, 1951.

Harris, E.J., and M. Maizels. Distribution of ions in suspensions of human erythrocytes. *J. Physiol.* 118:40–53, 1952.

Harris, J. The reversible nature of the potassium loss from erythrocytes during storage of blood at 2–5 °C. *Biol. Bull.* 79:373, 1940.

Harris, J.E. The influence of the metabolism of human erythrocytes on their potassium content. *J. Biol. Chem.* 141:579–595, 1941.

Harris, R.A., J.T. Penniston, J. Asai, and D.E. Green. The conformational basis of energy conservation in membrane systems. *Proc. Natl. Acad. Sci. USA* 59:830–837, 1968.

Hartshorne, R.P., and W.A. Catterall. Purification of the saxitoxin receptor of the sodium channel from rat brain. *Proc. Natl. Acad. Sci. USA* 78:4620–4624, 1981.

Hartshorne, R.P., B.U. Keller, J.A. Talvenheimo, W.A. Catterall, and M. Montal. Functional reconstitution of the purified brain sodium channel in planar lipid bilayers. *Proc. Natl. Acad. Sci. USA* 82:240–244, 1985.

Harvey, E.N., and J.F. Danielli. Properties of the cell surface. *Biol. Rev.* 13:319–341, 1938.

Hasselbach, W., and M. Makinose. Die Calciumpumpe der "Erschlaffungsgrana" des Muskels und ihre Abhängigkeit von der ATP-Spaltung. *Biochem. Z.* 333:518–528, 1961.

Hasselbach, W., and M. Makinose. ATP and active transport. *Biochem. Biophys. Res. Commun.* 7:132–136, 1962.

Hasselbach, W., and M. Makinose. Über den Mechanismus des Calciumtransportes durch die Membranen des sarkoplasmatischen Reticulums. *Biochem. Z.* 339:94–111, 1963.

Hasselbach, W., and A. Weber. Models for the study of the contraction of muscle and of cell protoplasm. *Pharmacol. Rev.* 7:97–117, 1957.

Hatefi, Y., A.G. Haavik, L.R. Fowler, and D.E. Griffiths. Studies on the electron transfer system. *J. Biol. Chem.* 237:2661–2669, 1962.

Haydon, D.A., and J. Taylor. The stability and properties of bimolecular lipid leaflets in aqueous solutions. *J. Theor. Biol.* 4:281–296, 1963.

Hebert, H., P.L. Jørgensen, E. Skriver, and A.B. Maunsbach. Crystallization patterns of membrane-bound $(Na^+ + K^+)$-ATPase. *Biochim. Biophys. Acta* 689:571–574, 1982.

Hebert, H., E. Skriver, R. Hegerl, and A.B. Maunsbach. Structure of two-dimensional crystals of membrane-bound Na,K-ATPase as analyzed by correlation averaging. *J. Ultrastruct. Res.* 92:28–35, 1985.

Hediger, M.A., M.J. Coady, T.S. Ikeda, and E.M. Wright. Expression cloning and cDNA sequencing of the Na^+/glucose co-transporter. *Nature* 330:379–381, 1987a.

Hediger, M.A., T. Ikeda, M. Coady, C.B. Gundersen, and E.M. Wright. Expression of size-selected mRNA encoding the intestinal Na/glucose cotransporter in *Xenopus laevis* oocytes. *Proc. Natl. Acad. Sci. USA* 84:2634–2637, 1987b.

Hegyvary, C., and R.L. Post. Binding of adenosine triphosphate to sodium and potassium ion-stimulated adenosine triphosphatase. *J. Biol. Chem.* 246:5234–5240, 1971.

Heilbrunn, L.V., and F.J. Wiercinski. The action of various cations on muscle protoplasm. *J. Cell. Comp. Physiol.* 29:15–32, 1947.

Heinz, E., P. Geck, and B. Pfeiffer. Energetic problems of the transport of amino acids in Ehrlich cells. *J. Membr. Biol.* 57:91–94, 1980.

Hempling, H.G. Potassium and sodium movements in the Ehrlich mouse ascites tumor cell. *J. Gen. Physiol.* 41:565–583, 1958.

Hempling, H.G., and D. Hare. The effect of glycine transport on potassium fluxes in the Ehrlich mouse ascites tumor cell. *J. Biol. Chem.* 236:2498–2502, 1961.

Henderson, R., J.M. Baldwin, T.A. Ceska, F. Zemlin, E. Beckmann, and K.H. Downing. Model for the structure of bacteriorhodopsin based on high-resolution electron cryo-microscopy. *J. Mol. Biol.* 213:899–929, 1990.

Henderson, R., and P.N.T. Unwin. Three-dimensional model of purple membrane obtained by electron microscopy. *Nature* 257:28–32, 1975.

Henriquez, V., and S.L. Ørskov. Untersuchungen über die Schwankungen des Kationenge-haltes der roten Blutkörperchen. *Skand. Arch. Physiol.* 74:78–85, 1936.

Heppel, L.A. The electrolytes of muscle and liver in potassium-depleted rats. *Am. J. Physiol.* 127:385–392, 1939.

Heppel, L.A. The diffusion of radioactive sodium into the muscles of potassium-deprived rats. *Am. J. Physiol.* 128:449–454, 1940.

Heppel, L.A., and C.L.A. Schmidt. Studies on the potassium metabolism of the rat during pregnancy, lactation, and growth. *Univ. Calif. Publ. Physiol.* 8:189–206, 1938.

Herbert, E. A study of the liberation of orthophosphate from adenosine triphosphate by the stromata of human erythrocytes. *J. Cell Comp. Physiol.* 47:11–36, 1956.

Hess, H.H. Effect of Na^+ and K^+ on Mg^{++}-stimulated adenosinetriphosphatase activity of brain. *J. Neurochem.* 9:613–621, 1962.

Hess, H.H., and A. Pope. Effect of metal cations on adenosinetriphosphatase activity of rat brain. *Federation Proc.* 16:196, 1957.

Hesse, J.E., L. Wieczorek, K. Altendorf, A.S. Reicin, E. Dorus, and W. Epstein. Sequence homology between two membrane transport ATPases, the Kdp-ATPase of *Escherichia coli* and the Ca^{2+}-ATPase of sarcoplasmic reticulum. *Proc. Natl. Acad. Sci. USA* 81:4746–4750, 1984.

Hevesy, G., and O. Rebbe. Rate of penetration of phosphate into muscle cells. *Acta Physiol. Scand.* 1:171–182, 1940.

Hilden, S., and L.E. Hokin. Active potassium transport coupled to active sodium transport in vesicles reconstituted from purified sodium and potassium ion-activated adenosine triphosphase from the rectal gland of *Squalus acanthias*. *J. Biol. Chem.* 250:6296–6303, 1975.

Hilden, S., and L. Hokin. Coupled Na^+-K^+ transport in vesicles containing a purified (NaK)-ATPase and only phosphatidyl choline. *Biochem. Biophys. Res. Commun.* 69:521–527, 1976.

Hilden, S., H.M. Rhee, and L.E. Hokin. Sodium transport by phospholipid vesicles contain-ing purified sodium and potassium ion-activated adenosine triphosphatase. *J. Biol. Chem.* 249:7432–7440, 1974.

Hille, B. The permeability of the sodium channel to organic cations in myelinated nerve. *J. Gen. Physiol.* 58:599–619, 1971.

Hille, B. The permeability of the sodium channel to metal cations in myelinated nerve. *J. Gen. Physiol.* 59:637–658, 1972.

Hille, B. Ionic selectivity, saturation, and block in sodium channels. *J. Gen. Physiol.* 66:535–560, 1975.

Hille, B. *Ion Channels of Excitable Membranes.* 2nd ed. Sunderland, MA: Sinauer Associates, 1992.

Hilpert, W., B. Schink, and P. Dimroth. Life by a new decarboxylation-dependent energy conservation mechanism with Na^+ as coupling ion. *EMBO J.* 3:1665–1670, 1984.

Hind, G., and A.T. Jagendorf. Separation of light and dark stages in photophosphorylation. *Proc. Natl. Acad. Sci. USA* 49:715–722, 1963.

Hind, G., and A.T. Jagendorf. Light scattering changes associated with the production of a possible intermediate in photophosphorylation. *J. Biol. Chem.* 240:3195–3201, 1965.

Hinke, J.A.M., J.P. Caillé, and D.C. Gayton. Distribution and state of monovalent ions in skeletal muscle based on ion electrode, isotope, and diffusion analyses. *Ann. N.Y. Acad. Sci.* 204:274–296, 1973.

Hinkle, P.C., J.J. Kim, and E. Racker. Ion transport and respiratory control in vesicles formed from cytochrome oxidase and phospholipids. *J. Biol. Chem.* 247:1338–1339, 1972.

Hinkle, P.C., M.A. Kumar, A. Resetar, and D.L. Harris. Mechanistic stoichiometry of mitochondrial oxidative phosphorylation. *Biochemistry* 30:3576–3582, 1991.

Höber, R. The permeability of red blood corpuscles to organic anions. *J. Cell. Comp. Physiol.* 7:367–391, 1936.

Höber, R. *Physical Chemistry of Cells and Tissues.* Philadelphia: Blakiston, 1945.

Höber, R. The membrane theory. *Ann. N.Y. Acad. Sci.* 47:381–394, 1946.

Hodgkin, A. *Chance and Design.* Cambridge: Cambridge University Press, 1992.

Hodgkin, A.L. Ionic exchange and electrical activity in nerve and muscle. *Arch. Sci. Physiol.* 3:151–163, 1949.

Hodgkin, A.L. The ionic basis of electrical activity in nerve and muscle. *Biol. Rev.* 26:339–409, 1951.

Hodgkin, A.L. Ionic movements and electrical activity in giant nerve fibres. *Proc. R. Soc. B* 148:1–37, 1958.

Hodgkin, A.L. Chance and design in electrophysiology: An informal account of certain experiments on nerve carried out between 1934 and 1952. *J. Physiol.* 263:1–21, 1976.

Hodgkin, A.L., and A.F. Huxley. Action potentials recorded from inside a nerve fibre. *Nature* 144:706–707, 1939.

Hodgkin, A.L., and A.F. Huxley. Resting and action potentials in single nerve fibres. *J. Physiol.* 104:176–195, 1945.

Hodgkin, A.L., and A.F. Huxley. Potassium leakage from an active nerve fibre. *Nature* 158:376–377, 1946.

Hodgkin, A.L., and A.F. Huxley. Potassium leakage from an active nerve fibre. *J. Physiol.* 106:341–367, 1947.

Hodgkin, A.L., and A.F. Huxley. Currents carried by sodium and potassium ions through the membrane of the giant axon of *Loligo. J. Physiol.* 116:449–472, 1952a.

Hodgkin, A.L., and A.F. Huxley. The components of membrane conductance in the giant axon of *Loligo. J. Physiol.* 116:473–496, 1952b.

Hodgkin, A.L., and A.F. Huxley. The dual effect of membrane potential on sodium conductance in the giant axon of *Loligo. J. Physiol.* 116:497–506, 1952c.

Hodgkin, A.L., and A.F. Huxley. A quantitative description of membrane current and its application to conduction and excitation in nerve. *J. Physiol.* 117:500–544, 1952d.

Hodgkin, A.L., A.F. Huxley, and B. Katz. Ionic currents underlying activity in the giant axon of the squid. *Arch. Sci. Physiol.* 3:129–150, 1949.

Hodgkin, A.L., A.F. Huxley, and B. Katz. Measurement of current-voltage relations in the membrane of the giant axon of *Loligo. J. Physiol.* 116:424–448, 1952.

Hodgkin, A.L., and B. Katz. The effect of sodium ions on the electrical activity of the giant axon of the squid. *J. Physiol.* 108:37–77, 1949.

Hodgkin, A.L., and R.D. Keynes. Sodium extrusion and potassium absorption in *Sepia* axons. *J. Physiol.* 120:46P–47P, 1953a.

Hodgkin, A.L., and R.D. Keynes. Metabolic inhibitors and sodium movements in giant axons. *J. Physiol.* 120:45P–46P, 1953b.

Hodgkin, A.L., and R.D. Keynes. The mobility and diffusion coefficient of potassium in giant axons from *Sepia*. *J. Physiol.* 119:513–528, 1953c.

Hodgkin, A.L., and R.D. Keynes. Movements of cations during recovery in nerve. *Symp. Soc. Exp. Biol.* 8:423–437, 1954.

Hodgkin, A.L., and R.D. Keynes. Active transport of cations in giant axons from *Sepia* and *Loligo*. *J. Physiol.* 128:28–60, 1955a.

Hodgkin, A.L., and R.D. Keynes. The potassium permeability of a giant nerve fibre. *J. Physiol.* 128:61–88, 1955b.

Hodgkin, A.L., and R.D. Keynes. Experiments on the injection of substances into squid giant axons by means of a microsyringe. *J. Physiol.* 131:592–616, 1956.

Hodgkin, A.L., and R.D. Keynes. Movements of labelled calcium in squid giant axons. *J. Physiol.* 138:253–281, 1957.

Hoffman, J.F. The link between metabolism and the active transport of Na in human red cell ghosts. *Federation Proc.* 19:127, 1960.

Hoffman, J.F. Molecular mechanism of active cation transport. *In* Shanes, A.M. (ed.) *Biophysics of Physiological and Pharmacological Actions.* Washington: Am. Assoc. Adv. Science, 1961, pp. 3–17.

Hoffman, J.F. The active transport of sodium by ghosts of human red blood cells. *J. Gen. Physiol.* 45:837–859, 1962a.

Hoffman, J.F. Cation transport and structure of the red-cell plasma membrane. *Circulation* 26:1201–1213, 1962b.

Hoffman, J.F. The link between metabolism and active transport of sodium in human red cell ghosts. *J. Membr. Biol.* 57:143–161, 1980.

Hoffman, J.F., D.C. Tosteson, and R. Whittam. Retention of potassium by human erythrocyte ghosts. *Nature* 185:186–187, 1960.

Hogeboom, G.H., W.C. Schneider, and G.E. Palade. Cytochemical studies of mammalian tissues. I. Isolation of intact mitochondria from rat liver; some biochemical properties of mitochondria and submicroscopic particulate material. *J. Biol. Chem.* 172:619–635, 1948.

Hokin, L.E. The synthesis and secretion of amylase by pigeon pancreas *in vitro*. *Biochem. J.* 48:320–326, 1951.

Hokin, L.E. Purification of the (sodium+potassium)-activated adenosinetriphosphatase and reconstitution of sodium transport. *Ann. N.Y. Acad. Sci.* 242:12–23, 1974.

Hokin, L.E. The road to the phosphoinositide-generated second messengers. *Trends Pharmacol. Sci.* 8:53–56, 1987.

Hokin, L.E., J.L. Dahl, J.D. Deupree, J.F. Dixon, J.H. Hackney, and J.F. Perdue. Studies on the characterization of the sodium-potassium transport adenosine triphosphatase. X. Purification of the enzyme from the rectal gland of *Squalus acanthias*. *J. Biol. Chem.* 248:2593–2605, 1973.

Hokin, L.E., and M.R. Hokin. Effects of acetylcholine on phosphate turnover in phospholipides of brain cortex *in vitro*. *Biochim. Biophys. Acta* 16:229–237, 1955a.

Hokin, L.E., and M.R. Hokin. Effects of acetylcholine on the turnover of phosphoryl units in individual phospholipids of pancreas slices and brain cortex slices. *Biochim. Biophys. Acta* 18:102–110, 1955b.

Hokin, L.E., and M.R. Hokin. The actions of pancreozymin in pancreas slices and the role of phospholipids in enzyme secretion. *J. Physiol.* 132:442–453, 1956.

Hokin, L.E., and M.R. Hokin. The mechanism of phosphate exchange in phosphatidic acid in response to acetylcholine. *J. Biol. Chem.* 234:1387–1390, 1959a.

Hokin, L.E., and M.R. Hokin. Evidence for phosphatidic acid as the sodium carrier. *Nature* 184:1068–1069, 1959b.

Hokin, L.E., and M.R. Hokin. Studies on the carrier function of phosphatidic acid in sodium transport. *J. Gen. Physiol.* 44:61–85, 1960a.

Hokin, L.E., and M.R. Hokin. The role of phosphoinositide in transmembrane transport elicited by acetylcholine and other humoral agents. *Int. Rev. Neurobiol.* 2:99–136, 1960b.

Hokin, L.E., and M.R. Hokin. Diglyceride kinase and phosphatidic acid phosphatase in erythrocyte membranes. *Nature* 189:836–837, 1961.

Hokin, L.E., and M.R. Hokin. The role of phosphatides in active transport with particular reference to sodium transport. In *Proceedings of the First International Pharmacological Meeting*. New York: Macmillan, 1963a, Vol. 4, pp. 23–40.

Hokin, L.E., and M.R. Hokin. Biological transport. *Annu. Rev. Biochem.* 32:553–578, 1963b.

Hokin, L.E., and M.R. Hokin. The incorporation of ^{32}P from [γ-^{32}P]adenosine triphosphate into polyphosphoinositides and phosphatidic acid in erythrocyte membranes. *Biochim. Biophys. Acta* 84:563–575, 1964.

Hokin, L.E., M.R. Hokin, and D. Mathison. Phosphatidic acid phosphatase in the erythrocyte membrane. *Biochim. Biophys. Acta* 67:485–497, 1963.

Hokin, L.E., and M. Hokin-Neaverson. Commentary. *Biochim. Biophys. Acta* 1000:465–469, 1989.

Hokin, L.E., P.S. Sastry, P.R. Galsworthy, and A. Yoda. Evidence that a phosphorylated intermediate in a brain transport adenosine triphosphatase is an acyl phosphate. *Proc. Natl. Acad. Sci. USA* 54:177–184, 1965.

Hokin, M.R., B.G. Benfey, and L.E. Hokin. Phospholipides and adrenaline secretion in guinea pig adrenal medulla. *J. Biol. Chem.* 233:814–817, 1958.

Hokin, M.R., and L.E. Hokin. Enzyme secretion and the incorporation of P^{32} into phospholipides of pancreas slices. *J. Biol. Chem.* 203:967–977, 1953.

Hokin, M.R., and L.E. Hokin. Effects of acetylcholine on phospholipides in the pancreas. *J. Biol. Chem.* 209:549–557, 1954.

Hokin, M.R., and L.E. Hokin. The synthesis of phosphatidic acid from diglyceride and adenosine triphosphate in extracts of brain microsomes. *J. Biol. Chem.* 234:1381–1386, 1959.

Hokin, M.R., and L.E. Hokin. Further evidence for phosphatidic acid as the sodium carrier. *Nature* 190:1016–1017, 1961.

Hokin, M.R., and L.E. Hokin. The synthesis of phosphatidic acid and protein-bound phosphorylserine in salt gland homogenates. *J. Biol. Chem.* 239:2116–2122, 1964a.

Hokin, M.R., and L.E. Hokin. Interconversions of phosphatidylinositol and phosphatidic acid involved in the response to acetylcholine in the salt gland. *In* Dawson, R.M.C., and D.N. Rhodes (eds.) *Metabolism and Physiological Significance of Lipids*. London: Wiley, 1964b, pp. 423–434.

Holm-Jensen, I., A. Krogh, and V. Wartiovaara. Some experiments on the exchange of potassium between single cells of Characeae and the bathing fluid. *Acta Bot. Fenn.* 36:1–22, 1944.

Hopfer, U., K. Nelson, J. Perrotto, and K.J. Isselbacher. Glucose transport in isolated brush border membrane from rat small intestine. *J. Biol. Chem.* 248:25–32, 1973.

Hosang, M., E.M. Gibbs, D.F. Diedrich, and G. Semenza. Photoaffinity labeling and identification of (a component of) the small-intestinal Na^+, D-glucose transporter using 4-azidophlorizin. *FEBS Lett.* 130:244–248, 1981.

Huang, C. Studies on phosphatidylcholine vesicles. *Biochemistry* 8:344–351, 1969.

Huf, E. Versuche über den Zusammenhang zwischen Stoffwechsel, Potentialbildung und Funktion der Froschaut. *Pflügers Arch.* 235:655–673, 1935.

Huf, E. Über aktiven Wasser- und Salztransport durch die Froschaut. *Pflügers Arch.* 237:143–166, 1936a.

Huf, E. Die Bedeutung der Atmungsvorgänge für die Resorptionsleistung und Potentialbildung bei der Froschhaut. *Biochem. Z.* 288:116–122, 1936b.

Huf, E.G., and J. Parrish. Nature of the electrolyte pump in surviving frog skin. *Am. J. Physiol.* 164:428–436, 1951.

Huf, E.G., and J. Wills. Influence of some inorganic cations on active salt and water uptake by isolated frog skin. *Am. J. Physiol.* 167:255–260, 1951.

Hunter, F.E., and L. Ford. Inactivation of oxidative and phosphorylative systems in mitochondria by preincubation with phosphate and other ions. *J. Biol. Chem.* 216:357–369, 1955.

Huxley, A. Kenneth Stewart Cole. *Biograph. Mem. Fellows R. Soc.* 38:99–110, 1992.

Huxley, A.F., and R. Niedergerke. Structural changes in muscle during contraction. *Nature* 173:971–973, 1954.

Huxley, A.F., and R. Stämpfli. Evidence for a saltatory conduction in peripheral myelinated nerve fibres. *J. Physiol.* 108:315–339, 1949.

Huxley, A.F., and R. Stämpfli. Direct determination of membrane resting potential and action potential in single myelinated nerve fibres. *J. Physiol.* 112:476–495, 1951a.

Huxley, A.F., and R. Stämpfli. Effect of potassium and sodium on resting and action potentials of single myelinated nerve fibres. *J. Physiol.* 112:496–508, 1951b.

Huxley, H., and J. Hanson. Changes in the cross-striations of muscle during contraction and stretch and their structural interpretation. *Nature* 173:973–976, 1954.

Inesi, G., J.J. Goodman, and S. Watanabe. Effect of diethyl ether on the adenosine triphosphatase activity and the calcium uptake of fragmented sarcoplasmic reticulum of rabbit skeletal muscle. *J. Biol. Chem.* 242:4637–4643, 1967.

Ivey, D.M., and T.A. Krulwich. Two unrelated alkaliphilic *Bacillus* species possess identical deviations in sequence from those of other prokaryotes in regions of F_o proposed to be involved in proton translocation through ATP synthase. *Res. Microbiol.* 143:467–470, 1992.

Iwata, S., C. Ostermeier, B. Ludwig, and H. Michel. Structure at 2.8 Å resolution of cytochrome c oxidase from *Paracoccus dentrificans*. *Nature* 376:660–669, 1995.

Jacobs, M.H. Permeability. *Annu. Rev. Biochem.* 4:1–16, 1935.

Jacobs, M.H. Permeability. *Annu. Rev. Physiol.* 1:1–20, 1939.

Jacobs, M.H. Early osmotic history of the plasma membrane. *Circulation* 26:1013–1021, 1962.

Jacobson, K., E.D. Sheets, and R. Simson. Revisiting the fluid mosaic model of membranes. *Science* 268:1441–1442, 1995.

Jagendorf, A.T., and E. Uribe. ATP formation caused by acid-base transition of spinach chloroplasts. *Proc. Natl. Acad. Sci. USA* 55:170–177, 1966.

Jagendorf, A.T., and E. Uribe. Photophosphorylation and the chemi-osmotic hypothesis. *Brookhaven Symp. Biol.* 19:215–245, 1967.

Jan, L.Y., and Y.N. Jan. Voltage-sensitive ion channels. *Cell* 56:13–25, 1989.

Jardetzky, O. Simple allosteric model for membrane pumps. *Nature* 211:969–970, 1966.

Järnefelt, J. The relative rates of adenosinetriphosphatase and phosphatidic acid synthesis in microsomes from rat brain. *Exp. Cell Res.* 25:211–213, 1961.

Jeanneney, G., L. Servantie, and G. Ringenbach. Les modifications du rapport potassium/sodium du plasma dans le sang citraté conservé a la glacière. *Compt. Rend. Soc. Biol.* 130:472–473, 1938.

Jencks, W.P. Binding energy, specificity, and enzymic catalysis: The Circe effect. *Adv. Enzymol.* 43:219–410, 1975.

Jensen, J., and J.G. Nørby. On the specificity of the ATP-binding site of $(Na^+ + K^+)$-activated ATPase from brain microsomes. *Biochim. Biophys. Acta* 233:395–403, 1971.

Johnson, L.N., and D.C. Phillips. Structure of some crystalline lysozyme-inhibitor complexes determined by X-ray analysis at 6 Å resolution. *Nature* 206:761–763, 1965.

Jones, D.H., and P.D. Boyer. The apparent absolute requirement of adenosine diphosphate for the inorganic phosphate ⇔ water exchange of oxidative phosphorylation. *J. Biol. Chem.* 244:5767–5772, 1969.

Jorgensen, C.B. The effect of adrenaline and related compounds on the permeability of isolated frog skin to ions. *Acta Physiol. Scand.* 14:213–219, 1947.

Jorgensen, C.B., H. Levi and H.H. Ussing. On the influence of the neurohypophyseal principles on the sodium metabolism in the axolotl (Amblystoma mexicanum). *Acta Physiol. Scand.* 12:350–371, 1947.

Jørgensen, P.L. Purification and characterization of $(Na^+ + K^+)$-ATPase III. Purification from the outer medulla of mammalian kidney after selective removal of membrane components by sodium dodecylsulfate. *Biochim. Biophys. Acta* 356:36–52, 1974a.

Jørgensen, P.L. Purification and characterization of $(Na^+ + K^+)$-ATPase. IV. Estimation of the purity and of the molecular weight and polypeptide content per enzyme unit in preparations from the outer medulla of rabbit kidney. *Biochim. Biophys. Acta* 356:53–67, 1974b.

Jørgensen, P.L. Purification and characterization of $(Na^+ + K^+)$-ATPase. V. Conformational changes in the enzyme. Transitions between the Na-form and the K-form studied with tryptic digestion as a tool. *Biochim. Biophys. Acta* 401:399–415, 1975.

Jørgensen, P.L. Purification and characterization of $(Na^+ + K^+)$-ATPase VI. Differential tryptic modification of catalytic functions of the purified enzyme in presence of NaCl and KCl. *Biochim. Biophys. Acta* 466:97–108, 1977.

Jørgensen, P.L. Mechanism of the Na^+, K^+ pump. Protein structure and conformations of the pure $(Na^+ + K^+)$-ATPase. *Biochim. Biophys. Acta* 694:27–68, 1982.

Jørgensen, P.L., and J.P. Andersen. Thermoinactivation and aggregation of $\alpha\beta$ units in soluble and membrane-bound (Na,K)-ATPase. *Biochemistry* 25:2889–2897, 1986.

Jørgensen, P.L., and J.H. Collins. Tryptic and chymotryptic cleavage sites in sequence of α-subunit of $(Na^+ + K^+)$-ATPase from outer medulla of mammalian kidney. *Biochim. Biophys. Acta* 860:570–576, 1986.

Jørgensen, P.L., and J.C. Skou. Preparation of highly active $(Na^+ + K^+)$-ATPase from the outer medulla of rabbit kidney. *Biochem. Biophys. Res. Commun.* 37:39–46, 1969.

Jørgensen, P.L., and J.C. Skou. Purification and characterization of $(Na^+ + K^+)$-ATPase. I. The influence of detergents on the activity of $(Na^+ + K^+)$-ATPase in preparations from the outer medulla of rabbit kidney. *Biochim. Biophys. Acta* 233:366–380, 1971.

Jørgensen, P.L., J.C. Skou, and L.P. Solomonson. Purification and characterization of $(Na^+ + K^+)$-ATPase. II. Preparation by zonal centrifugation of highly active $(Na^+ + K^+)$-ATPase from the outer medulla of rabbit kidneys. *Biochim. Biophys. Acta* 233:381–394, 1971.

Josephson, L., and L.C. Cantley. Isolation of a potent (Na-K)ATPase inhibitor from striated muscle. *Biochemistry* 16:4572–4578, 1977.

Joyce, C.R.B., and M. Weatherall. Cardiac glycosides and potassium exchange of human erythrocytes. *J. Physiol.* 127:33P, 1954.

Judah, J.D., and K. Ahmed. The biochemistry of sodium transport. *Biol. Rev.* 39:160–193, 1964.

Kaczorowski, G.J., and H.R. Kaback. Mechanism of lactose translocation in membrane vesicles from *Escherichia coli*. *Biochemistry* 18:3691–3697, 1979.

Kagawa, Y., and E. Racker. Partial resolution of the enzymes catalyzing oxidative phosphorylation. XXV. Reconstitution of vesicles catalyzing $^{32}P_i$-adenosine triphosphate exchange. *J. Biol. Chem.* 246:5477–5487, 1971.

Kahlenberg, A., P.R. Galsworthy, and L.E. Hokin. Sodium-potassium adenosine triphosphatase: Acyl phosphate "intermediate" shown to be L-glutamate-γ-phosphate. *Science* 157:434–436, 1967.

Kahlenberg, A., P.R. Galsworthy, and L.E. Hokin. Studies on the characterization of the sodium-potassium transport adenosinetriphosphatase. II. Characterization of the acyl phosphate intermediate as an L-glutamyl-γ-phosphate residue. *Arch. Biochem. Biophys.* 126:331–342, 1968.

Kalckar, H.M. *Biological Phosphorylations.* Englewood Cliffs, NJ: Prentice-Hall, 1969.

Kanai, Y., and M.A. Hediger. Primary structure and functional characterization of a high-affinity glutamate transporter. *Nature* 360:467–474, 1992.

Kanazawa, H., K. Mabuchi, T. Kayano, F. Tamura, and M. Futai. Nucleotide sequence of genes coding for dicyclohexylcarbodiimide-binding protein and the α subunit of proton-translocating ATPase of *Escherichia coli. Biochem. Biophys. Res. Commun.* 100:219–225, 1981.

Kanazawa, T., M. Saito, and Y. Tonomura. Formation and decomposition of a phosphorylated intermediate in the reaction of Na^+-K^+ dependent ATPase. *J. Biochem.* 67:693–711, 1970.

Kandpal, R.P., and P.D. Boyer. *Escherichia coli* F_1 ATPase is reversibly inhibited by intra- and intersubunit crosslinking. An approach to assessing rotational catalysis. *Biochim. Biophys. Acta* 890:97–105, 1987.

Kapakos, J.G., and M. Steinberg. Fluorescent labeling of $(Na^+ + K^+)$-ATPase by 5-iodoacetamidofluorescein. *Biochim. Biophys. Acta* 693:493–496, 1982.

Kapakos, J.G., and M. Steinberg. Ligand binding to (Na,K)-ATPase labeled with 5-iodoacetamidofluorescein. *J. Biol. Chem.* 261:2084–2089, 1986a.

Kapakos, J.G., and M. Steinberg. 5-iodoacetamidofluorescein-labeled (Na,K)-ATPase. Steady-state fluorescence during turnover. *J. Biol. Chem.* 261:2090–2095, 1986b.

Karlish, S.J.D. Characterization of conformational changes in (Na,K) ATPase labeled with fluorescein at the active site. *J. Bioenerg. Biomembr.* 12:111–136, 1980.

Karlish, S.J.D., R. Goldshleger, and W.D. Stein. A 19-kDa C-terminal tryptic fragment of the α chain of Na/K-ATPase is essential for occlusion and transport of cations. *Proc. Natl. Acad. Sci. USA* 87:4566–4570, 1990.

Karlish, S.J.D., and D.W. Yates. Tryptophan fluorescence of $(Na^+ + K^+)$-ATPase as a tool for study of the enzyme mechanism. *Biochim. Biophys. Acta* 527:115–130, 1978.

Kasahara, M., and P.C. Hinkle. Reconstitution and purification of the D-glucose-transporter from human erythrocytes. *J. Biol. Chem.* 252:7384–7390, 1977.

Kasahara, M., and H.S. Penefsky. High affinity binding of monovalent P_i by beef heart mitochondrial adenosine triphosphatase. *J. Biol. Chem.* 253:4180–4187, 1978.

Katz, B. The effect of electrolyte deficiency on the rate of conduction in a single nerve fibre. *J. Physiol.* 106:411–417, 1947.

Katz, B., and R. Miledi. The statistical nature of the acetylcholine potential and its molecular components. *J. Physiol.* 224:665–699, 1972.

Katzin, L.I. The ionic permeability of frog skin as determined with the aid of radioactive indicators. *Biol. Bull.* 77:302–303, 1939.

Katzin, L.I. The use of radioactive tracers in the determination of irreciprocal permeability of biological membranes. *Biol. Bull.* 79:342, 1940.

Kauzmann, W. Some factors in the interpretation of protein denaturation. *Adv. Prot. Res.* 14:1–63, 1959.

Kawakami, K., S. Noguchi, M. Noda, H. Takahashi, T. Ohta, M. Kawamura, H. Nojima, K. Nagano, T. Hirose, S. Inayama, H. Hayashida, T. Miyata, and S. Numa. Primary structure of the α-subunit of *Torpedo californica* $(Na^+ + K^+)$ ATPase deduced from cDNA sequence. *Nature* 316:733–736, 1985.

Kawakami, K., H. Nojima, T. Ohta, and K. Nagano. Molecular cloning and sequence analysis of human Na,K-ATPase β-subunit. *Nucl. Acids Res.* 14:2833–2844, 1986.

Kayalar, C., J. Rosing, and P.D. Boyer. 2,4-Dinitrophenol causes a marked increase in the apparent K_m of P_i and of ADP for oxidative phosphorylation. *Biochem. Biophys. Res. Commun.* 72:1153–1159, 1976.

Kayalar, C., J. Rosing, and P.D. Boyer. An alternating site sequence for oxidative phosphorylation suggested by measuring substrate binding patterns and exchange reaction inhibitions. *J. Biol. Chem.* 252:2486–2491, 1977.

Kearney, E.B. Studies on succinic dehydrogenase. *J. Biol. Chem.* 229:363–375, 1957.

Kendrew, J.C., G. Bodo, H.M. Dintzis, R.G. Parrish, H.M. Wyckoff, and D.C. Phillips. A three-dimensional model of the myoglobin molecule obtained by X-ray analysis. *Nature* 181:662–666, 1958.

Kendrew, J.C., R.E. Dickerson, B.E. Strandberg, R.G. Hart, D.R. Davies, D.C. Phillips, and V.C. Shore. Structure of myoglobin. *Nature* 185:422–427, 1960.

Kennedy, E.P., and A.L. Lehninger. Oxidation of fatty acids and tricarboxylic acid cycle intermediates by isolated rat liver mitochondria. *J. Biol. Chem.* 179:957–972, 1949.

Kerr, S.E. Studies on the inorganic composition of blood. IV. The relationship of potassium to the acid-soluble phosphorus fractions. *J. Biol. Chem.* 117:227–235, 1937.

Keynes, R.D. The leakage of radioactive potassium from stimulated nerve. *J. Physiol.* 107:35P–36P, 1948.

Keynes, R.D. The movement of radioactive sodium during nervous activity. *J. Physiol.* 109:13P, 1949a.

Keynes, R.D. The movements of radioactive ions in resting and stimulated nerve. *Arch. Sci. Physiol.* 3:165–175, 1949b.

Keynes, R.D. The leakage of radioactive potassium from stimulated nerve. *J. Physiol.* 113:99–114, 1951a.

Keynes, R.D. The ionic movements during nervous activity. *J. Physiol.* 114:119–150, 1951b.

Keynes, R.D. The ionic fluxes in frog muscle. *Proc. R. Soc. Lond. B* 142:359–382, 1954.

Keynes, R.D. Forty years of exploring the sodium channel: An autobiographical account. *In Mélanges Monnier.* Paris: Privately published, 1989, pp. 171–178.

Keynes, R.D., and P.R. Lewis. The resting exchange of radioactive potassium in crab nerve. *J. Physiol.* 113:73–98, 1951a.

Keynes, R.D., and P.R. Lewis. The sodium and potassium content of cephalopod nerve fibres. *J. Physiol.* 114:151–182, 1951b.

Keynes, R.D., and P.R. Lewis. The intracellular calcium content of some invertebrate nerves. *J. Physiol.* 134:399–407, 1956.

Keynes, R.D., and G.W. Maisel. The energy requirement for sodium extrusion from a frog muscle. *Proc. R. Soc. Lond. B* 142:383–392, 1954.

Keynes, R.D., and R.C. Swan. The effect of external sodium concentration on the sodium fluxes in frog skeletal muscle. *J. Physiol.* 147:591–625, 1959.

Khorana, H.G., G.E. Gerber, W.C. Herlihy, C.P. Gray, R.J. Anderegg, K. Nihei, and K. Biemann. Amino acid sequence of bacteriorhodopsin. *Proc. Natl. Acad. Sci. USA* 76:5046–5050, 1979.

Kielley, W.W., and O. Meyerhof. Studies on adenosinetriphosphatase of muscle. II. A new magnesium-activated adenosinetriphosphatase. *J. Biol. Chem.* 76:591–601, 1948.

Kielley, W.W., and O. Meyerhof. Studies on adenosinetriphosphatase of muscle. III. The lipoprotein nature of the magnesium-activated adenosinetriphosphatase. *J. Biol. Chem.* 183:391–401, 1950.

Kimura, T.E., and K.P. DuBois. Inhibition of the enzymatic hydrolysis of ATP by certain cardiac drugs. *Science* 106:370–371, 1947.

Kincaid, H. Molecular biology and the unity of science. *Phil. Sci.* 57:575–593, 1990.

Kirley, T.L., E.T. Wallick, and L.K. Lane. The amino acid sequence of the fluorescein isothiocyanate reactive site of lamb and rat kidney Na^+- and K^+-dependent ATPase. *Biochem. Biophys. Res. Commun.* 125:767–773, 1984.

Kirschner, L.B. The cation content of phospholipides from swine erythrocytes. *J. Gen. Physiol.* 42:231–241, 1958.

Kirschner, L.B., and J. Barker. Turnover of phosphatidic acid and sodium extrusion from mammalian erythrocytes. *J. Gen. Physiol.* 47:1061–1078, 1964.

Kirshner, N. Uptake of catecholamines by a particulate fraction of the adrenal medulla. *J. Biol. Chem.* 237:2311–2317, 1962.

Kitcher, P. *The Advancement of Science.* New York: Oxford University Press, 1993.

Kleinzeller, A. Exploring the cell membrane: Conceptual developments. *Comp. Biochem.* 39:1–359, 1995.

Kleinzeller, A., and A. Knotková. The effect of ouabain on the electrolyte and water transport in kidney cortex and liver slices. *J. Physiol.* 175:172–192, 1964.

Klingenberg, M. The ADP-ATP translocation in mitochondria, a membrane potential controlled transport. *J. Membr. Biol.* 56:97–105, 1980.

Klingenberg, M., and P. Schollmeyer. Zur Reversibilität der oxydativen Phosphorylierung. *Biochem. Z.* 333:335–350, 1960.

Klodos, I., and J.C. Skou. The effect of Mg^{2+} and chelating agents on intermediary steps of the reaction of Na^+, K^+-activated ATPase. *Biochim. Biophys. Acta* 391:474–485, 1975.

Knauf, P.A., F. Proverbio, and J.F. Hoffman. Electrophoretic separation of different phosphoproteins associated with Ca-ATPase and Na,K-ATPase in human red cell ghosts. *J. Gen. Physiol.* 63:324–336, 1974.

Knauf, P.A., and A. Rothstein. Chemical modification of membranes. *J. Gen. Physiol.* 58:190–210, 1971.

Koefoed-Johnsen, V. The effect of g-strophanthin (ouabain) on the active transport of sodium through the isolated frog skin. *Acta Physiol. Scand. Suppl.* 145:87–88, 1957.

Koefoed-Johnsen, V., H. Levi, and H.H. Ussing. The mode of passage of chloride ions through the isolated frog skin. *Acta Physiol. Scand.* 25:150–163, 1952a.

Koefoed-Johnsen, V., and H.H. Ussing. The nature of the frog skin potential. *Acta Physiol. Scand.* 42:298–308, 1958.

Koefoed-Johnsen, V., H.H. Ussing, and K. Zerahn. The origin of the short-circuit current in the adrenaline stimulated frog skin. *Acta Physiol. Scand.* 27:38–48, 1952b.

Koepsell, H., K. Korn, D. Ferguson, H. Menuhr, D. Ollig, and W. Haase. Reconstitution and partial purification of several Na^+ cotransport systems from renal brush-border membranes. *J. Biol. Chem.* 259:6548–6558, 1984.

Kohler, R.E. The enzyme theory and the origin of biochemistry. *Isis* 64:181–196, 1973.

Komai, H., D.R. Hunter, J.H. Southard, R.A. Haworth, and D.E. Green. Energy coupling in lysolecithin-treated submitochondrial particles. *Biochem. Biophys. Res. Commun.* 69:695–704, 1976.

Kopito, R.R., and H.F. Lodish. Primary structure and transmembrane orientation of the murine anion exchange protein. *Nature* 316:234–238, 1985.

Koshland, D.E. Application of a theory of enzyme specificity to protein synthesis. *Proc. Natl. Acad. Sci. USA* 44:98–104, 1958.

Koshland, D.E., G. Némethy, and D. Filmer. Comparison of experimental binding data and theoretical models in proteins containing subunits. *Biochemistry* 5:365–385, 1966.

Krebs, E.G., and E.H. Fischer. The phosphorylase b to a converting enzyme of rabbit skeletal muscle. *Biochim. Biophys. Acta* 20:150–157, 1956.

Krogh, A. Osmotic regulation in the frog (*R. esculenta*) by active absorption of chloride ions. *Skand. Arch. Physiol.* 76:60–74, 1937.

Krogh, A. The active and passive exchanges of inorganic ions through the surfaces of living cells and through living membranes generally. *Proc. R. Soc. Lond. B* 133:140–200, 1946.

Kromphardt, H., H. Grobecker, K. Ring, and E. Heinz. Über den Einfluss von Alkali-Ionen auf den Glycintransport in Ehrlich-Ascites-Tumorzellen. *Biochim. Biophys. Acta* 74:549–551, 1963.

Kuhn, T.S. *The Structure of Scientific Revolutions*, 2nd ed. Chicago: University of Chicago Press, 1970.

Kuhn, T.S. *The Essential Tension*. Chicago: University of Chicago Press, 1977.

Kumagai, H., S. Ebashi, and F. Takeda. Essential relaxing factor in muscle other than myokinase and creatine phosphokinase. *Nature* 176:166, 1955.

Kunz, H.A., and F. Sulser. Über die Hemmung des aktiven Kationentransportes durch Herzglykoside. *Experientia* 13:365–367, 1957.

Kyte, J. Purification of the sodium- and potassium-dependent adenosine triphosphatase from canine renal medulla. *J. Biol. Chem.* 246:4157–4165, 1971a.

Kyte, J. Phosphorylation of a purified (Na$^+$ + K$^+$) adenosine triphosphatase. *Biochem. Biophys. Res. Commun.* 43:1259–1265, 1971b.

Kyte, J. Properties of the two polypeptides of sodium- and potassium-dependent adenosine triphosphatase. *J. Biol. Chem.* 247:7642–7649, 1972.

Kyte, J. Structural studies of sodium and potassium ion-activated adenosine triphosphatase. *J. Biol. Chem.* 250:7443–7449, 1975.

Kyte, J., and R.F. Doolittle. A simple method for displaying the hydropathic character of a protein. *J. Mol. Biol.* 157:105–132, 1982.

Laimins, L., D.B. Rhoads, K. Altendorf, and W. Epstein. Identification of the structural proteins of an ATP-driven potassium transport system in *Escherichia coli*. *Proc. Natl. Acad. Sci. USA* 75:3216–3219, 1978.

Lane, L.K., J.H. Copenhaver, Jr., G.E. Lindenmayer, and A. Schwartz. Purification and characterization of, and [^{3}H]ouabain binding to the transport adenosine triphosphatase from outer medulla of canine kidney. *J. Biol. Chem.* 248:7197–7200, 1973.

Langer, G.A. Effects of digitalis on myocardial ionic exchange. *Circulation* 46:180–187, 1972.

Lanyi, J. Bacteriorhodopsin as a model for proton pumps. *Nature* 375:461–463, 1995.

Lardy, H.A., and C.A. Elvehjem. Biological oxidations and reductions. *Annu. Rev. Biochem.* 14:1–29, 1945.

Lardy, H.A., D. Johnson, and W.C. McMurray. Antibiotics as tools for metabolic studies. *Arch. Biochem. Biophys.* 78:587–597, 1958.

Lardy, H.A., and H. Wellman. Oxidative phosphorylations: Role of inorganic phosphate and acceptor systems in control of metabolic rates. *J. Biol. Chem.* 195:215–224, 1952.

Laris, P.C., D.P. Bahr, and R.R.J. Chaffee. Membrane potentials in mitochondrial preparations as measured by means of a cyanine dye. *Biochim. Biophys. Acta* 376:415–425, 1975.

Larsen, S.H., J. Adler, J.J. Gargus, and R.W. Hogg. Chemomechanical coupling without ATP: The source of energy for motility and chemotaxis in bacteria. *Proc. Natl. Acad. Sci. USA* 71:1239–1243, 1974.

Larsen, T.M., T. Laughlin, H.M. Holden, I. Rayment, and G.H. Reed. Structure of rabbit muscle pyruvate kinase complexed with Mn^{2+}, K$^+$, and pyruvate. *Biochemistry* 33:6301–6309, 1994.

Laubinger, W., G. Deckers-Hebestreit, K. Altendorf, and P. Dimroth. A hybrid adenosinetriphosphatase composed of F_1 of *Escherichia coli* and F_o of *Propionigenium modestum* is a functional sodium pump. *Biochemistry* 29:5458–5463, 1990.

Laubinger, W., and P. Dimroth. Characterization of the Na$^+$-stimulated ATPase of *Propionigenium modestum* as an enzyme of the F_1F_o type. *Eur. J. Biochem.* 168:475–480, 1987.

Laubinger, W., and P. Dimroth. Characterization of the ATP synthase of *Propionigenium modestum* as a primary sodium pump. *Biochemistry* 27:7531–7537, 1988.

Laudan, L. *Progress and its Problems*. Berkeley: University of California Press, 1977.

Laudan, L. *Beyond Positivism and Relativism*. Boulder: Westview Press, 1996.

Läuger, P. Ion transport through pores: A rate-theory analysis. *Biochim. Biophys. Acta* 311:423–441, 1973.

Lawford, H.G., and P.B. Garland. Proton translocation coupled to quinol oxidation in ox heart mitochondria. *Biochem. J.* 136:711–720, 1973.

Leaf, A., and A. Renshaw. Ion transport and respiration of isolated frog skin. *Biochem. J.* 65:82–93, 1957.

Lee, C.-P., G.F. Azzone, and L. Ernster. Evidence for energy-coupling in non-phosphorylating electron transport particles from beef-heart mitochondria. *Nature* 201:152–155, 1964.

Lee, J., G. Simpson, and P. Scholes. An ATPase from dog gastric mucosa: Changes of outer pH in suspensions of membrane vesicles accompanying ATP hydrolysis. *Biochem. Biophys. Res. Commun.* 60:825–832, 1974.

Lee, K.H., and R. Blostein. Red cell sodium fluxes catalysed by the sodium pump in the absence of K^+ and ADP. *Nature* 285:338–339, 1980.

Lehninger, A.L. Phosphorylation coupled to oxidation of dihydrodiphosphopyridine nucleotide. *J. Biol. Chem.* 190:345–359, 1951.

Lehninger, A.L. Oxidative phosphorylation. *Harvey Lect.* 49:176–215, 1954.

Lehninger, A.L. Oxidative phosphorylation in submitochondrial systems. *Federation Proc.* 19:952–962, 1960.

Lehninger, A.L. *The Mitochondrion.* New York: Benjamin, 1964.

Lehninger, A.L. Introduction. *Federation Proc.* 26:1333–1334, 1967.

Lehninger, A.L., and C.L. Wadkins. Oxidative phosphorylation. *Annu. Rev. Biochem.* 31:47–78, 1962.

Leicester, H.M. *Development of Biochemical Concepts from Ancient to Modern Times.* Cambridge: Harvard University Press, 1974.

Lenard, J., and S.J. Singer. Protein conformation in cell membrane preparations as studied by optical rotatory dispersion and circular dichroism. *Proc. Natl. Acad. Sci. USA* 56:1828–1835, 1966.

Leung, K.H., and P.C. Hinkle. Reconstitution of ion transport and respiratory control in vesicles formed from reduced coenzyme Q-cytochrome c reductase and phospholipids. *J. Biol. Chem.* 250:8467–8471, 1975.

Levi, H., and H.H. Ussing. The exchange of sodium and chloride ions across the fibre membrane of the isolated frog sartorius. *Acta Physiol. Scand.* 16:232–249, 1948.

Levi, H., and H.H. Ussing. Resting potential and ion movements in the frog skin. *Nature* 164:928–929, 1949.

Lew, V.L., I.M. Glynn, and J.C. Ellory. Net synthesis of ATP by reversal of the sodium pump. *Nature* 225:865–866, 1970.

Lew, V.L., M.A. Hardy, and J.C. Ellory. The uncoupled extrusion of Na^+ through the Na^+ pump. *Biochim. Biophys. Acta* 323:251–266, 1973.

Libet, B. Adenosinetriphosphatase (ATP-ase) in nerve. *Federation Proc.* 7:72, 1948.

Lindberg, O., J.J. Duffy, A.W. Norman, and P.D. Boyer. Characteristics of bound phosphohistidine labeling in mitochondria. *J. Biol. Chem.* 240:2850–2854, 1965.

Lindenmayer, G.E., A.H. Laughter, and A. Schwartz. Incorporation of inorganic phosphate-32 into a Na^+, K^+-ATPase preparation: Stimulation by ouabain. *Arch. Biochem. Biophys.* 127:187–192, 1968.

Linderholm, H. Active transport of ions through frog skin with special reference to the action of certain diuretics. *Acta Physiol. Scand.* 27 (Suppl. 97):3–144, 1952.

Linderholm, H. On the behavior of the "sodium pump" in frog skin at various concentrations of ions in the solution on the epithelial side. *Acta Physiol. Scand.* 31:36–61, 1954.

Ling, G.N. The role of phosphate in the maintenance of the resting potential and selective ionic accumulation in frog muscle cells. *In* McElroy, W.D., and B. Glass (eds.) *Phosphorus Metabolism.* Baltimore: Johns Hopkins University Press, 1952, Vol. 2, pp. 748–797.

Ling, G.N. The interpretation of selective ionic permeability and cellular potentials in terms of the fixed charge-induction hypothesis. *J. Gen. Physiol.* 43:149–174, 1960.

Ling, G.N. *A Physical Theory of the Living State.* New York: Blaisdell, 1962.

Ling, G.N. The association-induction hypothesis. *Tex. Rep. Biol. Med.* 2:244–265, 1964.

Ling, G.N. The membrane theory and other views for solute permeability, distribution, and transport in living cells. *Perspect. Biol. Med.* 9:87–106, 1965.

Ling, G.N. A new model for the living cell. *Int. Rev. Cytol.* 26:1–61, 1969.

Ling, G.N. Potassium accumulation in frog muscle. *Science* 198:1281–1283, 1977.

Ling, G., and R.W. Gerard. The normal membrane potential of frog sartorius fibers. *J. Cell. Comp. Physiol.* 34:383–396, 1949a.

Ling, G., and R.W. Gerard. The membrane potential and metabolism of muscle fibers. *J. Cell. Comp. Physiol.* 34:413–438, 1949b.

Ling, G.N., and W. Negendank. Do isolated membranes and purified vesicles pump sodium? A critical review and reinterpretation. *Perspect. Biol. Med.* 23:215–239, 1980.

Ling, G.N., and M.M. Ochsenfeld. Studies on ion accumulation in muscle cells. *J. Gen. Physiol.* 49:819–843, 1966.

Ling, G.N., and M.M. Ochsenfeld. Control of cooperative adsorption of solutes and water in living cells by hormones, drugs, and metabolic products. *Ann. N.Y. Acad. Sci.* 204:325–336, 1973.

Lingrel, J., and T. Kuntzweiler. Na$^+$, K$^+$-ATPase. *J. Biol. Chem.* 269:19659–19662, 1994.

Lipmann, F. Metabolic generation and utilization of phosphate bond energy. *Adv. Enzymol.* 1:99–162, 1941.

Lipmann, F. Metabolic process patterns. *In* Green, D.E., (ed.) *Currents in Biochemical Research.* New York: Interscience, 1946, pp. 137–148.

Loomis, W.F., and F. Lipmann. Reversible inhibition of the coupling between phosphorylation and oxidation. *J. Biol. Chem.* 173:807–808, 1948.

Lorente de Nó, R. Effects of choline and acetylcholine chloride upon peripheral nerve fibers. *J. Cell. Comp. Physiol.* 24:85–97, 1944.

Lorente de Nó, R. *A Study of Nerve Physiology.* New York: Studies from the Rockefeller Institute, 1947, Vols. 131, 132.

Lovatt Evans, C. *Principles of Human Physiology,* 11th ed. Philadelphia, Lea and Febiger, 1952.

Lowry, O.H., N.R. Roberts, M.-L. Wu, W.S. Hixon, and E.J. Crawford. The quantitative histochemistry of brain. II. Enzyme measurements. *J. Biol. Chem.* 207:19–37, 1954.

Lunardi, J., and P.V. Vignais. Studies of the nucleotide-binding sites on the mitochondrial F$_1$-ATPase through the use of a photoactivatable derivative of adenylyl imidodiphosphate. *Biochim. Biophys. Acta* 682:124–134, 1982.

Lund, E.J. The electrical polarity of Obelia and frog's skin, and its reversible inhibition by cyanide, ether, and cholorform. *J. Exp. Zool.* 44:383–396, 1926.

Lund, E.J. Relation between continuous bio-electric currents and cell respiration. III. Effects of concentration of oxygen on cell polarity in the frog skin. *J. Exp. Zool.* 51:291–307, 1928.

Lund, E.J., and J.B. Morrman. Electric polarity and velocity of cell oxidation as functions of temperature. *J. Exp. Zool.* 60:249–267, 1931.

Lundegårdh, H. An electro-chemical theory of salt absorption and respiration. *Nature* 143:203–204, 1939.

Lundegårdh, H. Salt absorption of plants. *Nature* 145:114–115, 1940.

Lünsdorf, H., K. Ehrig, P. Friedl, and H.U. Schairer. Use of monoclonal antibodies in immuno-electron microscopy for the determination of subunit stoichiometry in oligomeric enzymes. *J. Mol. Biol.* 173:131–136, 1984.

Luzzati, V., and F. Huson. The structure of the liquid-crystalline phases of lipid-water systems. *J. Cell Biol.* 12:207–219, 1962.

MacLennan, D.H. Purification and properties of an adenosine triphosphatase from sarcoplasmic reticulum. *J. Biol. Chem.* 245:4508–4518, 1970.

MacLennan, D.H., C.J. Brandl, B. Korczak, and N.M. Green. Amino-acid sequence of a Ca^{2+} + Mg^{2+}-dependent ATPase from rabbit muscle sarcoplasmic reticulum, deduced from its complementary DNA sequence. *Nature* 316:696–700, 1985.

MacLennan, D.H., P. Seeman, G.H. Iles, and C.C. Yip. Membrane formation by the adenosine triphosphatase of sarcoplasmic reticulum. *J. Biol. Chem.* 246:2702–2710, 1971.

MacLennan, D.H., and A. Tzagoloff. Studies on the mitochondrial adenosine triphosphatase system. *Biochemistry* 7:1603–1610, 1968.

Maddy, A.H. The chemical organization of the plasma membrane of animal cells. *Int. Rev. Cytol.* 20:1–65, 1966.

Maddy, A.H., and B.R. Malcolm. Protein conformations in the plasma membrane. *Science* 150:1616–1618, 1965.

Maguire, M.E. MgtA and MgtB: Prokaryotic P-type ATPases that mediate Mg^{2+} influx. *J. Bioenerg. Biomembr.* 24:319–328, 1992.

Mahler, H.R., and E.H. Cordes. *Biological Chemistry.* New York: Harper & Row, 1966.

Maizels, M. The permeation of human erythrocytes by anions and cations. *Trans. Faraday Soc.* 33:959–966, 1937.

Maizels, M. Phosphate, base and haemolysis in stored blood. *Q. J. Exp. Physiol.* 32:143–181, 1943.

Maizels, M. Cation control in human erythrocytes. *J. Physiol.* 108:247–263, 1949.

Maizels, M. Factors in the active transport of cations. *J. Physiol* 112:59–83, 1951.

Maizels, M. Active cation transport in erythrocytes. *Symp. Soc. Exp. Biol.* 8:202–227, 1954.

Maizels, M. Edward Joseph Conway. *Biograph. Mem. Fellows R. Soc.* 15:69–81, 1969.

Maizels, M., and J.H. Paterson. Survival of stored blood after transfusion. *Lancet* 417–420, 1940.

Makinose, M. The phosphorylation of the membranal protein of the sarcoplasmic vesicles during active calcium transport. *Eur. J. Biochem.* 10:74–82, 1969.

Makinose, M. Calcium efflux dependent formation of ATP from ADP and orthophosphate by the membranes of the sarcoplasmic vesicles. *FEBS Lett.* 12:269–270, 1971.

Makinose, M., and W. Hasselbach. ATP synthesis by the reverse of the sarcoplasmic calcium pump. *FEBS Lett.* 12:271–272, 1971.

Manery, J.F., and W.F. Bale. The penetration of radioactive sodium and phosphorus into the extra- and intracellular phases of tissues. *Am. J. Physiol.* 132:215–231, 1941.

Manery, J.F., I.S. Danielson, and A.B. Hastings. Connective tissue electrolytes. *J. Biol. Chem.* 124:359–375, 1938.

Manery, J.F., and A.B. Hastings. The distribution of electrolytes in mammalian tissues. *J. Biol. Chem.* 127:657–676, 1939.

Manson, M.D., P. Tedesco, H.C. Berg, F.M. Harold, and C. van der Drift. A protonmotive force drives bacterial flagella. *Proc. Natl. Acad. Sci. USA* 74:3060–3064, 1977.

Marchesi, V.T., and E. Steers. Selective solubilization of a protein component of the red cell membrane. *Science* 159:203–204, 1968.

Marmont, G. Studies on the axon membrane. *J. Cell. Comp. Physiol.* 34:351–382, 1949.

Marsh, B.B. A factor modifying muscle fibre synaeresis. *Nature* 167:1065–1066, 1951.

Marsh, B.B. The effects of adenosine triphosphate on the fibre volume of a muscle homogenate. *Biochim. Biophys. Acta* 9:247–260, 1952.

Martin, A.J.P., and R.L.M. Synge. A new form of chromatogram employing two liquid phases. *Biochem. J.* 35:1358–1368, 1941.

Martin, W.H., A.D. Beavis, and K.D. Garlid. Identification of an 82,000-dalton protein responsible for K^+/H^+ antiport in rat liver mitochondria. *J. Biol. Chem.* 259:2062–2065, 1984.

Martin-Vasallo, P., W. Dackowski, J.R. Emanuel, and R. Levenson. Identification of a putative isoform of the Na,K-ATPase β subunit. *J. Biol. Chem.* 264:4613–4618, 1989.

Martonosi, A. The role of phospholipids in the ATP-ase activity of skeletal muscle microsomes. *Biochem. Biophys. Res. Commun.* 29:753–757, 1967.

Masuda, H., and L. deMeis. Phosphorylation of the sarcoplasmic reticulum membrane by orthophosphate. Inhibition by calcium ions. *Biochemistry* 12:4581–4585, 1973.

Matchett, P.A., and J.A. Johnson. Inhibition of sodium and potassium transport in frog sartorii in the presence of ouabain. *Federation Proc.* 13:384, 1954.

Matsuno-Yagi, A., T. Yagi, and Y. Hatefi. Studies on the mechanism of oxidative phosphorylation. *Proc. Natl. Acad. Sci. USA* 82:7550–7554, 1985.

Matsuura, S., J. Shioi, and Y. Imae. Motility in *Bacillus subtilis* driven by an artificial protonmotive force. *FEBS Lett.* 82:187–190, 1977.

Matthews, B.W., P.B. Sigler, R. Henderson, and D.M. Blow. Three-dimensional structure of tosyl-α-chymotrypsin. *Nature* 214:652–656, 1967.

Maxam, A.M., and W. Gilbert. A new method for sequencing DNA. *Proc. Natl. Acad. Sci. USA* 74:560–564, 1977.

Mayr, E. *The Growth of Biological Thought.* Cambridge: Harvard University Press, 1982.

Mayr, E. *Toward a New Philosophy of Biology.* Cambridge: Harvard University Press, 1988.

McCarty, R.E., and E. Racker. Effects of an antiserum to the chloroplast coupling factor on phosphorylation and related processes. *Federation Proc.* 25:226, 1966.

McDougal, D.B., K.D. Little, and R.K. Crane. Studies on the mechanism of intestinal absorption of sugars. *Biochim. Biophys. Acta* 45:483–489, 1960.

McGilvery, R.W. *Biochemistry.* Philadelphia: W.B. Saunders, 1970.

McIlwain, H. Phosphate and nucleotides of the central nervous system. *Biochem. Soc. Symp.* 8:27–43, 1952a.

McIlwain, H. Phosphates of brain during *in vitro* metabolism: Effects of oxygen, glucose, glutamate, glutamine, and calcium and potassium salts. *Biochem. J.* 52:289–295, 1952b.

Mercer, R.W., J.W. Schneider, A. Savitz, J. Emanuel, E.J. Benz, Jr., and R. Levenson. Rat brain β-chain gene: Primary structure, tissue specific expression, and amplification in ouabain-resistant HeLa C^+ cells. *Mol. Cell. Biol.* 6:3884–3890, 1986.

Michell, R.H. Inositol phospholipids and cell surface receptor function. *Biochim. Biophys. Acta* 415:81–147, 1975.

Michell, R.H., C.J. Kirk, L.M. Jones, C.P. Downes, and J.A. Creba. The stimulation of inositol lipid metabolism that accompanies calcium mobilization in stimulated cells: Defined characteristics and unanswered questions. *Phil. Trans. R. Soc. B* 296:123–137, 1981.

Miller, C., and E. Racker. Ca^{++}-induced fusion of fragmented sarcoplasmic reticulum with artificial planar bilayers. *J. Membr. Biol.* 30:283–300, 1976.

Mirsky, A.E., and L. Pauling. On the structure of native, denatured, and coagulated proteins. *Proc. Natl. Acad. Sci. USA* 22:439–447, 1936.

Mitchell, P. A general theory of membrane transport from studies of bacteria. *Nature* 180:134–136, 1957a.

Mitchell, P. Structure and function of subcellular components. *Nature* 179:661–662, 1957b.

Mitchell, P. Structure and function in microorganisms. *Biochem. Soc. Symp.* 16:73–93, 1959.

Mitchell, P. Approaches to the analysis of specific membrane transport. In Goodson, T.W., and O. Lindberg (eds.) *Biological Structure and Function.* London, Academic Press, 1961a, Vol. 2, pp. 581–603.

Mitchell, P. Coupling of phosphorylation to electron and hydrogen transfer by a chemiosmotic type of mechanism. *Nature* 191:144–148, 1961b.

Mitchell, P. Conduction of protons through the membranes of mitochondria and bacteria by uncouplers of oxidative phosphorylation. *Biochem. J.* 81:24P, 1961c.

Mitchell, P. Biological transport phenomena and the spatially anisotropic characteristics of enzyme systems causing a vector component of metabolism. In Keinzeller, A., and A.

Kotyk (eds.) *Membrane Transport and Metabolism.* New York: Academic Press, 1961d, pp. 22–34.

Mitchell, P. Metabolism, transport, and morphogenesis: Which drives which? *J. Gen. Microbiol.* 29:25–37, 1962.

Mitchell, P. Molecule, group and electron translocation through natural membranes. *Biochem. Soc. Symp.* 22:142–169, 1963.

Mitchell, P. *Chemiosmotic Coupling in Oxidative and Photosynthetic Phosphorylation.* Bodmin: Glynn Research Laboratories, 1966.

Mitchell, P. Proton-translocation phosphorylation in mitochondria, chloroplasts and bacteria: Natural fuel cells and solar cells. *Federation Proc.* 26:1370–1379, 1967.

Mitchell, P. *Chemiosmotic Coupling and Energy Transduction.* Bodmin: Glynn Research Laboratories, 1968.

Mitchell, P. The chemical and electrical components of the electrochemical potential of H^+ ions across the mitochondrial cristae membrane. *FEBS Symp.* 17:219–232, 1969.

Mitchell, P. Self-electrophoretic locomotion in microorganisms: Bacterial flagella as giant ionophores. *FEBS Lett.* 28:1–4, 1972.

Mitchell, P. Proton translocation mechanisms and energy transduction by adenosine triphosphatases: An answer to criticisms. *FEBS Lett.* 50:95–97, 1975.

Mitchell, P. Possible molecular mechanisms of the proton motive function of cytochrome systems. *J. Theor. Biol.* 62:327–367, 1976.

Mitchell, P. Bioenergetic aspects of unity in biochemistry. *In* Semenza, G., (ed.) *Of Oxygen, Fuels, and Living Matter.* Chichester: Wiley, 1981, pp. 1–160.

Mitchell, P., and W.H. Koppenol. Chemiosmotic ATPase mechanisms. *Ann. N.Y. Acad. Sci.* 402:584–601, 1982.

Mitchell, P., R. Mitchell, J. Moody, I. West, H. Baum, and J. Wrigglesworth. Chemiosmotic coupling in cytochrome oxidase. *FEBS Lett.* 188:1–7, 1985.

Mitchell, P., and J. Moyle. Group-translocation: A consequence of enzyme-catalysed group-transfer. *Nature* 182:372–373, 1958.

Mitchell, P., and J. Moyle. Stoichiometry of proton translocation through the respiratory chain and adenosine triphosphatase systems of rat liver mitochondria. *Nature* 208:147–151, 1965.

Mitchell, P., and J. Moyle. Acid-base titration across the membrane system of rat-liver mitochondria. *Biochem. J.* 104:588–600, 1967a.

Mitchell, P., and J. Moyle. Respiration-driven proton translocation in rat liver mitochondria. *Biochem. J.* 105:1147–1162, 1967b.

Mitchell, P., and J. Moyle. Proton translocation coupled to ATP hydrolysis in rat liver mitochondria. *Eur. J. Biochem.* 4:530–539, 1968.

Mitchell, P., and J. Moyle. Estimation of membrane potential and pH difference across the cristae membrane of rat liver mitochondria. *Eur. J. Biochem.* 7:471–484, 1969.

Mitchell, P., J. Moyle, and L. Smith. Bromthymol blue as a pH indicator in mitochondrial suspensions. *Eur. J. Biochem.* 4:9–19, 1968.

Mitchell, R.A., L.G. Butler, and P.D. Boyer. The association of readily-soluble bound phosphohistidine from mitochondria with succinate thiokinase. *Biochem. Biophys. Res. Commun.* 16:545–550, 1964.

Mitchell, R.A., R.D. Hill, and P.D. Boyer. Mechanistic implications of Mg^{++}, adenine nucleotide, and inhibitor effects on energy-linked reactions of submitochondrial particles. *J. Biol. Chem.* 242:1793–1801, 1967.

Moczydlowski, E.C., and P.A.G. Fortes. Inhibition of sodium and potassium adenosine triphosphatase by 2',3'-O-(2,4,6-trinitrocyclohexadienylidene) adenine nucleotides. Implications for the structure and mechanism of the Na:K pump. *J. Biol. Chem.* 256:2357–2366, 1981.

Mond, R., and H. Netter. Über die Regulation des Natriums durch den Muskel. *Pflügers Arch.* 230:42–69, 1932.

Monod, J., J.-P. Changeux, and F. Jacob. Allosteric proteins and cellular control systems. *J. Mol. Biol.* 6:306–329, 1963.

Monod, J., J. Wyman, and J.-P. Changeux. On the nature of allosteric transitions: A plausible model. *J. Mol. Biol.* 12:88–118, 1965.

Moore, C., and B.C. Pressman. Mechanism of action of valinomycin on mitochondria. *Biochem. Biophys. Res. Commun.* 15:562–567, 1964.

Morus, I.R. Book review. *Br. J. Hist. Sci.* 26:92–93, 1993.

Mueckler, M., C. Caruso, S.A. Baldwin, M. Panico, I. Blench, H.R. Morris, W.J. Allard, G.E. Lienhard, and H.F. Lodish. Sequence and structure of a human glucose transporter. *Science* 229:941–945, 1985.

Mueller, P., and D.O. Rudin. Development of K^+-Na^+ discrimination in experimental bimolecular lipid membranes by macrocyclic antibiotics. *Biochem. Biophys. Res. Commun.* 26:398–404, 1967.

Mueller, P., D.O. Rudin, H.T. Tien, and W.C. Wescott. Reconstitution of cell membrane structure *in vitro* and its transformation into an excitable system. *Nature* 194:979–980, 1962.

Muirhead, H., and M.F. Perutz. Structure of reduced human hemoglobin. *Cold Spring Harb. Symp. Quant. Biol.* 28:451–459, 1963.

Mullins, L.J. An analysis of pore size in excitable membranes. *J. Gen. Physiol.* 43:105–117, 1960.

Mullins, L.J., and F.J. Brinley, Jr. Potassium fluxes in dialyzed squid axons. *J. Gen. Physiol.* 53:704–740, 1969.

Mullins, L.J., W.O. Fenn, T.R. Noonan, and L. Haege. Permeability of erythrocytes to radioactive potassium. *Am. J. Physiol.* 135:93–101, 1941.

Muntz, J.A., and J. Hurwitz. Effect of potassium and ammonium ions upon glycolysis catalyzed by an extract of rat brain. *Arch. Biochem. Biophys.* 32:124–136, 1951.

Murer, M., and U. Hopfer. Demonstration of electrogenic Na^+-dependent D-glucose transport in intestinal brush border membranes. *Proc. Natl. Acad. Sci. USA* 71:484–488, 1974.

Mushkin, S.J. *Biomedical Research: Costs and Benefits.* Cambridge: Ballinger, 1979.

Nagai, T., M. Makinose, and W. Hasselbach. Der physiologische Erschlaffungsfaktor und die Muskelgrana. *Biochim. Biophys. Acta* 43:223–238, 1960.

Nagano, K., T. Kanazawa, N. Mizuno, Y. Tashima, T. Nakao, and M. Nakao. Some acyl phosphate-like properties of P^{32}-labeled sodium-potassium-activated adenosine triphosphatase. *Biochem. Biophys. Res. Commun.* 19:759–764, 1965.

Nakao, M., T. Nakao, Y. Hara, F. Nagai, S. Yagasaki, M. Koi, A. Nakagawa, and K. Kawai. Purification and properties of Na, K-ATPase from pig brain. *Ann. N.Y. Acad. Sci.* 242:24–33, 1974.

Narahashi, T. Chemicals as tools in the study of excitable membranes. *Physiol. Rev.* 54:813–889, 1974.

Nastuk, W.L., and A.L. Hodgkin. The electrical activity of single muscle fibers. *J. Cell. Comp. Physiol.* 35:39–73, 1950.

Needham, D.M. *Machina Carnis.* Cambridge: Cambridge University Press, 1971.

Negrin, R.S., D.L. Foster, and R.H. Fillingame. Energy-transducing H^+-ATPase of *Escherichia coli. J. Biol. Chem.* 255:5643–5648, 1980.

Neher, E., and B. Sakmann. Single-channel currents recorded from membrane of denervated frog muscle fibres. *Nature 260:799–802, 1976.*

Neumann, J., and A.T. Jagendorf. Light-induced pH changes related to phosphorylation by chloroplasts. *Arch. Biochem. Biophys.* 107:109–119, 1964.

Neville, D.M. The isolation of a cell membrane fraction from rat liver. *J. Biophys. Biochem. Cytol.* 8:413–422, 1960.

Newman, M.J., D.L. Foster, T.H. Wilson, and H.R. Kaback. Purification and reconstitution of functional lactose carrier from *Escherichia coli*. *J. Biol. Chem.* 256:11804–11808, 1981.

Nicholls, D., J. Kanfer, and E. Titus. The effect of ouabain on the incorporation of inorganic P^{32} into phospholipid. *J. Biol. Chem.* 237:1043–1049, 1962.

Nicholls, P. Cytochromes—a survey. *Enzymes* 8:3–40, 1963.

Nicoll, D.A., S. Longoni, and K.D. Philipson. Molecular cloning and functional expression of the cardiac sarcolemmal Na^+-Ca^{2+} exchanger. *Science* 250:562–565, 1990.

Nielsen, J., F.G. Hansen, J. Hoppe, P. Friedl, and K. von Meyenburg. The nucleotide sequence of the *atp* genes coding for the F_o subunits a, b, c and the F_1 subunit δ of the membrane bound ATP synthase of *Escherichia coli*. *Mol. Gen. Genet.* 184:33–39, 1981.

Niggli, V., J.T. Penniston, and E. Carafoli. Purification of the $(Ca^{2+}$-$Mg^{2+})$-ATPase from human erythrocyte membranes using a calmodulin affinity column. *J. Biol. Chem.* 254:9955–9958, 1979.

Niggli, V., E. Sigel, and E. Carafoli. The purified Ca^{2+} pump of human erythrocyte membranes catalyzes an electroneutral Ca^{2+}-H^+ exchange in reconstituted liposomal systems. *J. Biol. Chem.* 257:2350–2356, 1982.

NIH Factbook. Chicago: Marquis Academic Media, 1976.

Nilius, B. Gating properties and modulation of Na channels. *News Physiol. Sci.* 4:225–230, 1989.

Noda, M., T. Ikeda, H. Suzuki, H. Takeshima, T. Takahashi, M. Kuno, and S. Numa. Expression of functional sodium channels from cloned cDNA. *Nature* 322:826–828, 1986.

Noda, M., S. Shimizu, T. Tanabe, T. Takai, T. Kayano, T. Ikeda, H. Takahashi, H. Nakayama, Y. Kanaoka, N. Minamino, K. Kangawa, H. Matsuo, M.A. Raftery, T. Hirose, S. Inayama, H. Hayashida, T. Miyata, and S. Numa. Primary structure of *Electrophorus electricus* sodium channel deduced from cDNA sequence. *Nature* 312:121–127, 1984.

Noguchi, S., M. Mishina, M. Kawamura, and S. Numa. Expression of functional $(Na^+ + K^+)$-ATPase from cloned cDNAs. *FEBS Lett.* 225:27–32, 1987.

Noguchi, S., M. Noda, H. Takahashi, K. Kawakami, T. Ohta, K. Nagano, T. Hirose, S. Inayama, M. Kawamura, and S. Numa. Primary structure of the β-subunit of *Torpedo californica* $(Na^+ + K^+)$-ATPase deduced from the cDNA sequence. *FEBS Lett.* 196:315–320, 1986.

Noonan, T.R., W.O. Fenn, and L. Haege. The distribution of injected radioactive potassium in rats. *Am. J. Physiol.* 132:474–488, 1941.

Nørby, J.G. Na, K-ATPase: Structure and kinetics. Comparison with other ion transport systems. *Chemica Scripta* 27B:119–129, 1987.

Nørby, J.G., and J. Jensen. Binding of ATP to brain microsomal ATPase. Determination of the ATP-binding capacity and the dissociation constant of the enzyme-ATP complex as a function of K^+ concentration. *Biochim. Biophys. Acta* 233:104–116, 1971.

Novikoff, A.B., L. Hecht, E. Podber, and J. Ryan. Phosphatases of rat liver. I. The dephosphorylation of adenosinetriphosphate. *J. Biol. Chem.* 194:153–170, 1952.

Ochoa, S. 'Coupling' of phosphorylation with oxidation of pyruvic acid in brain. *J. Biol. Chem.* 138:751–773, 1941.

Ochoa, S. Efficiency of aerobic phosphorylation in cell-free heart extracts. *J. Biol. Chem.* 151:493–505, 1943.

Oesterhelt, D., and W. Stoeckenius. Rhodopsin-like protein from the purple membrane of *Halobacterium halobium*. *Nature New Biol.* 233:149–152, 1971.

Oesterhelt, D., and W. Stoeckenius. Functions of a new photoreceptor membrane. *Proc. Natl. Acad. Sci. USA* 70:2853–2857, 1973.

Ohnishi, M., and D.W. Urry. Solution conformation of valinomycin-potassium ion complex. *Science* 168:1091–1092, 1970.

Ohtsubo, M., S. Noguchi, K. Takeda, M. Morohashi, and M. Kawamura. Site-directed mutagenesis of Asp-376, the catalytic phosphorylation site, and Lys-507, the putative ATP-binding site, of the α subunit of *Torpedo californica* Na$^+$/K$^+$-ATPase. *Biochim. Biophys. Acta* 1021:157–160, 1990.

Okamoto, C.T., J.M. Karpilow, A. Smolka, and J.G. Forte. Isolation and characterization of gastric microsomal glycoproteins. Evidence for a glycosylated β-subunit of the H$^+$/K$^+$-ATPase. *Biochim. Biophys. Acta* 1037:360–372, 1990.

Okamoto, H., N. Sone, H. Hirata, M. Yoshida, and Y. Kagawa. Purified proton conductor in proton translocating adenosine triphosphatase of a thermophilic bacterium. *J. Biol. Chem.* 252:6125–6131, 1977.

Ovchinnikov, Y.A., N.G. Abdulaev, M.Y. Feigina, A.V. Kiselev, and N.A. Lobanov. The structural basis of the functioning of bacteriorhodopsin. *FEBS Lett.* 100:219–224, 1979.

Ovchinnikov, Y.A., N.N. Modyanov, N.E. Broude, K.E. Petrukhin, A.V. Grishin, N.M. Arzamazova, N.A. Aldanova, G.S. Monastyrskaya, and E.D. Sverdlov. Pig kidney Na$^+$, K$^+$-ATPase. *FEBS Lett.* 201:237–245, 1986.

Painter, A.A., and F.E. Hunter. Phosphorylation coupled to oxidation of thiol groups (GSH) by cytochrome c with disulfide (GSSG) as an essential catalyst. *Biochem. Biophys. Res. Commun.* 40:360–395, 1970.

Palade, G.E. The fine structure of mitochondria *Anat. Rec.* 114:427–451, 1952.

Palade, G.E. An electron microscope study of the mitochondrial structure. *J. Histochem. Cytochem.* 1:188–211, 1953.

Palmer, L.G., and J. Gulati. Potassium accumulation in muscle: A test of the binding hypothesis. *Science* 194:521–523, 1976.

Pandit-Hovenkamp, H.G. The P$_i$-incorporating factor of *Alcaligenes faecalis*. *Biochim. Biophys. Acta.* 99:552–555, 1965.

Parker, J.C. Solute and water transport in dog and cat red blood cells. *In* Ellory, J.C., and V.L. Lew (eds.) *Membrane Transport in Red Cells.* London: Academic Press, 1977, pp. 427–465.

Parpart, A.K., and A.J. Dziemian. The chemical composition of the red cell membrane. *Cold Spring Harb. Symp. Quant. Biol.* 8:17–22, 1940.

Parsons, D.F. Mitochondrial structure: Two types of subunits on negatively stained mitochondrial membranes. *Science* 140:985–987, 1963.

Pauling, L., R.B. Corey, and H.R. Branson. The structure of proteins: Two hydrogen-bonded helical configurations of the polypeptide chain. *Proc. Natl. Acad. Sci. USA.* 37:205–211, 1951.

Pearce, B.E., and E.M. Wright. Sodium-induced conformational changes in the glucose transporter of intestinal brush borders. *J. Biol. Chem.* 259:14105–14112, 1984.

Pearce, B.E., and E.M. Wright. Evidence for tyrosyl residus at the Na$^+$ site on the intestinal Na$^+$/glucose cotransporter. *J. Biol. Chem.* 260:6026–6031, 1985.

Pedemonte, C.H., and J.H. Kaplan. Chemical modification as an approach to elucidation of sodium pump structure-function relations. *Am. J. Physiol.* 258:C1–C23, 1990.

Pedersen, P.L., and E. Carafoli. Ion motive ATPases. *Trends Biochem. Sci.* 12:146–150, 1987.

Penefsky, H.S. Mechanism of inhibition of mitochondrial adenosine triphosphatase by dicyclohexylcarbodiimide and oligomycin. *Proc. Natl. Acad. Sci. USA* 82:1589–1593, 1985a.

Penefsky, H.S. Reaction mechanism of the membrane-bound ATPase of submitochondrial particles from beef heart. *J. Biol. Chem.* 260:13728–13734, 1985b.

Penefsky, H.S. Energy-dependent dissociation of ATP from high affinity catalytic sites of beef heart mitochondrial adenosine triphosphatase. *J. Biol. Chem.* 260:13735–13741, 1985c.

Penefsky, H.S., M.E. Pullman, A. Datta, and E. Racker. Partial resolution of the enzymes catalyzing oxidative phosphorylation. *J. Biol. Chem.* 235:3330–3336, 1960.

Penefsky, H.S., and R.C. Warner. Partial resolution of the enzymes catalyzing oxidative phosphorylation. VI. Studies on the mechanism of cold inactivation of mitochondrial adenosine triphosphatase. *J. Biol. Chem.* 240:4694–4702, 1965.

Peracchia, C. *Handbook of Membrane Channels.* San Diego: Academic Press, 1994.

Perlin, D.S., K. Kasamo, R.J. Brooker, and C.W. Slayman. Electrogenic H^+ translocation by the plasma membrane ATPase of *Neurospora. J. Biol. Chem.* 259:7884–7892, 1984.

Perutz, M.F. Stereochemistry of cooperative effects in haemoglobin. *Nature* 228:726–739, 1970.

Perutz, M.F., J.C. Kendrew, and H.C. Watson. Structure and function of hemoglobin. *J. Mol. Biol.* 13:669–678, 1965.

Perutz, M.F., M.G. Rossmann, A.F. Cullis, H. Muirhead, G. Will, and A.C.T. North. Structure of haemoglobin. *Nature* 185:416–422, 1960.

Peter, J.B., D.E. Hultquist, M. DeLuca, G. Kriel, and P.D. Boyer. Bound phosphohistidine as an intermediate in a phosphorylation reaction of oxidative phosphorylation catalyzed by mitochondrial extracts. *J. Biol. Chem.* 238:PC1182–PC1184, 1963.

Peters, J.P. Transfers of water and solutes in the body. *Bull. N.Y. Acad. Med.* 14:299–324, 1938.

Peters, J.P., and D.D. Van Slyke. *Quantitative Clinical Chemistry.* Vol. I. *Interpretations.* Baltimore: Williams and Wilkins, 1931.

Phillips, D.C. The hen egg-white lysozyme molecule. *Proc. Natl. Acad. Sci. USA* 57:484–495, 1967.

Philo, R.D., and A.A. Eddy. The membrane potential of mouse ascites-tumour cells studied with the fluorescent probe 3,3′-dipropyloxadicarbocyanine. *Biochem. J.* 174:801–810, 1978.

Pikula, S., N. Mullner, L. Dux, and A. Martonosi. Stabilization and crystallization of Ca^{2+}-ATPase in detergent-solubilized sarcoplasmic reticulum. *J. Biol. Chem.* 263:5277–5286, 1988.

Pinchot, G.B. A high-energy intermediate of oxidative phosphorylation. *Proc. Natl. Acad. Sci. USA* 46:929–938, 1960.

Pinchot, G.B. The first phosphorylation site—observations on mechanism. *Federation Proc.* 22:1076–1079, 1963.

Pinchot, G.B., and M. Hormanski. Characterization of a high energy intermediate of oxidative phosphorylation. *Proc. Natl. Acad. Sci. USA* 48:1970–1977, 1962.

Pinchot, G.B., and B.J. Salmon. An intermediate of oxidative phosphorylation from *Alcaligenes faecalis. Arch. Biochem. Biophys.* 115:345–359, 1966.

Pinto da Silva, P., and D. Branton. Membrane splitting in freeze-etching. *J. Cell Biol.* 45:598–605, 1970.

Pitts, B.J.R. Stoichiometry of sodium-calcium exchange in cardiac sarcolemmal vesicles. *J. Biol. Chem.* 254:6232–6235, 1979.

Ponder, E. Accumulation of potassium by human red cells. *J. Gen. Physiol.* 33:745–757, 1950.

Ponder, E., and J. Macleod. The potential and respiration of frog skin. *J. Gen. Physiol.* 20:433–447, 1937.

Porter, K.R., and G.E. Palade. Studies on the endoplasmic reticulum. III. Its form and distribution in striated muscle cells. *J. Biophys. Biochem. Cytol.* 3:269–299, 1957.

Portzehl, H. Die Bindung des Erschaffungsfaktors von Marsh an die Muskelgrana. *Biochim. Biophys. Acta* 26:373–377, 1957.

Post, R.L. Relationship of an ATPase in human erythrocyte membranes to the active transport of sodium and potassium. *Federation Proc.* 18:121, 1959.

Post, R.L. A reminiscence about sodium, potassium-ATPase. *Ann. N.Y. Acad. Sci.* 242:6–11, 1974.

Post, R.L. Seeds of sodium, potassium ATPase. *Annu. Rev. Physiol.* 51:1–15, 1989.

Post, R.L., and C.D. Albright. Membrane adenosine triphosphatase system as a part of a system for active sodium and potassium transport. *In* Kleinzeller, A., and A. Kotyk (eds.) *Membrane Transport and Metabolism.* New York: Academic Press, 1961, pp. 219–227.

Post, R.L., C. Hegyvary, and S. Kume. Activation by adenosine triphosphate in the phosphorylation kinetics of sodium and potassium ion transport adenosine triphosphatase. *J. Biol. Chem.* 247:6530–6540, 1972.

Post, R.L., and P.C. Jolly. The linkage of sodium, potassium, and ammonium active transport across the human erythrocyte membrane. *Biochim. Biophys. Acta* 25:118–128, 1957.

Post, R.L., and S. Kume. Evidence for an aspartyl phosphate residue at the active site of sodium and potassium ion transport adenosine triphosphatase. *J. Biol. Chem.* 248:6993–7000, 1973.

Post, R.L., S. Kume, and F.N. Rogers. Alternating paths of phosphorylation of the sodium and potassium ion pump of plasma membranes. *In* Azzone, G.F., L. Ernster, S. Papa, E. Quagliariello, and N. Siliprandi (eds.) *Mechanisms in Bioenergetics.* New York: Academic Press, 1973, pp. 203–218.

Post, R.L., S. Kume, T. Tobin, B. Orcutt, and A.K. Sen. Flexibility of an active center in sodium-plus-potassium adenosine triphosphatase. *J. Gen. Physiol.* 54:306S–326S, 1969.

Post, R.L., C.R. Merritt, C.R. Kinsolving, and C.D. Albright. Membrane adenosine triphosphatase as a participant in the active transport of sodium and potassium in the human erythrocyte. *J. Biol. Chem.* 235:1796–1802, 1960.

Post, R.L., and A.K. Sen. An enzymatic mechanism of active sodium and potassium transport. *J. Histochem. Cytochem.* 13:105–112, 1965.

Post, R.L., A.K. Sen, and A.S. Rosenthal. A phosphorylated intermediate in adenosine triphosphate-dependent sodium and potassium transport across kidney membranes. *J. Biol. Chem.* 240:1437–1445, 1965.

Post, R.L., G. Toda, and F.N. Rogers. Phosphorylation by inorganic phosphate of sodium plus potassium ion transport adenosine triphosphate. *J. Biol. Chem.* 250:691–701, 1975.

Pratap, P.R., J.D. Robinson, and M.I. Steinberg. The reaction sequence of the Na^+/K^+-ATPase: Rapid kinetic measurements distinguish between alternative schemes. *Biochim. Biophys. Acta* 1069:288–298, 1991.

Prebble, J.N. Successful theory development in biology. *Biosci. Rep.* 16:207–215, 1996.

Pressman, B.C. Critique of the chemiosmotic hypothesis. *In* Azzone, G.F., E. Carafoli, A.L. Lehninger, E. Quagliariello, and N. Siliprandi (eds.) *Biochemistry and Biophysics of Mitochondrial Membranes.* New York: Academic Press, 1972, pp. 591–602.

Pressman, B.C., E.J. Harris, W.S. Jagger, and J.H. Johnson. Antibiotic-mediated transport of alkalai ions across lipid barriers. *Proc. Natl. Acad. Sci. USA* 58:1949–1956, 1967.

Pullman, M.E., H.S. Penefsky, A. Datta, and E. Racker. Partial resolution of the enzymes catalyzing oxidative phosphorylation. *J. Biol. Chem.* 235:3322–3329, 1960.

Raaflaub, J. Die Schwellung isolierter Leberzellmitochondrien und ihre physikalisch-chemische Beeinflussbarkeit. *Helv. Physiol. Acta* 11:142–156, 1953.

Rabon, E., R.D. Gunther, A. Soumarmon, S. Bassilian, M. Lewin, and G. Sachs. Solubilization and reconstitution of the gastric H,K-ATPase. *J. Biol. Chem.* 260:10200–10207, 1985.

Racker, E. A mitochondrial factor conferring oligomycin sensitivity on soluble mitochondrial ATPase. *Biochem. Biophys. Res. Commun.* 10:435–439, 1963.

Racker, E. Reconstitution of a calcium pump with phospholipids and a purified Ca^{++}-

adenosine triphosphatase from sarcoplasmic reticulum. *J. Biol. Chem.* 247:8198–8200, 1972.

Racker, E. A new procedure for the reconstitution of biologically active phospholipid vesicles. *Biochem. Biophys. Res. Commun.* 55:224–230, 1973.

Racker, E., and L.L. Horstman. Mechanism and control of oxidative phosphorylation. *In* Mehlman, M.A., and R.W. Hanson (eds.) *Energy Metabolism and the Regulation of Metabolic Processes in Mitochondria.* New York: Academic Press, 1972, pp. 1–25.

Racker, E., and I. Krimsky. The mechanism of oxidation of aldehydes by glyceraldehyde-3-phosphate dehydrogenase. *J. Biol. Chem.* 198:731–743, 1952.

Racker, E., and W. Stoeckenius. Reconstitution of purple membrane vesicles catalyzing light-driven proton uptake and adenosine triphosphate formation. *J. Biol. Chem.* 249:662–663, 1974.

Racker, E., D.D. Tyler, R.W. Estabrook, T.E. Conover, D.F. Parsons, and B. Chance. Correlations between electron-transport activity, ATPase, and morphology of submitochondrial particles. *In* King, T.E., H.S. Mason, and M. Morrison (eds.) *Oxidases and Related Redox Systems.* New York: John Wiley, 1965, Vol. 2, pp. 1077–1101.

Ragan, C.I., and P.C. Hinkle. Ion transport and respiratory control in vesicles formed from reduced nicotinamide adenine dinucleotide coenzyme Q reductase and phospholipids. *J. Biol. Chem.* 250:8472–8476, 1975.

Raker, J.W., I.M. Taylor, J.M. Weller, and A.B. Hastings. Rate of potassium exchange of the human erythrocyte. *J. Gen. Physiol.* 33:691–702, 1950.

Rakowski, R.F., D.C. Gadsby, and P. DeWeer. Stoichiometry and voltage dependence of the sodium pump in voltage-clamped, internally dialyzed squid giant axon. *J. Gen. Physiol.* 93:903–941, 1989.

Rall, T.W., E.W. Sutherland, and J. Berthet. The relationship of epinephrine and glucagon to liver phosphorylase. *J. Biol. Chem.* 224:463–475, 1957.

Rasmussen, H. Cell communication, calcium ion and cyclic adenosine monophosphate. *Science* 170:404–412, 1970.

Ray, T.K., and J.G. Forte. Studies on the phosphorylated intermediates of a K^+-stimulated ATPase from gastric mucosa. *Biochim. Biophys. Acta* 443:451–467, 1976.

Reeves, J.P., and J.L. Sutko. Sodium-calcium ion exchange in cardiac membrane vesicles. *Proc. Natl. Acad. Sci. USA* 76:590–594, 1979.

Rega, A.F., and P.J. Garrahan. Calcium ion-dependent phosphorylation of human erythrocyte membranes. *J. Membr. Biol.* 22:313–327, 1975.

Rega, A.F., and P.J. Garrahan. Calcium ion-dependent dephosphorylation of the Ca^{2+}-ATPase of human red cells by ADP. *Biochim. Biophys. Acta* 507:182–184, 1978.

Reid, R.A., J. Moyle, and P. Mitchell. Synthesis of adenosine triphosphate by a protonmotive force in rat liver mitochondria. *Nature* 212:257–258, 1966.

Reizer, J., A. Reizer, and M.H. Saier, Jr. A functional superfamily of sodium/solute symporters. *Biochim. Biophys. Acta* 1197:133–166, 1994.

Repke, K. Metabolism of cardiac glycosides. *In* Uvnä, B., (ed.) *Proceedings of the First International Pharmacological Meeting.* New York: Macmillan, 1963, Vol. 3, pp. 47–73.

Repke, K.R.H. Über den biochemischen Wirkungsmodus von Digitalis. *Klin. Wochenschr.* 42:157–165, 1964.

Repke, K.R.H., and R. Schön. Flip-flop model of (Na,K)-ATPase function. *Acta Biol. Med. Ger.* 31:K19–K30, 1973.

Reuben, M.A., L.S. Lasater, and G. Sachs. Characterization of a β subunit of the gastric H^+/K^+-transporting ATPase. *Proc. Natl. Acad. Sci. USA* 87:6767–6771, 1990.

Reuter, H., and N. Seitz. The dependence of calcium efflux from cardiac muscle on temperature and external ion composition. *J. Physiol.* 195:451–470, 1968.

Rhoads, D.B., and W. Epstein. Energy coupling to net K^+ transport in *Escherichia coli* K-12. *J. Biol. Chem.* 252:1394–1401, 1977.

Rhoads, D.B., F.B. Waters, and W. Epstein. Cation transport in *Escherichia coli.* VIII. Potassium transport mutants. *J. Gen. Physiol.* 67:325–341, 1976.

Richards, D.E., A.F. Rega, and P.J. Garrahan. Two classes of sites for ATP in the Ca^{2+}-ATPase from human red cell membranes. *Biochim. Biophys. Acta* 511:194–201, 1978.

Richardson, J.S. The anatomy and taxonomy of protein structure. *Adv. Prot. Chem.* 34:167–339, 1981.

Riggs, T.R., L.M. Walker, and H.N. Christensen. Potassium migration and amino acid transport. *J. Biol. Chem.* 233:1479–1484, 1958.

Riklis, E., and J.H. Quastel. Effects of cations on sugar absorption by isolated surviving guinea pig intestine. *Can. J. Biochem. Physiol.* 36:347–361, 1958.

Ritchie, J.M., and R.W. Straub. The hyperpolarization which follows activity in mammalian non-medulated fibres. *J. Physiol.* 136:80–97, 1957.

Robertson, J.D. The ultrastructure of cell membranes and their derivatives. *Biochem. Soc. Symp.* 16:3–43, 1959.

Robertson, J.D. Unit membranes: A review with recent new studies. *Symp. Soc. Study Dev. Growth* 22:1–81, 1964.

Robertson, R.N. Ion transport and respiration. *Biol. Rev.* 35:231–264, 1960.

Robertson, R.N., and M. Wilkins. Quantitative relation between salt accumulation and salt respiration in plant cells. *Nature* 161:101, 1948.

Robinson, J.D. Kinetic studies on a brain microsomal adenosine triphosphatase. Evidence suggesting conformational changes. *Biochemistry* 6:3250–3258, 1967.

Robinson, J.D. Kinetic studies on a brain microsomal adenosine triphosphatase. II. Potassium-dependent phosphatase activity. *Biochemistry* 8:3348–3355, 1969.

Robinson, J.D. Nucleotide and divalent cation interactions with the $(Na^+ + K^+)$-dependent ATPase. *Biochim. Biophys. Acta* 341:232–247, 1974a.

Robinson, J.D. Free Mg^{2+} and proposed isomerizations of the $(Na^+ + K^+)$-dependent ATPase. *FEBS Lett.* 47:352–355, 1974b.

Robinson, J.D. Substrate sites of the $(Na^+ + K^+)$-dependent ATPase. *Biochim. Biophys. Acta* 429:1006–1019, 1976.

Robinson, J.D. The chemiosmotic hypothesis of energy coupling and the path of scientific opportunity. *Perspect. Biol. Med.* 27:367–383, 1984.

Robinson, J.D. Reduction, explanation, and the quests of biological research. *Phil. Sci.* 53:333–353, 1986a.

Robinson, J.D. Appreciating key experiments. *Br. J. Hist. Sci.* 19:51–56, 1986b.

Robinson, J.D. Aims and achievements of the reductionist approach in biochemistry/molecular biology/cell biology: A response to Kincaid. *Phil. Sci.* 59:465–470, 1992.

Robinson, J.D., R.L. Davis, and M. Steinberg. Fluoride and beryllium interact with the (Na + K)-dependent ATPase as analogs of phosphate. *J. Bioenerg. Biomembr.* 18:521–531, 1986.

Robinson, J.D., and M.S. Flashner. The $(Na^+ + K^+)$-activated ATPase. Enzymatic and transport properties. *Biochim. Biophys. Acta* 549:145–176, 1979.

Robinson, J.D., and P.R. Pratap. Indicators of conformational changes in the Na^+/K^+-ATPase and their interpretation. *Biochim. Biophys. Acta* 1154:83–104, 1993.

Rose, R.C., and S.G. Schultz. Studies on the electrical potential profile across rabbit ileum. *J. Gen. Physiol.* 57:639–663, 1971.

Rosenberg, R.L., S.A. Tomiko, and W.S. Agnew. Single-channel properties of the reconstituted voltage-regulated Na channel isolated from the electroplax of *Electrophorus electricus. Proc. Natl. Acad. Sci. USA* 81:5594–5598, 1984.

Rosenberg, T. On accumulation and active transport in biological systems. *Acta Chem. Scand.* 2:14–33, 1948.

Rosenberg, T., and W. Wilbrant. Uphill transport induced by counterflow. *J. Gen. Physiol.* 41:289–296, 1957.

Rosing, J., C. Kayalar, and P.D. Boyer. Evidence for energy-dependent change in phosphate binding for mitochondrial oxidative phosphorylation based on measurements of medium and intermediate phosphate-water exchanges. *J. Biol. Chem.* 252:2478–2485, 1977.

Rossi, J.P.F.C., P.J. Garrahan, and A.F. Rega. Reversal of the calcium pump in human red cells. *J. Membr. Biol.* 44:37–46, 1978.

Rothenberg, M.A. Studies on permeability in relation to nerve function. II. Ionic movements across axonal membranes. *Biochim. Biophys. Acta* 4:96–114, 1950.

Rothenberg, M.A., and E.A. Feld. Rate of penetration of electrolytes into nerve-fibers. *J. Biol. Chem.* 172:345–346, 1948.

Rottenberg, H., and C.-P. Lee. Energy-dependent hydrogen ion accumulation in submitochondrial particles. *Biochemistry* 14:2675–2680, 1975.

Rowen, L.S. *Normative Epistemology and Scientific Research: Reflection on the "Ox-phos" Controversy.* Ph.D. thesis, Vanderbilt University, 1986.

Ruoho, A., and J. Kyte. Photoaffinity labeling of the ouabain-binding site on $(Na^+ + K^+)$ adenosine triphosphatase. *Proc. Natl. Acad. Sci. USA* 71:2352–2356, 1974.

Rushton, W.A.H. Initiation of the propagated disturbance. *Proc. R. Soc. Lond. B* 124:210–243, 1937.

Saccomani, G., G. Shah, J.G. Spenney, and G. Sachs. Characterization of gastric mucosal membranes. *J. Biol. Chem.* 250:4802–4809, 1975.

Saccomani, G., H.B. Stewart, D. Shaw, M. Lewin, and G. Sachs. Characterization of gastric mucosal membranes. *Biochim. Biophys. Acta* 465:311–330, 1977.

Sachs, J.R. Potassium-potassium exchange as part of the over-all mechanism of the sodium pump of the human red blood cell. *J. Physiol.* 374:221–244, 1986.

Salmon, W. *The Foundations of Scientific Inference.* Pittsburg: University of Pittsburgh Press, 1966.

Sanadi, D.R. Energy linked reactions in mitochondria. *Annu. Rev. Biochem.* 34:21–48, 1965.

Sandow, A. Excitation-contraction coupling in muscular response. *Yale J. Biol. Med.* 25:189–201, 1952.

Sanger, F., S. Nicklen, and A.R. Coulson. DNA sequencing with chain-terminating inhibitors. *Proc. Natl. Acad. Sci. USA* 74:5463–5467, 1977.

Sanger, F., and H. Tuppy. The amino-acid sequence in the phenylalanyl chain of insulin. 1. The identification of lower peptides from partial hydrolysates. *Biochem. J.* 49:463–481, 1951a.

Sanger, F., and H. Tuppy. The amino-acid sequence in the phenylalanyl chain of insulin. 2. The investigation of peptides from enzymic hydrolysates. *Biochem. J.* 49:481–490, 1951b.

Sanui, H., and N. Pace. Sodium and potassium binding by rat liver cell microsomes. *J. Gen. Physiol.* 42:1325–1345, 1959.

Scarborough, G.A. Isolation and characterization of *Neurospora crassa* plasma membranes. *J. Biol. Chem.* 250:1106–1111, 1975.

Scarborough, G.A. The *Neurospora* plasma membrane ATPase is an electrogenic pump. *Proc. Natl. Acad. Sci. USA* 73:1485–1488, 1976.

Scarborough, G.A. Properties of the *Neurospora crassa* plasma membrane ATPase. *Arch. Biochem. Biophys.* 180:384–393, 1977.

Scarborough, G.A. Proton translocation catalyzed by the electrogenic ATPase in the plasma membrane of *Neurospora. Biochemistry* 19:2925–2931, 1980.

Scarborough, G.A., and R. Addison. On the subunit composition of the *Neurospora* plasma membrane H^+-ATPase. *J. Biol. Chem.* 259:9109–9114, 1984.

Schatz, G., and J. Saltzgaber. The presence of denatured mitochondrial adenosine triphosphatase in "structural protein" from beef-heart mitochondria. *In* Chance, B., C.-P. Lee, and J.K. Blaisie (eds.) *Probes of Structure and Function of Macromolecules and Membranes.* New York: Academic Press, 1971, pp. 437–444.

Schatzmann, H.J. Herzglykoside als Hemmstoffe für den aktiven Kalium- und Natriumtransport durch die Erythrocytenmembran. *Helv. Physiol. Acta* 11:346–354, 1953.

Schatzmann, H.J. ATP-dependent Ca^{++} extrusion from human red cells. *Experientia* 22:364–365, 1966.

Schatzmann, H.J. Dependence on calcium concentration and stoichiometry of the calcium pump in human red cells. *J. Physiol.* 235:551–569, 1973.

Schatzmann, H.J. Amarcord. *Annu. Rev. Physiol.* 57:1–18, 1995.

Scheiner-Bobis, G., K. Fahlbusch, and W. Schoner. Demonstration of cooperating α subunits in working $(Na^+ + K^+)$-ATPase by the use of the MgATP complex analogue cobalt terammine ATP. *Eur. J. Biochem.* 168:123–131, 1987.

Schmidt, U.M., B. Eddy, C.M. Fraser, J.C. Venter, and G. Semenza. Isolation of (a subunit of) the Na^+/D-glucose cotransporter(s) of rabbit intestinal brush border membranes using monoclonal antibodies. *FEBS Lett.* 161:279–283, 1983.

Schmidt-Nielsen, B. *August and Marie Krogh.* New York: American Physiological Society, 1995.

Schmitt, F.O., R.S. Bear, and E. Ponder. Optical properties of the red cell membrane. *J. Cell. Comp. Physiol.* 9:89–92, 1936.

Schneider, W.C. Intracellular distribution of enzymes. III. The oxidation of octanoic acid by rat liver fractions. *J. Biol. Chem.* 176:259–266, 1948.

Schoenheimer, R. *The Dynamic State of Body Constituents.* Cambridge: Harvard University Press, 1942.

Schramm, M., H.-G. Klieber, and J. Daut. The energy expenditure of actomyosin-ATPase, Ca^{2+}-ATPase and Na^+, K^+-ATPase in guinea-pig cardiac ventricular muscle. *J. Physiol.* 481:647–662, 1994.

Schuldinger, S., H. Rottenberg, and M. Avron. Determination of ΔpH in chloroplasts. *Eur. J. Biochem.* 25:64–70, 1972.

Schultz, S.G., and P.F. Curran. Coupled transport of sodium and organic solutes. *Physiol. Rev.* 50:637–718, 1970.

Schultz, S.G., and R. Zalusky. The interaction between active sodium transport and active sugar transport in the isolated rabbit ileum. *Biochim. Biophys. Acta* 71:503–505, 1963.

Schultz, S.G., and R. Zalusky. Ion transport in isolated rabbit ileum. I. Short-circuit current and Na fluxes. *J. Gen. Physiol.* 47:567–584, 1964a.

Schultz, S.G., and R. Zalusky. Ion transport in isolated rabbit ileum. II. The interaction between active sodium and active sugar transport. *J. Gen. Physiol.* 47:1043–1059, 1964b.

Schultz, S.G., and R. Zalusky. Interactions between active sodium transport and active amino-acid transport in isolated rabbit ileum. *Nature* 205:292–294, 1965.

Schuurmans Stekhoven, F.M.A.H., H.G.P. Swarts, J.J.H.H.M. De Pont, and S.L. Bonting. Studies on $(Na^+ + K^+)$-activated ATPase. XLV. Magnesium induces two low-affinity non-phosphorylating nucleotide binding sites per molecule. *Biochim. Biophys. Acta* 649:533–540, 1981.

Schwartz, A., J.C. Allen, and S. Harigaya. Possible involvement of cardiac Na^+, K^+-adenosine triphosphatase in the mechanism of action of cardiac glycosides. *J. Pharmacol. Exp. Ther.* 168:31–41, 1969.

Schwartz, A., H.S. Bachelard, and H. McIlwain. The sodium-stimulated adenosine-triphosphatase activity and other properties of cerebral microsomal fractions and subfractions. *Biochem. J.* 84:626–637, 1962.

Schwartz, A., H. Matsui, and A.H. Laughter. Tritiated digoxin binding to $(Na^+ + K^+)$-

activated adenosine triphosphatase: Possible allosteric sites. *Science* 160:323–325, 1968.

Sebald, W., W. Machleidt, and E. Wachter. N,N'-dicyclohexylcarbodiimide binds specifically to a single glutamyl residue of the proteolipid subunit of the mitochondrial adenosine-triphosphatases from *Neurospora crassa* and *Saccharomyces cerevisiae*. *Proc. Natl. Acad. Sci. USA* 77:785–789, 1980.

Sen, A.K., and R.L. Post. Stoichiometry and localization of adenosine triphosphate-dependent sodium and potassium transport in the erythrocyte. *J. Biol. Chem.* 239:345–352, 1964.

Sen, A.K., T. Tobin, and R.L. Post. A cycle for ouabain inhibition of sodium- and potassium-dependent adenosine triphosphatase. *J. Biol. Chem.* 244:6596–6604, 1969.

Senior, A.E. Mitochondrial adenosine triphosphatase. *Biochemistry* 14:660–664, 1975.

Senior, A.E., and J.C. Brooks. The subunit composition of the mitochondrial oligomycin-insensitive ATPase. *FEBS Lett.* 17:327–329, 1971.

Senior, A.E., and J.G. Wise. The proton-ATPase of bacteria and mitochondria. *J. Membr. Biol.* 73:105–124, 1983.

Serrano, R. Plasma membrane ATPase of fungi and plants as a novel type of proton pump. *Curr. Top. Cell. Regul.* 23:87–126, 1984.

Serrano, R., M.C. Kielland-Brandt, and G.R. Fink. Yeast plasma membrane ATPase is essential for growth and has homology with $(Na^+ + K^+)$, K^+-, Ca^{2+}-ATPase. *Nature* 319:689–693, 1986.

Shanes, A.M. Electrochemical aspects of physiological and pharmacological action in excitable cells. *Pharmacol. Rev.* 10:59–164, 1958.

Shanes, A.M., and M.D. Berman. Penetration of the desheathed toad sciatic nerve by ions and molecules. *J. Gen. Physiol.* 45:177–197, 1955.

Shapiro, A.B., and R.E. McCarty. Alteration of the nucleotide-binding site asymmetry of chloroplast coupling factor 1 by catalysis. *J. Biol. Chem.* 263:14160–14165, 1988.

Shaw, T.I. Potassium movements in washed erythrocytes. *J. Physiol.* 129:464–475, 1955.

Sheppard, C.W., and W.R. Martin. Cation exchange between cells and plasma of mammalian blood. I. Methods and application to potassium exchange in human blood. *J. Gen. Physiol.* 33:703–722, 1950.

Sheppard, C.W., W.R. Martin, and G. Beyl. Cation exchange between cells and plasma of mammalian blood. II. Sodium and potassium exchange in the sheep, dog, cow, and man and the effect of varying the plasma potassium concentration. *J. Gen. Physiol.* 34:411–429, 1951.

Shortland, M. Book review. *Br. J. Hist. Sci.* 21:264–267, 1988.

Shull, G.E. cDNA cloning of the β-subunit of the rat gastric H,K-ATPase. *J. Biol. Chem.* 265:12123–12126, 1990.

Shull, G.E., and J. Greeb. Molecular cloning of two isoforms of the plasma membrane Ca^{2+}-transporting ATPase from rat brain. *J. Biol. Chem.* 263:8646–8657, 1988.

Shull, G.E., J. Greeb, and J.B. Lingrel. Molecular cloning of three distinct forms of the Na^+, K^+-ATPase α-subunit from rat brain. *Biochemistry* 25:8125–8132, 1986a.

Shull, G.E., L.K. Lane, and J.B. Lingrel. Amino-acid sequence of the β-subunit of the $(Na^+ + K^+)$ ATPase deduced from a cDNA. *Nature* 321:429–431, 1986b.

Shull, G.E., and J.B. Lingrel. Molecular cloning of the rat stomach $(H^+ + K^+)$-ATPase. *J. Biol. Chem.* 261:16788–16791, 1986.

Shull, G.E., A. Schwartz, and J.E. Lingrel. Amino-acid sequence of the catalytic subunit of the $(Na^+ + K^+)$ATPase deduced from a complementary DNA. *Nature* 316:691–695, 1985.

Siebers, A., and K. Altendorf. Characterization of the phosphorylated intermediate of the K^+-translocating Kdp-ATPase from *Escherichia coli*. *J. Biol. Chem.* 264:5831–5838, 1989.

Siegel, G.J., and R.W. Albers. Sodium-potassium-activated adenosine triphosphatase of *Electrophorus* electric organ. IV. Modification of response to sodium and potassiuim by arsenite plus 2,3-dimercaptopropanol. *J. Biol. Chem.* 242:4972–4979, 1967.

Siegel, G.J., G.J. Koval, and R.W. Albers. Sodium-potassium-activated adenosine triphosphatase. VI. Characterization of the phosphoprotein formed from orthophosphate in the presence of ouabain. *J. Biol. Chem.* 244:3264–3269, 1969.

Silverman, M., and M. Simon. Flagellar rotation and the mechanism of bacterial motility. *Nature* 249:73–74, 1974.

Simons, T.J.B. Potassium: potassium exchange catalysed by the sodium pump in human red cells. *J. Physiol.* 237:123–155, 1974.

Simons, T.J.B. The interaction of ATP analogues possessing a blocked γ-phosphate group with the sodium pump in human red cells. *J. Physiol.* 244:731–739, 1975.

Singer, S.J. The molecular organization of biological membranes. *In* Rothfield, L.I. (ed.) *Structure and Function of Biological Membranes*. New York: Academic Press, 1971, pp. 145–222.

Singer, S.J., and G.L. Nicolson. The fluid mosaic model of the structure of cell membranes. *Science* 175:720–731, 1972.

Sjöstrand, F.J. Electron microscopy of mitochondria and cytoplasmic double membranes. *Nature* 171:30–32, 1953.

Sjöstrand, F.J. A new repeat structural element of mitochondrial and certain cytoplasmic membranes. *Nature* 199:1262–1264, 1963.

Skou, J.C. The influence of some cations on an adenosine triphosphatase from peripheral nerves. *Biochim. Biophys. Acta* 23:394–401, 1957a.

Skou, J.C. Discussion. *In* Richter, D., (ed.) *Metabolism of the Nervous System*. London: Pergamon Press, 1957b, p. 173.

Skou, J.C. Further investigations of Mg^{++} + Na^+-activated adenosinetriphosphatase, possibly related to the active, linked transport of Na^+ and K^+ across the nerve membrane. *Biochim. Biophys. Acta* 42:6–23, 1960.

Skou, J.C. The relationship of a (Mg^{2+} + Na^+)-activated, K^+-stimulated enzyme or enzyme system to the active, linked transport of Na^+ and K^+ across the cell membrane. *In* Kleinzeller, A., and A. Kotyk (eds.) *Membrane Transport and Metabolism*. New York: Academic Press, 1961, pp. 228–236.

Skou, J.C. Preparation from mammalian brain and kidney of the enzyme system involved in active transport of Na^+ and K^+. *Biochim. Biophys. Acta* 58:314–325, 1962.

Skou, J.C. The identification of the sodium-pump as the membrane-bound Na^+/K^+-ATPase: A commentary. *Biochim. Biophys. Acta* 1000:435–438, 1989.

Skou, J.C., and M. Esmann. Effects of ATP and protons on the Na:K selectivity of the (Na^+ + K^+)-ATPase studied by ligand effects on intrinsic and extrinsic fluorescence. *Biochim. Biophys. Acta* 601:386–402, 1980.

Skou, J.C., and M. Esmann. Eosin, a fluorescent probe of ATP binding to the (Na^+ + K^+)-ATPase. *Biochim. Biophys. Acta* 647:232–240, 1981.

Skulachev, V.P. Transmembrane electrochemical H^+-potential as a convertible energy source for the living cell. *FEBS Lett.* 74:1–9, 1977.

Skulachev, V.P. Bacterial Na^+ energetics. *FEBS Lett.* 250:106–114, 1989.

Skulachev, V.P. The laws of cell energetics. *Eur. J. Biochem.* 208:203–209, 1992.

Slater, E.C. Mechanism of phosphorylation in the respiratory chain. *Nature* 172:975–978, 1953.

Slater, E.C. Oxidative phosphorylation. *Comp. Biochem.* 14:327–396, 1966.

Slater, E.C. The coupling between energy-yielding and energy-utilizing reactions in mitochondria. *Q. Rev. Biophys.* 4:35–71, 1971.

Slater, E.C. A short history of the biochemistry of mitochondria. *In* Lee, C.-P., G. Schatz, and

G. Dallner (eds.) *Mitochondria and Microsomes.* Reading, MA: Addison-Wesley, 1981, pp. 15–43.

Slater, E.C., and K.W. Cleland. The effect of calcium on the respiratory and phosphorylative activities of heart-muscle sarcosomes. *Biochem. J.* 55:566–580, 1953.

Slater, E.C., and A. Kemp. Rate of labeling of mitochondrial phosphohistidine by radioactive inorganic phosphate. *Nature* 204:1268–1271, 1964.

Slater, E.C., A. Kemp, and J.M. Tager. Utilization of high-energy intermediates of oxidative phosphorylation and the proposed function of imidazole phosphate as an intermediate in oxidative phosphorylation. *Nature* 201:781–784, 1964.

Slater, E.C., J. Rosing, and A. Mol. The phosphorylation potential generated by respiring mitochondria. *Biochim. Biophys. Acta* 292:534–553, 1973.

Slayman, C.L. Electrical properties of *Neurospora crassa.* Effects of external cations on the intracellular potential. *J. Gen. Physiol.* 49:69–92, 1965a.

Slayman, C.L. Electrical properties of *Neurospora crassa.* Respiration and the intracellular potential. *J. Gen. Physiol.* 49:93–116. 1965b.

Slayman, C.L. Movements of ions and electrogenesis in microorganisms. *Am. Zool.* 10:377–392, 1970.

Slayman, C.L., C.Y.-H. Lu, and L. Shane. Correlated changes in membrane potential and ATP concentrations in *Neurospora. Nature* 226:274–276, 1970.

Slayman, C.L., and C.W. Slayman. Measurement of membrane potentials in *Neurospora. Science* 136:876–877, 1962.

Slayman, C.L., and C.W. Slayman. Net uptake of potassium by *Neurospora. J. Gen. Physiol.* 52:424–443, 1968.

Slayman, C.W., and E.L. Tatum. Potassium transport in *Neurospora.* III. Isolation of a transport mutant. *Biochim. Biophys. Acta* 109:184–193, 1965.

Smith, H.W. The plasma membrane, with notes on the history of botany. *Circulation* 26:987–1012, 1962.

Smith, R.L., K. Zinn, and L.C. Cantley. A study of the vanadate-trapped state of the (Na,K)-ATPase. Evidence against interacting nucleotide site models. *J. Biol. Chem.* 255:9852–9859, 1980.

Solomon, A.K. The permeability of the human erythrocyte to sodium and potassium. *J. Gen. Physiol.* 36:57–110, 1952.

Solomon, A.K. Transport pathways: Water movement across cell membranes. *In* Tosteson, D.C., (ed.) *Membrane Transport: People and Ideas.* Bethesda: American Physiological Society, 1989, pp. 125–153.

Solomon, A.K., T.J. Gill, and G.L. Gold. The kinetics of cardiac glycoside inhibition of potassium transport in human erythrocytes. *J. Gen. Physiol.* 40:327–350, 1956a.

Solomon, A.K., F. Lionetti and P.F. Curran. Possible cation-carrier substances in blood. *Nature* 178:582–583, 1956b.

Sone, N., M. Yoshida, H. Hirata, and Y. Kagawa. Purification and properties of a dicyclohexylcarbodiimide-sensitive adenosine triphosphatase from a thermophilic bacterium. *J. Biol. Chem.* 250:7917–7923, 1975.

Spector, W.G. Electrolyte flux in isolated mitochondria. *Proc. R. Soc. Lond.* B 141:268–279, 1953.

Srinivasan, P.R., J.S. Fruton, and J.T. Edsall. The origins of modern biochemistry. *Ann. N.Y. Acad. Sci.* 325:1–373, 1979.

Stasny, J.T., and F.L. Crane. The effect of sonic oscillation on the structure and function of beef heart mitochondria. *J. Cell. Biol.* 22:49–62, 1964.

Steck, T.L., G. Fairbanks, and D.F.H. Wallach. Disposition of the major proteins in the isolated erythrocyte membrane. *Biochemistry* 10:2617–2624, 1971.

Stein, W.D. James Frederic Danielli. *Biograph. Mem. Fellows R. Soc.* 32:117–135, 1986.

Stein, W.D., and J.F. Danielli. Structure and function in red cell permeability. *Disc. Faraday Soc.* 21:238–251, 1956.

Steinbach, H.B. The electrical potential difference across living frog skin. *J. Cell. Comp. Physiol.* 3:1–27, 1933.

Steinbach, H.B. Potassium in frog skin. *J. Cell. Comp. Physiol.* 10:51–60, 1937.

Steinbach, H.B. Sodium and potassium in frog muscle. *J. Biol. Chem.* 133:695–701, 1940a.

Steinbach, H.B. Electrolyte balance of animal cells. *Cold Spring Harb. Symp. Quant. Biol.* 8:242–254, 1940b.

Steinbach, H.B. Chloride in the giant axons of the squid. *J. Cell. Comp. Physiol.* 17:57–64, 1941.

Steinbach, H.B. The osmotic behavior of frog sartorius muscles. *J. Cell Comp. Physiol.* 24:291–297, 1944.

Steinbach, H.B. Intracellular inorganic ions and muscle action. *Ann. N.Y. Acad. Sci.* 47:849–874, 1947.

Steinbach, H.B. Calcium and apyrase system of muscle. *Arch. Biochem.* 22:328–330, 1949.

Steinbach, H.B. Sodium extrusion from isolated frog muscle. *Am. J. Physiol.* 167:284–287, 1951.

Steinbach, H.B. On the sodium and potassium balance of isolated frog muscles. *Proc. Natl. Acad. Sci. USA* 38:451–455, 1952.

Steinbach, H.B. The regulation of sodium and potassium in muscle fibres. *Symp. Soc. Exp. Biol.* 8:438–452, 1954.

Steinbach, H.B., and S. Spiegelman. The sodium and potassium balance in squid nerve axoplasm. *J. Cell. Comp. Physiol.* 22:187–196, 1943.

Steinberg, M., and S.J.D. Karlish. Studies on conformational changes in Na,K-ATPase labeled with 5-iodoacetamidofluorescein. *J. Biol. Chem.* 264:2726–2734, 1989.

Stekhoven, F.S., R.F. Waitus, and H.T.B. van Moerkerk. Identification of the dicyclohexyl-carbodiimide binding protein in the oligomycin-sensitive adenosine triphosphatase from bovine heart mitochondria. *Biochemistry* 11:1144–1150, 1972.

Sternweis, P.C., and A.G. Gilman. Aluminum: A requirement for activation of the regulatory component of adenylate cyclase by fluoride. *Proc. Natl. Acad. Sci. USA* 79:4888–4891, 1982.

Stewart, B., B. Wallmark, and G. Sachs. The interaction of H^+ and K^+ with the partial reactions of gastric $(H^+ + K^+)$-ATPase. *J. Biol. Chem.* 256:2682–2690, 1981.

Stewart, J.M.M., and C.M. Grisham. ^1H nuclear magnetic resonance studies of the conformation of an ATP analog at the active site of Na,K-ATPase from kidney medulla. *Biochemistry* 27:4840–4848, 1988.

Stiehler, R.D., and L.B. Flexner. A mechanism of secretion in the choroid plexus. The conversion of oxidation-reduction energy into work. *J. Biol. Chem.* 126:603–617, 1938.

Stoeckenius, W. From membrane structure to bacteriorhodopsin. *J. Membr. Biol.* 139:139–148, 1994.

Stokes, D.L., and N.M. Green. Structure of CaATPase: Electron microscopy of frozen-hydrated crystals at 6 Å resolution in projection. *J. Mol. Biol.* 213:529–538, 1990.

Straub, F.B. Muscle. *Annu. Rev. Biochem.* 19:371–388, 1950.

Straub, F.B. Über die Akkumulation der Kaliumionen durch menschliche Blutkörperchen. *Acta. Physiol. Hung.* 4:235–240, 1953.

Streb, H., R.F. Irvine, M.J. Berridge, and I. Schultz. Release of Ca^{2+} from a nonmitochondrial intracellular store in pancreatic acinar cells by inositol-1,4,5-trisphosphate. *Nature* 306:67–69, 1983.

Streeten, D.H.P., and A.K. Solomon. The effect of ACTH and adrenal steroids on K transport in human erythrocytes. *J. Gen. Physiol.* 37:643–661, 1954.

Strickland, S.P. *Politics, Science, and Dread Disease.* Cambridge: Harvard University Press, 1972.

Stürmer, W., H.-J. Apell, I. Wuddel, and P. Läuger. Conformational transitions and charge translocation by the Na,K-pump: Comparison of optical and electrical transients elicited by ATP-concentration jumps. *J. Membr. Biol.* 110:67–86, 1989.

Suelter, C.H., M. DeLuca, J.B. Peter, and P.D. Boyer. Detection of a possible intermediate in oxidative phosphorylation. *Nature* 192:43–47, 1961.

Sutherland, E.W., I. Øye, and R.W. Butcher. The action of epinephrine and the role of the adenyl cyclase system in hormone action. *Recent Prog. Horm. Res.* 21:623–646, 1965.

Sutherland, E.W., and T.W. Rall. Fractionation and characterization of a cyclic adenine ribonucleotide formed by tissue particles. *J. Biol. Chem.* 232:1077–1091, 1958.

Sweadner, K.J. Two molecular forms of (Na$^+$ + K$^+$)-stimulated ATPase in brain. *J. Biol. Chem.* 254:6060–6067, 1979.

Sweadner, K.J., and S.M. Goldin. Reconstitution of active ion transport by the sodium and potassium ion-stimulated adenosine triphosphatase from canine brain. *J. Biol. Chem.* 250:4022–4024, 1975.

Szent-Györgyi, A. *Chemistry of Muscular Contraction.* New York: Academic Press, 1947.

Tager, J.M., R.D. Veldsema-Currie, and E.C. Slater. Chemi-osmotic theory of oxidative phosphorylation. *Nature* 212:376–379, 1966.

Takai, Y., A. Kishimoto, Y. Iwasa, Y. Kawahara, T. Mori, and Y. Nishizuka. Calcium-dependent activation of a multifunctional protein kinase by membrane phospholipids. *J. Biol. Chem.* 254:3692–3695, 1979a.

Takai, Y., A. Kishimoto, U. Kikkawa, T. Mori, and Y. Nishizuka. Unsaturated diacylglycerol as a possible messenger for the activation of calcium-activated, phospholipid-dependent protein kinase system. *Biochem. Biophys. Res. Commun.* 91:1218–1224, 1979b.

Takeuchi, M., and H. Terayama. Preparation and chemical composition of rat liver cell membranes. *Exp. Cell Res.* 40:32–44, 1965.

Takisawa, H., and M. Makinose. Occluded bound calcium on the phosphorylated sarcoplasmic transport ATPase. *Nature* 290:271–273, 1981.

Talvenheimo, J.A., M.M. Tamkun, and W.A. Catterall. Reconstitution of neurotoxin-stimulated sodium transport by the voltage-sensitive sodium channel purified from rat brain. *J. Biol. Chem.* 257:11868–11871, 1982.

Tanford, C. Contribution of hydrophobic interactions to the stability of the globular conformation of proteins. *J. Am. Chem. Soc.* 84:4240–4247, 1962.

Taniguchi, K., and R.L. Post. Synthesis of adenosine triphosphate and exchange between inorganic phosphate and adenosine triphosphate in sodium and potassium ion transport adenosine triphosphatase. *J. Biol. Chem.* 250:3010–3018, 1975.

Tanisawa, A.S., and J.G. Forte. Phosphorylated-intermediate of microsomal ATPase from rabbit gastric mucosa. *Arch. Biochem. Biophys.* 147:165–175, 1971.

Tasaki, I. Conduction of the nerve impulse, *In* Field, J., H.W. Magoun, and V.E. Hall (eds.) *Handbook of Physiology. Section 1, Neurophysiology.* Washington: American Physiological Society, 1959, pp. 75–121.

Taugner, G. The membrane of catecholamine storage vesicles of adrenal medulla. *Naunyn-Schmiedebergs Arch. Pharmacol.* 270:392–406, 1971.

Teich, M. *A Documentary History of Biochemistry 1770–1940.* Rutherford, NJ: Fairleigh Dickinson University Press, 1992.

Thayer, W.S., and P.C. Hinkle. Kinetics of adenosine triphosphate synthesis in bovine heart submitochondrial particles. *J. Biol. Chem.* 250:5336–5342, 1975.

Thesleff, S., and K. Schmidt-Nielsen. An electrophysiological study of the salt gland of the herring gull. *Am. J. Physiol.* 202:597–600, 1962.

Thomas, R.C. Membrane current and intracellular sodium changes in a snail neurone during extrusion of injected sodium. *J. Physiol.* 201:495–514, 1969.

Tobias, J.M. Biophysical aspects of conduction and transmission in the nervous system. *Annu. Rev. Physiol.* 21:299–324, 1959.

Toh, B.-H., P.A. Gleeson, R.J. Simpson, R.L. Moritz, J.M. Callaghan, I. Goldkorn, C.M. Jones, T.M. Martinelli, F.-T. Mu, D.C. Humphris, J.M. Pettitt, Y. Mori, T. Masuda, P. Sobieszczuk, J. Weinstock, T. Mantamadiotis, and G.S. Baldwin. The 60- to 90-kDa parietal cell autoantigen associated with autoimmune gastritis is a β subunit of the gastric H^+/K^+-ATPase (proton pump). *Proc. Natl. Acad. Sci. USA* 87:6418–6422, 1990.

Tokuda, H., and T. Unemoto. Characterization of the respiration-dependent Na^+ pump in the marine bacterium *Vibrio alginolyticus*. *J. Biol. Chem.* 257:10007–10014, 1982.

Tomita, M., and V.T. Marchesi. Amino-acid sequence and oligosaccharide attachement sites of human erythrocyte glycophorin. *Proc. Natl. Acad. Sci. USA* 72:2964–2968, 1975.

Tosteson, D.C. Sodium and potassium transport in red blood cells. *In* Shanes, A.M., (ed.) *Electrolytes in Biological Systems*. Washington: American Physiological Society, 1955, pp. 123–156.

Tosteson, D.C., and J.F. Hoffman. Regulation of cell volume by active cation transport in high and low potassium sheep red cells. *J. Gen. Physiol.* 44:169–194, 1960.

Tosteson, D.C., R.H. Moulton, and M. Blaustein. An enzymatic basis for the difference in active cation transport in two genetic types of sheep red cells. *Federation Proc.* 19:128, 1960.

Troshin, A.S. *Problems of Cell Permeability*. Oxford: Pergamon Press, 1966.

Trumpower, B.L. The protonmotive Q cycle. *J. Biol. Chem.* 265:11409–11412, 1990.

Tyson, P.A., M. Steinberg, E.T. Wallick, and T.L. Kirley. Identification of the 5-iodoacetamido-fluorescein reporter site on the Na,K-ATP. *J. Biol. Chem.* 264:726–734, 1989.

Uesugi, S., N.C. Dulak, J.F. Dixon, T.D. Hexum, J.L. Dahl, J.F. Perdue, and L.E. Hokin. Studies on the characterization of the sodium-potassium transport adenosine triphosphatase. VI. Large scale partial purification and properties of a Lubrol-solubilized bovine brain enzyme. *J. Biol. Chem.* 246:531–543, 1971.

Ussing, H.H. Interpretation of the exchange of radio-sodium in isolated muscle. *Nature* 160:262–263, 1947.

Ussing, H.H. The use of tracers in the study of active ion transport across animal membranes. *Cold Spring Harb. Symp. Quant. Biol.* 13:193–200, 1948.

Ussing, H.H. The distinction by means of tracers between active transport and diffusion. *Acta Physiol. Scand.* 19:43–56, 1949a.

Ussing, H.H. The active ion transport through the isolated frog skin in the light of tracer studies. *Acta Physiol. Scand.* 17:1–37, 1949b.

Ussing, H.H. Transport of ions across cellular membranes. *Physiol. Rev.* 29:127–155, 1949c.

Ussing, H.H. Some aspects of the applicability of tracers in permeability studies. *Adv. Enzymol.* 13:21–65, 1952.

Ussing, H.H. Ion transport across biological membranes. *In* Clarke, H.T., (ed.) *Ion Transport across Membranes*. New York: Academic Press, 1954, pp. 3–22.

Ussing, H.H. General principles and theories of membrane transport. *In* Murphy, Q.R., (ed.) *Metabolic Aspects of Transport across Cell Membranes*. Madison: University of Wisconsin Press, 1957, pp. 39–71.

Ussing, H.H. The frog skin potential. *J. Gen. Physiol.* 43 Suppl.:135–147, 1960.

Ussing, H.H. Life with tracers. *Annu. Rev. Physiol.* 42:1–16, 1980.

Ussing, H.H. Does active transport exist? *J. Membr. Biol.* 137:91–98, 1994.

Ussing, H.H., and K. Zerahn. Active transport of sodium as the source of electric current in the short-circuited isolated frog skin. *Acta Physiol. Scand.* 23:110–127, 1951.

Utter, M.F. Mechanism of inhibition of anaerobic glycolysis of brain by sodium ions. *J. Biol. Chem.* 185:499–517, 1950.

Van Winkle, W.B., L.K. Lane, and A. Schwartz. The subunit fine structure of isolated, purified Na$^+$,K$^+$-adenosine triphosphatase. *Exp. Cell. Res.* 100:291–296, 1976.

Vasington, F.D., and J.V. Murphy. Ca^{++} uptake by rat kidney mitochondria and its dependence on respiration and phosphorylation. *J. Biol. Chem.* 237:2670–2677, 1962.

Veatch, W., and L. Stryer. The dimeric nature of the gramicidin A transmembrane channel. *J. Mol. Biol.* 113:89–102, 1977.

Verma, A.K., A.G. Filoteo, D.R. Stanford, E.D. Wieben, J.T. Penniston, E.E. Strehler, R. Fischer, R. Heim, G. Vogel, S. Mathews, M.-A. Strehler-Page, P. James, T. Vorherr, J. Krebs, and E. Carafoli. Complete primary structure of a human plasma membrane Ca^{2+} pump. *J. Biol. Chem.* 263:14152–14159, 1988.

Vidaver, G. Glycine transport by hemolyzed and restored pigeon red cells. *Biochemistry* 3:795–799, 1964.

Vincenzi, F.F., and H.J. Schatzmann. Some properties of Ca-activated ATPase in human red cell membranes. *Helv. Physiol. Pharmacol. Acta* 25:233–234, 1967.

Vogel, F., H.W. Meyer, R. Grosse, and K.R.H. Repke. Electron microscopic visualization of the arrangement of the two protein components of (Na$^+$ + K$^+$)-ATPase. *Biochim. Biophys. Acta* 470:497–502, 1977.

Vogel, G., and R. Steinhart. ATPase of *Escherichia coli. Biochemistry* 15:208–216, 1976.

Walderhaug, M.O., R.L. Post, G. Saccomani, R.T. Leonard, and D.P. Briskin. Structural relatedness of three ion-transport adenosine triphosphatases around their active sites of phosphorylation. *J. Biol. Chem.* 260:3852–3859, 1985.

Walker, J.E., I.M. Fearnley, N.J. Gay, B.W. Gibson, F.D. Northrop, S.J. Powell, M.J. Runswick, M. Saraste, and V.L.J. Tybulewicz. Primary structure and subunit stoichiometry of F$_1$-ATPase from bovine mitochondria. *J. Mol. Biol.* 184:677–701, 1985.

Walker, J.E., M. Sarasate, and N.J. Gay. E. coli F$_1$-ATPase interacts with a membrane protein component of a proton channel. *Nature* 298:867–869, 1982a.

Walker, J.E., M. Sarasate, M.J. Runswick, and N.J. Gay. Distantly related sequences in the α- and β-subunits of ATP synthase, myosin, kinases and other ATP-requiring enzymes and a common nucleotide binding fold. *EMBO J.* 1:945–951, 1982b.

Wallach, D.F.H., and P.H. Zahler. Protein conformation in cellular membranes. *Proc. Natl. Acad. Sci. USA* 56:1552–1559, 1966.

Warncke, J., and C.L. Slayman. Metabolic modulation of stoichiometry in a proton pump. *Biochim. Biophys. Acta* 591:224–233, 1980.

Watson, J.D., and F.H.C. Crick. A structure for deoxyribose nucleic acid. *Nature* 171:737–738, 1953.

Webb, D.A., and J.Z. Young. Electrolyte content and action potential of the giant nerve fibres of *Loligo. J. Physiol.* 98:299–313, 1940.

Webb, M.R., C. Grubmeyer, H.S. Penefsky, and D.R. Trentham. The stereochemical course of phosphoric residue transfer catalyzed by beef heart mitochondrial ATPase. *J. Biol. Chem.* 255:11637–11639, 1980.

Weber, A. On the role of calcium in the activity of adenosine 5'-triphosphate hydrolysis by actomyosin. *J. Biol. Chem.* 234:2764–2769, 1959.

Weber, B.H. Glynn and the conceptual development of the chemiosmotic theory. *Biosci. Rep.* 6:577–617, 1991.

Webster, G., A.L. Smith, and M. Hansen. Intermediate reactions in the phosphorylation coupled to the oxidation of reduced cytochrome c. *Proc. Natl. Acad. Sci. USA* 49:259–266, 1963.

Weidmann, S. Transport of ions across cardiac membranes. *In* Murphy, Q.R. (ed.) *Metabolic Aspects of Transport Across Cell Membranes.* Madison: University of Wisconsin Press, 1955, pp. 115–125.

Weigele, J.B., and R.L. Barchi. Functional reconstitution of the purified sodium channel protein from rat sarcolemma. *Proc. Natl. Acad. Sci. USA* 79:3651–3655, 1982.

West, I.C. Lactose transport coupled to proton movements in *Escherichia coli*. *Biochem. Biophys. Res. Commun.* 41:655–661, 1970.

Whittam, R. Potassium movements and ATP in human red cells. *J. Physiol.* 140:479–497, 1958.

Whittam, R. Active cation transport as a pace-maker of respiration. *Nature* 191:603–604, 1961.

Whittam, R. Spatial asymmetry in the stimulation of a membrane adenosine triphosphatase. *Biochem. J.* 83:29P–30P, 1962a.

Whittam, R. The asymmetrical stimulation of a membrane adenosine triphosphatase in relation to active cation transport. *Biochem. J.* 84:110–118, 1962b.

Whittam, R., M.E. Ager, and J.S. Wiley. Control of lactate production by membrane adenosine triphosphatase activity in human erythrocytes. *Nature* 202:1111–1112, 1964.

Wickner, W.T., and H.F. Lodish. Multiple mechanisms of protein insertion into and across membranes. *Science* 230:400–407, 1985.

Wikström, M.K.F. Proton pump coupled to cytochrome c oxidase in mitochondria. *Nature* 266:271–273, 1977.

Wilbrandt, W. A relation between the permeability of the red cell and its metabolism. *Trans. Faraday Soc.* 33:956–959, 1937.

Wilbrandt, W. Die Abhängigkeit der Ionenpermeabilität der Erythrocyten vom glycolytischen Stoffwechsel. *Pflügers Arch.* 243:519–536, 1940.

Wilbrandt, W. Permeability. *Annu. Rev. Physiol.* 9:581–604, 1947.

Wilbrandt, W., and T. Rosenberg. The concept of carrier transport and its corollaries in pharmacology. *Pharmacol. Rev.* 13:109–183, 1961.

Wilde, W.S. The chloride equilibrium in muscle. *Am. J. Physiol.* 143:666–676. 1945.

Williams, R.J.P. Possible functions of chains of catalysts. *J. Theoret. Biol.* 1:1–17, 1961.

Williams, R.J.P. Possible functions of chains of catalysts II. *J. Theoret. Biol.* 3:209–229, 1962.

Williams, R.J.P. Proton-driven phosphorylation reactions in mitochondrial and chloroplast membranes. *FEBS Lett.* 53:123–125, 1975.

Williams, R.J.P. The nature of local chemical potentials. *FEBS Lett.* 150:1–3, 1982.

Williams, R.J.P. The history of proton-driven ATP formation. *Biosci. Rep.* 13:191–212, 1993.

Wilson, D.F., P.L. Dutton, M. Erecinska, J.G. Lindsay, and N. Sato. Mitochondrial electron transport and energy conservation. *Acct. Chem. Res.* 5:234–241, 1972.

Woelders, H., W.J. van der Zande, A.-M.A.F. Colen, R.J.A. Wanders, and K. van Dam. The phosphate potential maintained by mitochondria in state 4 is proportional to the proton-motive force. *FEBS Lett.* 179:278–282, 1985.

Woodbury, L.A., J.W. Woodbury, and H.H. Hecht. Membrane resting and action potentials of single cardiac muscle fibers. *Circulation* 1:264–266, 1950.

Woodward, J., and D. Goodstein. Conduct, misconduct and the structure of science. *Am. Sci.* 84:479–490, 1996.

Xie, X.-S., D.K. Stone, and E. Racker. Activation and partial purification of the ATPase of clathrin-coated vesicles and reconstitution of the proton pump. *J. Biol. Chem.* 259:11676–11678, 1984.

Xue, Z., J.-M. Zhou, T. Melese, R.L. Cross, and P.D. Boyer. Chloroplast F_1 ATPase has more than three nucleotide sites, and 2-azido-ADP or 2-azido-ATP at both catalytic and noncatalytic sites labels the β subunit. *Biochemistry* 26:3749–3753, 1987.

Yamamoto, T., and Y. Tonomura. Reaction mechanism of the Ca^{++}-dependent ATPase of sarcoplasmic reticulum from skeletal muscle. I. Kinetic studies. *J. Biochem.* 62:558–575, 1967.

Yamamoto, T., and Y. Tonomura. Reaction mechanism of the Ca^{++}-dependent ATPase of sarcoplasmic reticulum from skeletal muscle. II. Intermediate formation of phosphoryl protein. *J. Biochem.* 64:137–145, 1968.

Yoshida, H., and H. Fujisawa. Influence of subcellular structures on the activity of Na$^+$, K$^+$-activated adenosine triphosphatase in brain. *Biochim. Biophys. Acta* 60:443–444, 1962.

Yoshida, H., H. Fujisawa, and T. Nukada. Increase in incorporation of inorganic phosphate-32 into phosphatidic acid in brain slices by ouabain, EDTA or a high concentration of potassium. *Jap. J. Pharmacol.* 12:156–162, 1962.

Yoshida, H., T. Nukada, and H. Fujisawa. Effect of ouabain on ion transport and metabolic turnover of phospholipid of brain slices. *Biochim. Biophys. Acta* 48:614–615, 1961.

Yoshida, M., N. Sone, H. Hirata, Y. Kagawa, Y. Takeuchi, and K. Ohno. ATP synthesis catalyzed by purified DCCD-sensitive ATPase incorporated into reconstituted purple membrane vesicles. *Biochem. Biophys. Res. Commun.* 67:1295–1300, 1975.

Young, J.Z. Structure of nerve fibres and synapses in some invertebrates. *Cold Spring Harb. Symp. Quant. Biol.* 4:1–6, 1936.

Zerahn, K. Oxygen consumption and active transport of sodium in the isolated, short-circuited frog skin. *Nature* 177:937–938, 1956a.

Zerahn, K. Oxygen consumption and active sodium transport in the isolated and short-circuited frog skin. *Acta Physiol. Scand.* 36:300–318, 1956b.

INDEX

Abderhalden, E., 19–21
Abood, L.G., 129
Action potential, 11, 68–82, 84, 86, 130, 228, 302. *See also* channels; Goldman-Hodgkin-Katz equation; Hodgkin-Huxley equation; Nernst equation; patch clamp; voltage clamp
Acyl phosphate, 137, 144, 162, 165–66, 174, 193, 195, 197–98, 202n.58. *See also* enzyme phosphorylation
Adair, G.S., 119
Addison, R., 197
Adenylate kinase, 118, 124n.73, 136, 157, 200, 251, 275
ADP/ATP exchange, 135–37, 140, 162, 168, 170, 191, 194, 258. *See also* enzyme phosphorylation/ dephosphorylation
Adrian, E.D., 85
Agnew, W.S., 230
Akera, T., 165
Albers, R.W., 136–37, 153, 162–65, 168, 178, 192, 227
Albers-Post scheme, 162–64, 170, 192, 194
Allison, W., 289, 293
Allosteric enzymes and kinetics, 125n.83, 151–52, 158
Altendorf, K., 197–98
Altman, K.I., 142
Anfinsen, C., 151, 153
Apell, H.-J., 177
Apps, D.K., 292
Armstrong, C.M., 228
Arrhenius, S., 68

Askari, A., 170–71, 177–78
Astbury, W.T., 19–20
ATP, 3, 5–12, 22–24, 57, 61, 93, 99–100, 106, 115–18, 121-22,126–44, 152, 185–97, 200, 212–13, 215–17, 222–26, 234,247–59, 261–80, 283–97, 300
ATPase, 7, 23, 99, 222–23, 227, 229–300. *See also* transport pumps
  bacterial Kdp, 197–98
  F-type, 292–93
  fungal, 196–97
  gastric H^+/K^+, 194–95, 198
  Kielley-Meyerhof, 99, 190, 192
  Libet, 99, 129, 192
  mitochondrial, 249, 254–56, 258, 261, 264–67, 270, 272, 283–97. *See also* F_1; F_oF_1
  myosin, 99, 190, 284, 286
  Na^+/K^+, 10–11, 126–44, 156–79, 183–88, 192–98, 205, 209–11, 214–16, 222, 226, 234-35, 245
  plasma membrane Ca^{2+}, 193–94, 197, 227
  P-type, 188–89, 194, 197–98, 277, 292–93
  sarcoplasmic reticulum Ca^{2+}, 189–93, 197, 214
  V-type, 277, 292–93
ATP hydrolysis, enzymatic. *See* ATPase
ATP synthase, 9, 187, 200, 283–97. *See also* F_oF_1; oxidative phosphorylation
Avery, O.T., 98
Avron, M., 296
Azzone, G.F., 276

Bacteriorhodopsin, 243, 270–71
Bader, H., 139
Bale, W., 37, 42
Ball, E.G., 233
Ballou, C.E., 211
Band 3 of red blood cell membrane proteins, 240–41, 244
Bangham, A.D., 149
Barchi, R., 231
Bäuerlein, E., 275
Beadle, G.W., 98
Beaugé, L.A., 138, 167
Beechey, R.B., 289
Behn equation, 97n.15
Belitser, V.A., 23, 99
Bendall, J.R., 190–91
Bergmann, M., 19–20
Bernal, J.D., 19–20
Bernstein, J., 68–71, 78, 80
Berridge, M.J., 211
Bezanilla, F., 228
Black lipid membrane. *See* planar bilayer
Blaustein, M.P., 226
Blostein, R., 158, 171–72, 174, 177
Blow, D.M., 241
Bodansky, M., 20–21
Bohr, C., 40
Bonting, S.L., 134
Boyer, P.D., 100, 192, 252–53, 257, 275, 283–87, 289, 291, 293, 296–97, 307
Boyle, P.J., 31, 34, 36–37, 39, 41, 71
Branton, D., 239
Bretscher, M.S., 240
Brinley, F.J., 158
Brockerhoff, H., 211
Brody, T.M., 165
Bunge, G., 20
Burgen, A.S.V., 119
Bush, V., 152

Cabantchik, Z.I., 244
Caldwell, P.C., 105, 118, 120
Cantley, L.C., 166
Carafoli, E., 194, 198, 292
Cardiotonic steroids, 103–107, 110, 117–18, 120, 122, 123n.3, 134–36, 141–42, 146, 165, 171, 175–77, 222–23. *See also* digoxin; Na$^+$/K$^+$-ATPase; Na$^+$/K$^+$ pump; ouabain; strophanthidin

Carlsberg Foundation, 95
Carriers, 62, 73, 75, 100, 103, 106, 111, 114, 118, 192, 207, 209, 220, 222, 225, 240, 244, 262, 265, 272, 289. *See also* transport; transport pumps
Cation occlusion, 139, 166–68, 179, 188, 192
Catterall, W.A., 230
Cavieres, J.D., 138, 157
cDNA techniques, 184–88, 192, 194–95, 197, 199–200, 220n.4, 225-27, 231, 234, 240, 242, 244–45, 289
Centrifugation, 12, 19, 23, 54, 101, 129–30, 149, 160, 179, 303
Chance, B., 254, 256–57, 270–72, 275, 278–80, 296, 307
Changeux, J.-P., 151
Channels, 10–11, 228–231, 238, 290, 293, 300, 302, 309. *See also* action potentials; pores
Chappell, J.B., 137, 266
Charnock, J.S., 136, 139
Chin, G., 158
Chloride/bicarbonate antiporter. *See* Band 3
Chloroplasts, 9, 13, 269–70, 272–73, 277
Christensen, H.V., 225–26
Christian, W., 22
Chromatography, 101, 167, 179, 198, 303
Chymotrypsin, 241–42
Claude, A., 101
Cohn, E.T., 54, 57, 60, 64, 306–307
Cohn, M., 283
Cohn, W.E., 54, 57, 60, 64, 306–307
Cole, K.S., 70–72, 74, 78, 83–85
Collins, H.M., 64, 121
Collins, J.H., 183–84
Communication. *See* journals; meetings
Complexes I-IV of respiratory chain, 9, 225, 268, 273–74, 276
Conway, E.J., 29–34, 36–37, 39–45, 48–49, 60, 71, 80–81, 106, 114–16, 121, 194, 196, 213, 220, 264, 307, 309
Corey, R.B., 98, 149
Cori, C., 233
Cox, G., 121
Cox, G.B., 291
Crane, F.L., 254
Crane, R.K., 221–24, 227, 233, 265, 277–78
Crane, S., 233
Crick, F.H.C., 151

Cross, R.L., 289–91, 293
Crowfoot, D., 19–20
Csáky, T., 222–23, 226
Curran, P.F., 226
Curtis, H.S., 70–72, 78, 84
Cyanide, 51, 91–92, 114–15, 118, 121, 196, 212–13, 225, 254
Cyclic AMP. *See* second messengers
Cytochromes, 23, 99, 247, 254–56, 276. *See also* respiratory chain

Danielli, J.F., 16–18, 38, 53, 59, 63–64, 98, 118–19, 148, 238, 258, 265, 277
Danowski, T., 58–61, 64–65, 307
Darrow, D.C., 44
Davies, R.E., 114, 194–95, 264–66, 278
Davson, H., 53, 59–60, 63–64, 118, 306
DCCD, 289–91
Dean, R.B., 39, 41, 43, 48, 55, 57, 59–60, 63–64
deMeis, L., 192
deVries, H., 14–15
DeWeer, P., 158
Diffusion
  simple, 10, 18, 220
  exchange, 47, 56, 81, 110, 112, 213. *See also* energy for active transport
  facilitated, 10, 220, 244. *See also* carriers; transport
Digoxin, 105–106, 123, 165. *See also* cardiotonic steroids
Dimroth, P., 292–93
Dinitrophenol, 93, 100, 115, 117, 121, 247, 251–52, 254–56, 262, 265–66, 285–86. *See also* oxidative phosphorylation, uncouplers
DiPolo, R., 226–27
DNA, 98, 151, 244. *See also* cDNA techniques
Donnan, F.G., 29
Donnan equilibrium, 29–32, 41, 49, 54, 60, 73, 81, 213
Draper, M.H., 82
duBois-Reymond, E., 68, 90
Dunham, E.T., 117, 120, 126–27, 132, 134, 138, 141–42, 153, 193
Dunham, P.B., 177
Durell, J., 211

Ebashi, S., 190–91
Eddy, A., 226

Edidin, M., 240
Edsall, J.T., 139
Eisenman, A.J., 55, 57, 59
Eisenman, G., 229
Electrogenic pumps. *See* transport pumps, electrogenic
Electron microscopy, 14, 100, 147, 149, 151, 187, 191–93, 201n.24, 201n.26, 239, 243, 247, 253, 257–58, 289, 303
Electrophoresis, 23, 136, 159, 162, 179, 240, 251, 303. *See also* SDS-PAGE
Elvehjem, C.A., 257
Emmelot, P., 149
Energy for active transport, 42–43, 46–47, 60, 62–63, 81, 113–18, 126, 212–14. *See also* diffusion, exchange
Engelhardt, W.A., 23
Enzyme phosphorylation/dephosphorylation, 135–37, 144, 146, 152, 157, 162–63, 165–68, 170, 172, 175, 179, 185, 188, 191–93, 195, 197–200, 202, 292. *See also* acyl phosphate; ADP/ATP exchange; P-type ATPases
Epithelial transport, 40, 89–96, 103, 113, 213, 220–225, 231, 234
Epstein, W., 197–98
Ernster, L., 255, 257, 275
Esmann, M., 140, 164, 167
Exchange diffusion. *See* diffusion, exchange
Eykman, C., 15

F_1, 254, 256, 258, 267, 283–97. *See also* ATPase, mitochondrial
F_o, 256, 267, 283–97
F_oF_1 (and F_oF_1 ATPase), 256, 271, 273, 283–97. *See also* ATPase, mitochondrial
Fairbanks, G., 240, 244
Feng, T.P., 78, 82
Fenn, W.O., 26–29, 35–37, 39, 45, 48–49, 63, 65, 102, 296, 307
Fernández-Morán, H., 253, 258
Fick's law, 220. *See also* diffusion, simple
Fillingame, R., 289, 293
Financial support. *See* funding of research
Findlay, M., 131
Finean, J.B., 147
Fischer, E., 19, 21
Fiske, C.H., 22
Flame photometer, 84, 101–102

Fluoride, 56–60, 107, 166, 181n.67
Flux ratio method, 92–93, 96, 97n.15
Fogarty, J., 152
Forbush, B., 167
Forgac, M., 158
Forte, J.G., 194–95
Francis, W.L., 107
Fruton, J.S., 22
Frye, L.D., 240
Fuhrman, F.A., 93
Fukushima, Y., 140
Fuller, S., 301
Funding of research, 85, 152–54, 304, 310.
    See also Carlsberg Foundation; Medical Research Council; National Science Foundation; NIH; Nuffield Foundation; Rockefeller Foundation
Futai, M., 289, 293

Galeotti, G., 91
Gárdos, G., 117
Garrahan, P.J., 138, 157, 193–94
Gatty, O., 107
Gerard, R.W., 78, 82, 85, 129
Gilbert, G.N., 295–96
Gilbert, W., 184, 200
Ghosts, red blood cell, 15, 117, 126, 132–33, 135, 142, 209, 211, 240
Glucose transporters, 220–25, 233–34, 243–44
Glycolysis, 4–8, 22–23, 56–57, 59–60, 63, 99–100, 104, 107, 117–18, 130, 212, 248, 293. *See also* substrate-level phosphorylation
Glycophorin, 240
Glycosylation. *See* protein glycosylation
Glynn, I.M., 105, 109, 111, 120, 126–27, 130, 132–35, 137–140, 142, 156–158, 165–167, 172, 177, 193, 226
Goffeau, A., 197
Goldacre, R.J., 118
Goldin, S.M., 161–62
Goldman, D.E., 73–74, 92
Goldman-Hodgkin-Katz equation, 74, 81, 92
Gore, M.B.R., 131, 141–42
Gorter, E., 16–17
Graham, J., 82
Gramicidin, 240, 254, 256. *See also* ionophores
Grantham, J.J., 138

Green, D.E., 147, 253–58, 284, 294
Green, N.M., 192–93, 197
Grendel, F., 16–17
Greville, G.D., 266, 275, 279, 308
Grisham, C.M., 187
Grubmeyer, C., 290
Grundfest, H., 73, 78
Gryns, G., 15
Gulati, J., 215

Hald, P.M., 100
Hamburger, H.J., 15
Hansen, O., 140
Hanson, J., 151
Hara, Y., 138
Hare, D., 226
Harris, E.J., 47, 62–64, 107, 109, 212–13
Harris, J.E., 56–64, 93, 104, 110, 306–307.
Hasselbach, W., 191–92
Hatefi, Y., 255, 268
Hediger, M., 226
Hedin, S.G., 15
Hegyvary, C., 140, 166–67
Heilbrunn, L.V., 189
Heinz, E., 226
Helmholtz, H. v., 68
Hemoglobin, 15, 113, 149–51, 254
Hemolysis, 15–16, 56, 117
Hempling, H.G., 110, 226
Henderson, R., 243
Heppel, L.A., 34–37, 40–43, 48, 63
Herbert, E., 126
Hess, H.H., 132, 141–42
Hevesy, G., 36, 49, 54–55, 63–64
Hilden, S., 160–61
Hill, A.V., 26, 47
Hill, L., 152
Hille, B., 229
Hinkle, P.C., 244, 271, 273–74, 276, 278, 296
HK and LK sheep red blood cells, 112, 134
Hüber, R., 60, 73, 239
Hodgkin, A.L., 70–85, 92, 107–108, 118, 138–39, 142, 153, 193, 226, 228–29, 233
Hodgkin-Huxley equation and model, 74–79, 85–86, 228, 233–34, 303
Hoffman, J.F., 112, 117, 120, 132, 134, 153, 157, 174, 177, 193
Hogeboom, G.H., 101

Hokin, L.E., 137, 153, 159–61, 163, 187, 205–207, 209–11, 216, 307
Hokin-Neaverson, M.R., 205–207, 209–11, 216, 307
Hopfer, U., 224
Huf, E.G., 91, 93
Hunter, F.E., 275, 308
Hurwitz, J., 130
Huxley, A.F., 70–78, 80–81, 83–85, 151, 228–29, 233
Huxley, H.E., 83, 151
Hydrophobic interactions in membranes, 4, 18, 73, 98, 118, 238
Hydrophobic interactions in proteins, 151, 183, 185, 200, 238, 240, 244, 291. *See also* protein structure
Hyperbolic kinetics. *See* saturable kinetics

Inesi, G., 192
Iodoacetate, 59–60, 107, 117, 165, 212–13
Ionophore, 196, 203n.92, 224, 240, 270, 272. *See also* gramicidin; valinomycin
Isselbacher, K., 224

Jacob, F., 151
Jacobs, M.H., 53, 64
Jagendorf, A.T., 269–70, 278–79, 296, 307
Järnefelt, J., 211
Jeanneney, G., 56
Jensen, J., 140, 146
Jørgensen, P.L., 140, 160, 163–64, 172–73, 177
Journals, 95, 117, 120, 154, 216, 277, 279, 305, 310

Kaback, H.R., 243
Kagawa, Y., 271, 273, 289, 293
Kalckar, H.M., 23, 293
Kanai, Y., 226
Karlish, S.J.D., 138, 158, 164, 167–68, 173–74, 177
Katz, B., 73–74, 83–84, 92
Katz, J., 20
Katzin, L.I., 91, 95
Kauzmann, W., 151
Kawamura, M., 185
Keilin, D., 23, 247, 254–56, 277
Kendrew, J.C., 149
Kennedy, E.P., 100
Kerr, S.E., 21
Keynes, R.D., 79–81, 84, 105, 107–108, 113, 115, 118, 121, 135, 142, 193, 213, 228
Kielley, W.W., 99, 190, 192
Kirshner, N., 292
Klingenberg, M., 276
Klodos, I., 140, 177
Knauf, P.A., 193, 244
Koefoed-Johnsen, V., 108
Koepsell, H., 226
Kopito, R.R., 240, 244
Koshland, D.E., 150–51, 153
Krebs, E.G., 152–53
Krebs, H., 22, 205–206, 216
Krebs cycle, 8, 22–23, 99–100, 253. *See also* substrate-level phosphorylation
Krogh, A., 21, 40–44, 48–49, 60, 63, 91
Kuhn, T.S., 304, 307
Kume, S., 140
Kunz, H.A., 106
Kyte, J., 159, 161, 165

Lac permease, 242–43
Lang, D., 184
Lardy, H.A., 99, 254, 257, 279, 293
Lasker, M., 152
Läuger, P., 229
Lawrence, E.O., 54
Leaf, A., 116, 135, 307
Lehninger, A.L., 100, 248, 257, 261, 264–66, 275–76, 279, 294
Levi, H., 44, 47, 213
Lew, V.L., 138, 157–58
Lewis, P.R., 79
Libet, B., 99, 129–30, 142, 192
Lillie, R.S., 81
Ling, G.N., 82, 84–85, 205, 212–17, 220
Ling-Gerard microelectrode, 82, 84, 212
Lingrel, J., 184–85, 187, 195, 198
Lipid bilayer, 4–5, 16, 85, 98, 147–49, 229, 239–40, 244–45, 266, 272. *See also* planar lipid bilayers; plasma membrane
Lipid vesicles. *See* liposomes; reconstitution in lipid vesicles
Lipmann, F., 61, 98, 100, 137, 190–91, 233, 293
Liposomes, 180n.30. *See also* reconstitution in lipid vesicles
Ljubimowa, M.N., 23
Lodish, H., 240, 244
Lohmann, K., 22, 175

Loomis, F.W., 100
Lorente de Nó, R., 78, 81, 84
Lovatt Evans, C., 63
Lowry, O.H., 131, 142, 168
Lund, E.J., 91, 113
Lundegårdh, H., 113, 264, 266
Lüthi, U., 138
Lysozyme, 150, 241

MacLennan, D.H., 192, 198
Maizels, M., 61–64, 93, 107, 109, 126
Makinose, M., 191–92
Manery, J.F., 37, 42
Marchesi, V.T., 240
Mårdh, S., 139
Marmont, G., 74, 78, 83–85
Marsh, B.B., 189, 191
Martin, A.J.P., 101
Martonosi, A., 192
Matteucci, C., 68
Maunsbach, A.B., 188
Maxam, A.M., 184
Mayr, E., 294, 309
McIlwain, H., 131, 141–42, 176, 205
Medical Research Council, 153
Meetings, 56, 95, 107, 116, 118, 120, 132, 134, 141, 174–75, 178–79, 216, 222, 227, 278–79
Membrane. *See* plasma membrane
Menten, M., 21
Meyerhof, O., 22, 99, 190, 192
Michaelis, L., 21, 101
Michell, R.H., 211
Microsomes, 101, 102n.29, 160, 195, 211
Mines, G., 189
Mirsky, A.E., 19
Mitchell, P., 119, 227, 242, 256–57, 261–80, 283, 286–87, 290, 293–96, 303, 308
Mitochondria, 4–5, 9–10, 13, 16, 100–101, 231, 247–259, 261–280, 283–297. *See also* oxidative phosphorylation; respiratory chain
Mond, R., 26, 28, 36, 39, 42
Monod, J., 151
Moudrianakis, E.N., 251, 307
Moyle, J., 266, 270, 275, 278
Mueller, P., 149
Mulkay, M., 295–96
Müller-Hill, B., 242

Mullins, L.J., 158, 229
Muntz, J.A., 130
Muscle contraction/relaxation, 23, 68, 83, 99, 115, 121, 151, 189–93, 284
Myoglobin, 19, 149–50

Nachmansohn, D., 129
NAD$^+$/NADH, 6–9, 100, 102n.18, 247, 250–51, 254–55, 267, 276, 280, 291. *See also* glycolysis; Krebs cycle; oxidative phosphorylation; respiratory chain
Nagai, T., 191
Nägeli, C., 14
Nakao, M., 137, 174, 177
Nastuk, W.L., 82, 84
National Science Foundation, 152, 217
Neher, E., 229
Nernst, W., 68
Nernst equation, 69, 74, 108, 193, 196, 261
Netter, H., 26, 28, 36, 39, 42
Neville, D.M., 148
Nicolson, G.L., 240
Niedergerke, R., 151
Niemann, C., 19–20
NIH, 48, 117, 119–20, 134, 136, 151–52, 170, 217
Nishizuka, Y., 211
Nørby, J.G., 140, 166, 174–75, 177–78
Novikoff, A.B., 130
Nuffield Foundation, 85, 153
Numa, S., 185, 198, 231

Occlusion. *See* cation occlusion
Ochoa, S., 99, 257
Oesterhelt, D., 243
Oligomycin, 162, 167–68, 179, 254–56, 285, 289–91
Ørskov, S.L., 140
Osmotic pressure and swelling, 10, 14–15, 28–31, 35, 60, 212, 265
Ostwald, W., 21
Ottolenghi, P., 140
Ouabain, 105–107, 122–23, 132–33, 136, 144–46, 159, 161, 165, 186, 210, 215, 224. *See also* cardiotonic steroids
Ovchinnikov, Y.A., 198
Overton, E., 14–15, 69, 82
Oxidative phosphorylation, 3, 7–9, 23, 99–100, 115, 247–59, 261–80, 283–

97, 307. *See also* mitochondria; P/O ratios; respiratory chain; substrate-level phosphorylation
inhibitors, 254
uncouplers, 93, 100, 115, 247–48, 252, 254, 266, 269–70, 285–86, 292. *See also* dinitrophenol

Palade, G.E., 100
Park, C.R., 139, 143, 177
Palmer, L.G., 215
Pandit-Hovenkamp, H.G., 251, 307
Parnas, J.K., 22
Parpart, A.K., 18
Pasteur, L., 142
Patch clamp, 229, 303. *See also* action potential; Ling-Gerard microelectrode; voltage clamp
Pauling, L., 19, 98, 149
Pedersen, P.L., 288, 292
Penefsky, H.S., 256, 290–91, 293
Perutz, M.F., 149
Peters, J.P., 21, 53, 59–60, 65
Pfeffer, W., 14–15
Philipson, K., 227
Phillips, D.C., 150, 153
Phillips, P.H., 100, 293
Phlorizin, 222, 224. *See also* glucose transport
Phospholipid vesicles. *See* liposomes; reconstitution in lipid vesicles
Phosphorylation. *See* enzyme phosphorylation; oxidative phosphorylation; substrate-level phosphorylation
Pinchot, G.B., 250–51, 280, 307
Pitts, B.J.R., 227
Planar lipid bilayers, 149, 154n.11, 229, 240
Plasma membrane, 3–5, 9–10, 14–18, 45, 68–69, 82, 85, 96, 102n.29, 147–49, 196, 242, 250, 265
  structure, 19–20, 98, 149–51, 183, 187, 240–42
Plesner, L., 140
Ponder, E., 62, 64
Pope, A., 132, 141–42
Popper, K., 279
P/O ratios, 23, 99, 116, 247, 255, 262, 274, 276, 293. *See also* oxidative phosphorylation

Pores, 17–18, 27, 73, 76, 85, 96, 119, 148, 228, 300. *See also* channels
Post, R.L., 109–110, 112–13, 120, 122, 132–136, 138–142, 153, 162–64, 166–67, 174, 177, 183, 192, 195, 209
Pressman, B.C., 271–73, 278, 296
Protein glycosylation, 159, 185, 240, 244–45
Protein structure, 19–20, 98, 149–51, 183, 187, 240–42. *See also* hydrophobic interactions in proteins; X-ray diffraction
  α-helix, 98, 149–50, 187, 241, 243
  β-sheet, 98, 150, 187, 241
Prout, W., 207
Publication. *See* journals
Pullman, M.E., 256
Pumps. *See* ATPases; redox pumps; transport pumps

Quastel, J.H., 206–207, 221, 225

Racker, E., 160, 192, 243, 248, 250, 254, 256–57, 270–71, 273, 275, 278–79, 292, 308
Radda, G., 292
Raftery, M., 230
Raker, J.W., 62, 110
Reconstitution in lipid vesicles, 140, 160–62, 179, 192, 195, 197, 215, 224, 226, 229–31, 234, 240, 243–44, 270–71, 289, 292, 303
Redox reactions and pumps, 7, 9–10, 23, 24n.34, 78, 87n.41, 91, 99, 113–16, 121, 194–96, 205, 247–48, 251, 254–56, 258–59, 262, 264, 266, 269–70, 275–77, 280, 283, 307
Reeves, J.P., 227
Rega, A.F., 193–94
Rehm, W., 206
Reid, E.W., 90
Renshaw, A., 116, 135, 307
Repke, K.R.H., 165, 175–76, 178, 187
Respiratory chain, 9, 23, 24n.35, 87n.41, 99–100, 121, 247–48, 250, 254–56, 258–59, 262, 265–67, 272–73, 275–76, 291, 294, 307. *See also* Complexes I–IV; mitochondria; oxidative phosphorylation
Reuter, H., 226
Richards, D.E., 138

Riggs, T., 225–26
Riklis, E., 221, 225
Ringer, S., 189
Ritchie, J.M., 122
RNA, 151, 184–85, 225, 231, 244
Robertson, J.D., 147–48, 238, 258, 265
Robertson, R.N., 113, 264
Robinson, J.D., 166
Rockefeller Foundation, 40, 85, 153
Rose, R.C., 224
Rosenberg, T., 90, 93
Rosenthal, A.S., 136
Rothenberg, M.A., 79
Rothstein, A., 244
Rottenberg, H., 272
Rudin, D., 149

Sachs, G., 195
Sachs, J.R., 158
Saier, M., 227
Sakmann, B., 229
Sanger, F., 98, 139, 183–84, 192
Saturable kinetics, 21, 108–109, 197, 220
Saxitoxin, 230–31
Scarborough, G.A., 196–97
Schatz, G., 292
Schatzmann, H.J., 103–106, 193–94
Schmidt-Nielsen, K., 209
Schmitt, F.O., 18
Schneider, W.C., 101
Scholes, P., 195
Schultz, S.G., 197, 222, 224, 226
Schwartz, A., 165, 175–78, 183–85, 188
SDS-PAGE, 159–60, 162, 179, 186–87,
    193, 195, 197–99, 225, 240, 244, 288,
    292. *See also* electrophoresis
Second messengers, 11, 152, 211–12, 231
Semenza, G., 224
Sen, A.K., 135, 139, 166, 176–78
Senior, A.E., 288, 293
Serrano, R., 197
Shanes, A.M., 110
Shannon, J., 152
Shaw, T.I., 109, 111, 120, 133
Sheppard, C.W., 62, 110
Shortcircuit current, 93, 95–96, 115–16,
    222, 224, 226
Shull, G.E., 184–85, 187, 192, 194–95
Simons, T.J.B., 157
Singer, S.J., 238–40
Sjöstrand, F.J., 147

Skou, J.C., 127–36, 138–42, 144, 153,
    160, 164, 167, 173, 176-77, 179, 186,
    192, 209
Skulachev, V.P., 273, 277, 296
Slater, E.C., 247–49, 251, 253, 255–57,
    259, 266, 270, 273, 275, 277–78, 284,
    286, 294, 296
Slayman, C. W., 138, 196–97
Slayman, C. L., 196–97
Solomon, A.K., 62–63, 83, 108, 110, 118,
    197
Spectrophotometers, 101, 257
Stämpfli, R., 81, 83
Steady-state, 38, 40, 50n.32, 65, 79, 85,
    92, 112, 167, 210, 255, 297
Steck, T.L., 240
Stein, W.D., 148, 172, 238
Steinbach, H.B., 37–43, 47–49, 91, 130,
    309
Steinberg, M., 138, 164–65, 177
Stiehler, R.D., 113
Stoeckenius, W., 243, 270–71
Straub, F.B., 99, 117
Straub, R.W., 122
Streeten, D.H.P., 108
Strophanthidin, 118, 123, 134, 201. *See
    also* cardiotonic steroids
Subbarow, Y., 22
Substrate-level phosphorylation, 7–8, 248,
    253, 259, 307. *See also* oxidative phos-
    phorylation
Sugar transporters. *See* glucose transport-
    ers; lac permease
Sulser, F., 106
Sutherland, E.W., 151–53
Sutko, J., 227
Suzuki, K., 140
Svedberg, T., 19
Sweadner, K.J., 161, 186
Synge, R.L.M., 101
Szent-Györgyi, A., 99

Tanford, C., 151, 153
Taniguchi, K., 140, 166
Tatum, E.L., 98, 196
Taugner, G., 292
Tetrodotoxin, 230, 234
Thomas, R.C., 158–59
Thunberg, T., 23
Tobias, J.M., 142
Toda, G., 140

Tomita, M., 240
Tonomura, Y., 166, 192
Tosteson, D.C., 108–110, 120, 132, 134, 153, 174
Transport, 9–10. *See also* ATPases; carriers; epithelial transport; redox pumps; transport pumps
active, 10, 29–30, 38, 41, 43, 53, 60–61, 63, 65, 80–81, 89–90, 92–93, 106, 117, 144, 193, 213, 221–22, 227, 233–34, 264, 292. *See also* energy for active transport; transport pumps
passive, 10, 96, 221
primary active, 10–11, 227, 236n.60, 309
secondary active, 10, 11, 227, 236n.60, 242–43, 265, 293, 309
Transport pumps, 10–12, 37, 39, 41–43, 49, 61–63, 65, 103–22, 126–44, 156–179, 183–200, 205, 207, 209–11, 213, 216–17, 220, 222, 243, 283, 300. *See also* ATPases; cation occlusion; redox pumps
electrogenic, 122, 158–59, 192, 196–97, 226
Troshin, A.S., 212
Tsybakova, E.T., 23, 99

Ubiquinone, 254–55, 273, 276. *See also* respiratory chain
Unwin, P.N.T., 243
Ussing, H.H., 36, 44–49, 56, 63, 91–96, 108, 112, 116, 118, 153, 213, 222
Utter, M.F., 130–31, 141

Valinomycin, 240, 270–73. *See also* ionophores
Vanadate, 166, 188
Van Slyke, D.D., 21
van't Hoff, J.H., 14

Vidaver, G.A., 226
Vignais, P., 289
Voltage clamp, 74–75, 84, 303. *See also* action potential; patch clamp
von Meyenberg, K., 289

Walker, J.E., 200, 289, 291, 293
Wallach, D.F.H., 238, 240
Wallick, E.T., 177
Warburg, O., 22, 23
Watson, J.D., 151
Weidmann, S., 82
Whipple, G., 35
Whittam, R., 106–107, 120, 135
Widdas, W.F., 177
Wieland, H., 23
Wiercinski, F.J., 189
Wikström, M.K.F., 276
Wilbrandt, W., 44, 59–60, 64, 103
Wilde, W.S., 44
Williams, R.J.P., 264–66, 278, 287, 290
Wilson, D.F., 278
Wilson, T.H., 243
Woodbury, L.A., 82
Wright, E.M., 225
Wrinch, D., 19–20

X-ray diffraction and crystallography, 17, 19, 100, 147, 149–51, 153, 187, 241, 276, 286, 291, 303. *See also* protein structure

Yamaguchi, M., 140
Young, J.Z., 70, 84

Zalusky, R., 222, 224, 226
Zerahn, K., 93–94, 115–16, 121, 135, 194, 307
Zervas, L., 19